MW01625927

Productivity Improvement for Construction and Engineering

Other Titles of Interest

Constructability Concepts and Practice, **edited by John A. Gambatese; James B. Pocock; Phillip S. Dunston** (ASCE Technical Report, 2007). Examines constructability and the integration of construction knowledge and experience in the planning, design, procurement, construction, operation, maintenance, and decommissioning phases of a project consistent with overall project objectives.

Managing Gigaprojects: Advice from Those Who've Been There, Done That, **edited by Patricia D. Galloway, Ph.D., P.E.; Kris R. Nielsen, Ph.D., J.D.; Jack L. Dignum** (ASCE Press, 2013). A stellar group of financial, legal, and construction professionals share lessons learned and best practices developed from working on the world's biggest infrastructure construction projects.

Project Administration for Design-Build Contracts: A Primer for Owners, Engineers, and Contractors, **by James E. Koch, Ph.D., P.E.; Douglas D. Gransberg, Ph.D., P.E.; Keith R. Molenaar, Ph.D.** (ASCE Press, 2010). Explains the basics of administering a design-build project after the contract has been awarded.

Public-Private Partnerships: Case Studies on Infrastructure Development, **by Sidney M. Levy** (ASCE Press, 2011). Demystifies public-private partnerships as an innovative solution to the challenges of designing, financing, building, and operating major infrastructure projects.

Quality in the Constructed Project: A Guide for Owners, Designers, and Constructors, **Third Edition, prepared by the Construction Institute of ASCE** (ASCE Manual of Practice, 2012). Provides information and recommendations on principles and procedures that are effective in enhancing the quality of constructed projects.

Productivity Improvement for Construction and Engineering

Implementing Programs That Save Money and Time

J. K. Yates, Ph.D.

ASCE PRESS

Library of Congress Cataloging-in-Publication Data

Yates, J. K., 1955–
Productivity improvement for construction and engineering : implementing programs that save money and time/J.K. Yates, Ph.D.
pages cm
Includes bibliographical references and index.
ISBN 978-0-7844-1346-3 (print : alk. paper) — ISBN 978-0-7844-7832-5 (ebook) 1. Building—Superintendence. 2. Construction industry—Management. 3. Industrial productivity. I. Title.
TH438.Y38 2014
620.0068'5—dc23

2014007427

Published by American Society of Civil Engineers
1801 Alexander Bell Drive
Reston, Virginia, 20191-4382
www.asce.org/bookstore | ascelibrary.org

Errata: Errata, if any, can be found at http://dx.doi.org/10.1061/9780784413463.

ISBN 978-0-7844-1346-3 (paper)
ISBN 978-0-7844-7832-5 (PDF)
Manufactured in the United States of America.

21 20 19 18 17 16 15 14 1 2 3 4 5

Instructors: A PDF manual providing worked answers for the problems presented in this book and sample test questions is available at no charge. Contact ASCE by e-mail (pubsful@asce.org) or telephone (1-800-548-2723, domestic U.S.; 1-703-296-6300, international). Ask for *Productivity Improvement for Engineering and Construction: Solution Manual and Test Questions*, ISBN 978-0-7844-7833-2, Stock No. 47833.

Dedication

To my former students who contributed to this book by providing examples, case studies, figures, and photographs. To the students who over the years continually inspired me to be a better professor and administrator. I am very proud of what all of my former students have accomplished since they graduated from college.

Contents

Preface

This book provides a contemporary synopsis of productivity improvement investigation methods and techniques. There are other publications that include information on productivity improvement, but this book focuses exclusively on productivity improvement studies and analysis techniques for implementing engineering and construction productivity improvement programs. If members of engineering and construction firms are able to improve production rates by even small percentages, profits increase for firms. On multimillion-dollar projects, small increases in productivity might lead to additional profits in the millions of dollars.

The purpose of this book is to introduce engineering and construction professionals to productivity improvement investigation methods and techniques and to provide them with knowledge on how to design and implement productivity improvement programs for the engineering and construction industry. In addition to covering traditional productivity improvement topics, this book also includes information on new and emerging areas, such as building information modeling (BIM), sustainability concepts, computer simulation modeling, and global productivity issues.

The goal of this book is to help engineering and construction industry personnel learn to critically analyze engineering and construction work processes and to design improved work processes that help increase production rates at engineering firms and construction jobsites. Effectively using resources during construction processes results not only in higher profits but in improvements to jobsite environments that result in safer and more efficient construction jobsites.

This book is unique in that it addresses productivity improvement from both the engineering and construction perspectives and it provides detailed case studies that help to illustrate how to implement productivity improvement programs. The information presented in this book is mainly for engineering and construction industry personnel, but the concepts it introduces are also applicable to other industries where personnel work on projects that include repetitive, discrete activities.

The first chapter provides an introduction to productivity improvement programs, and it defines productivity improvement, productivity management, and productivity in terms of inputs and outputs. This chapter also explains the elements of performance and the role that productivity improvement teams have in analyzing productivity improvement. Productivity measures at the national, industrial, corporate, and personal levels are discussed, and the effects of productivity fluctuations on the economy are explained using examples of U.S. productivity outputs for the nonfarm business sectors and the construction industry.

The following chapters discuss productivity improvement techniques that could be applied in the construction industry, methods for improving productivity rates, and productivity improvement programs.

Chapter 2 provides information about the construction industry and its environment. This chapter discusses how productivity improvement programs are integrated into firms and the planning strategies of firms. It also includes information on strategic management approaches and provides a review of successful approaches to productivity improvement. Lean construction, quality control, and quality assurance procedures and prefabrication and modularization are also discussed in Chapter 2.

Chapter 3 discusses different methodologies and approaches for implementing productivity improvement programs. This chapter provides background information about the construction industry, the phases of construction, and the contract administration models that are used in the industry. Information is also provided on construction team members, the different construction sectors, construction project planning, the construction industry as it compares to the manufacturing industry, and project life cycle.

Chapter 4 discusses how the human component and jobsite working conditions affect productivity rates and how providing safe and well-organized jobsite environments contributes to the successful completion of construction projects. Additionally, understanding and accommodating the physical limitations of workers helps create working environments that reduce the incidence of accidents and that contribute to increasing production rates. This chapter provides suggestions on how to address physical limitations, inclement weather conditions, the monotony of repetitive tasks, fatigue, safety issues, and worker motivation. Information is also provided on safety issues, including case studies from legal cases and construction failures.

Chapter 5 discusses the responsibilities of productivity improvement team members as they are preparing productivity improvement studies. This chapter also includes techniques for preparing interview guides and discusses how learning curves are used in productivity improvement studies. The scope of productivity improvement studies is discussed along with analysis techniques and ways to prepare for conducting studies. Data-collection techniques are presented along with a discussion of work sampling and learning curves.

Chapter 6 discusses different work-sampling data analysis techniques that could be used to quantify work-process operations at construction jobsites. The work-sampling techniques presented in this chapter are activity sampling, field rating, 5-min rating, crew balance charts, process charts, and process diagrams. In addition, productivity measurement methods are introduced, and an explanation of the types of construction activities that could be analyzed using each of the productivity measurement methods is provided in this chapter. Systems engineering concepts are discussed in order to demonstrate the similarities and differences between systems engineering and productivity improvement methods.

Chapter 7 provides information on evaluating and prioritizing work-process alternatives, and it includes four detailed case studies that demonstrate how to collect and analyze productivity improvement data and provide recommendations for improving production rates.

Time-management studies are introduced in Chapter 8 to provide a method for evaluating the productivity of engineers and project and construction management personnel. Once engineers and project and construction management personnel have an understanding of how they waste time during work, they are able to start applying some of the suggestions provided in this chapter for improving their own productivity.

Chapter 9 provides background information on the computer hardware, software applications, and simulation models used to collect and analyze data for productivity improvement studies at construction jobsites. Computer software programs that could be used during productivity studies to help analyze specific work processes are discussed and include management information systems, databases, computer-aided design and drafting, building information modeling (BIM), and simulation software, such as *MicroCYCLONE, WebCYCLONE, State and Resource Based Simulation of Construction Processes* (*Stroboscope*), and the *General Purpose Simulation System* (*GPSS*).

Chapter 10 discusses how computer simulation modeling software programs and the queuing theory are used to model construction operations. In addition, GPSS is used to demonstrate how simulation software programs could be used to model heavy construction equipment operations and to determine optimal fleet configurations that maximize productivity rates. An example simulation model is presented to demonstrate that different combinations of heavy construction equipment could be tested using simulation software in order to determine the most cost-effective heavy construction equipment fleet. This chapter also provides information about sustainable alternatives to heavy construction equipment and hybrid-electric heavy construction equipment.

In the global engineering and construction marketplace, the one essential component of the bid estimate preparation process that provides a competitive advantage for firms is determining labor costs, which requires accurate productivity rates. In Chapter 11, productivity indices are provided for individual countries along with an explanation on how the indices could be used when preparing bid estimates. In addition, this chapter discusses some of the issues that affect the productivity of workers in foreign countries. Chapter 11 also provides statistical methods that might be used to assess the impact of several factors on productivity rates in newly emerging countries, as contrasted with industrialized countries, using a comparison of Nigeria and the United States as a case study. A second case study discusses the results of a productivity improvement study that was conducted in Saudi Arabia on construction projects that ranged from 10 million dollars to billions of dollars.

Chapter 12 provides background information on sustainability and sustainable development practices. It includes a discussion of why sustainable development practices are being integrated into engineering and construction projects and also covers life-cycle environmental and cost analysis, energy consumption, pollution prevention, material recycling, and the Leadership in Energy and Environmental Design (LEED) program and other sustainability certification systems and organizations.

Chapter 13 discusses the different types of sustainable construction materials being used to create green structures. It provides information on painting products; steel production and products; concrete production and products; formwork; masonry products; asphalt pavement; fiber-reinforced polymeric composite materials; wood products and wood substitutes; polyvinyl chloride and thermoplastic products; and mineral, mining and metal products.

Appendix A is a stone-panel curtain-wall installation case study. This case study summarizes the strategies used and the results obtained during a productivity improvement research investigation that evaluated the installation processes for stone-panel curtain walls on a tall office building. The researchers analyzed existing work processes and developed recommendations for improving work processes for the installation of stone panels on the curtain wall. Numerous photos are included in Appendix A that help illustrate the project task that was being evaluated and proposed alternatives that might lead to improved productivity.

This book helps engineering and construction personnel learn how to develop productivity improvement analysis techniques and design productivity improvement programs that could be implemented in the engineering and construction industry. Each small improvement in efficiency helps reduce the amount of time and resources required to complete construction projects, thus resulting in cost and time savings to owners and increased profits for construction firms.

Acknowledgments

The author would like to thank the members of the American Society of Civil Engineers for providing me with the opportunity to write this book, which contributes to the body of knowledge in the engineering and construction professions. Special thanks to Betsy Kulamer, who approved the original book proposal and worked with the author during the writing of the manuscript and during the production phase of the book. The contribution to the completion of the book by the Manager of Book Production Sharada Gilkey and the copyeditor Julie Kimmel is also acknowledged by the author, and I thank them for their assistance.

A special note of appreciation and a thank you to all of the construction management personnel and construction workers who allowed their jobsite activities to be observed and recorded for the case studies in this book. Their willingness to share their construction work environment with the authors and the research assistants was greatly appreciated by the author and the research team members.

The research assistants who conducted the case studies were former graduate students at Iowa State University, the University of Colorado, Polytechnic School of Engineering at New York University, and San José State University. The author would like to acknowledge the contributions of the following former graduate students:

Chapter 7, Case Study 1—Dr. Basel Mohamed Abulfateh, the Assistant Undersecretary for Services for the Royal Court, West Reffa, Kingdom of Bahrain.

Chapter 7, Case Study 2, and several figures in Chapter 6—Dr. Fred Rahbar, project consultant to Aramco Overseas Company (AOC), Dhahran, Saudi Arabia.

Chapter 7, Case Study 3—Alan Page, Matt Frechette, Meenakshi Grandhi, Nanik Yusuf, Anwar Beg Mirza, and Hasan Bas.

Appendix A Case Study 4—Guang Jiang, Amrita Mukherjee, Chakrit Raksamata, Iqbal Jeet Singh, Sorawut Srisakorn, and Milkias Lemma Tefera.

Several of the figures in Chapter 6 are from a study called *Productivity Improvement Study for Structural Steel Erection on the North Concourse Project* at the Norman Y. Mineta San José International Airport. The figures that were used from this study are Figures 6-3 to 6-9 and 6-11 to 6-15. The authors would like to thank the following former San José State University graduate students for contributing these figures—Neil Ogimachi, Joshua Spangrud, Christopher Untrauer, Viney Champaneria, Bahman Bina, and D. Alluri.

The author would also like to acknowledge the research conducted by Nancy Torres Terrero, who is a project manager for Jacobson and Company. The results of her research are included in Chapter 4 in Section 4.6, Causes of Construction Worker Injuries, and Section 4.13, Construction Accident Case Studies. She conducted a study on safety issues in construction while she was a graduate student at the Polytechnic School of Engineering at New York University (formerly Polytechnic University) in Brooklyn, New York. Dr. Edward Lockley, who was formerly a professor at City College of New York, contributed the material in Chapter 4 in Section 4.7, Construction Failures, that was part of his Ph.D. dissertation on construction failure investigations, which he completed at the Polytechnic School of Engineering at New York University.

The computer simulation model case study example in Chapter 10 was developed based on the work of former graduate student Khalid Al-Senani, who is the Performance Adviser for the Ministry of Petroleum and Minerals in Saudi Arabia. He is thanked for contributing his study on productivity improvement computer simulations.

Section 10.10, Sustainable Heavy Construction Equipment, and Section 10.11, Hybrid-Electric Heavy Construction Equipment, were written by Dr. Eric Asa, Associate Professor in the Department of Construction Management and Engineering at North Dakota State University.

In Chapter 13, Section 13.11, Fiber-Reinforced Polymeric Composite Materials, was written by Dr. Robert Steffen, Associate Professor in the Department of Construction Management at Western Carolina University. Also, the information in Chapter 13 in Section 13.12, Wood Products, on chromated copper arsenate was provided by Ken Nowak, a former graduate student at San José State University.

I would like to thank Dr. Rashad Zakieh for permitting the author to use material from his research investigation, which was conducted in Saudi Arabia and is cited in the case study in Chapter 11. His research provides insight into the types of productivity issues faced on major global construction projects. The case study in Chapter 11 in Section 11.6 on Nigeria and United States and Sections 11.4 and 11.5 were provided by Swagata Guhathakurta who was formerly a graduate student at Iowa State University.

Rebecca (Becky) Gill, who is the administrative coordinator for the Institute for Corrosion and Multiphase Technology at Ohio University, assisted with the typing of the first draft of the manuscript, and Dr. Jianfeng Qin, a former graduate student at Ohio University, who is now a project manager for Mcsam Downtown LLC in New York City, helped by drawing the figures used in some of the case studies and chapters. He also helped type preliminary draft editorial corrections. His assistance with drawing the figures was crucial. The assistance of both Rebecca Gill and Jianfeng Qin was greatly appreciated by the author.

Dr. Gary Smith, dean of the College of Engineering at North Dakota State University, provided one month of summer support to the author while the author was working on this book, and he is thanked for this contribution.

Professor Mario Baratta, an adjunct professor at San José State University, is acknowledged for his help in initiating this book.

CHAPTER 1

Introduction

When **productivity improvement** techniques are successfully implemented, work tasks are completed more efficiently and this might lead to reductions in the amount of time and money required to complete projects. Members of both private- and public-sector owner organizations, engineering firms, and construction companies are able to improve their operations by implementing productivity improvement programs. Private firms benefit from productivity improvements through the maximization of profits, and the public benefits when the cost of public projects are minimized because projects are funded by private citizens in the form of taxes, bonds, and levies. Even though the concepts introduced in this book are specific to the engineering and construction (E&C) industry, they are also applicable to other industries.

The level of commitment of engineers and construction management personnel influences the success of productivity improvement programs because they are responsible for designing and managing construction operations and are involved in the development and implementation of productivity improvement programs. Engineering and construction management personnel assist productivity improvement team members in analyzing existing operations, exploring methods for improving physical processes, designing training sessions for workers that demonstrate how to incorporate more productive techniques, and motivating workers to improve the efficiency of their work tasks and processes.

In the United States, worker **productivity rates** fluctuate when the delivery of information, materials, or equipment is delayed; while workers wait for others to finish their tasks; and when managers make ineffective decisions, provide unclear instructions or inadequate clarification of directions, and provide confusing job assignments. Some of the causes of delays at construction jobsites include inefficient site layout, overcrowded work areas, equipment operators waiting for others to move equipment to its proper location, worker interference, ineffective work crew configurations, and improper fabrication of materials.

The operations that occur at construction jobsites consist of **discrete activities**, each with a distinct beginning, end, and unique cycle. The discrete components of work tasks are shown in Figure 1-1. Productivity improvement studies examine the components of the work tasks shown in Figure 1-1 to determine whether altering the work processes could reduce the time required to complete work tasks. Potential

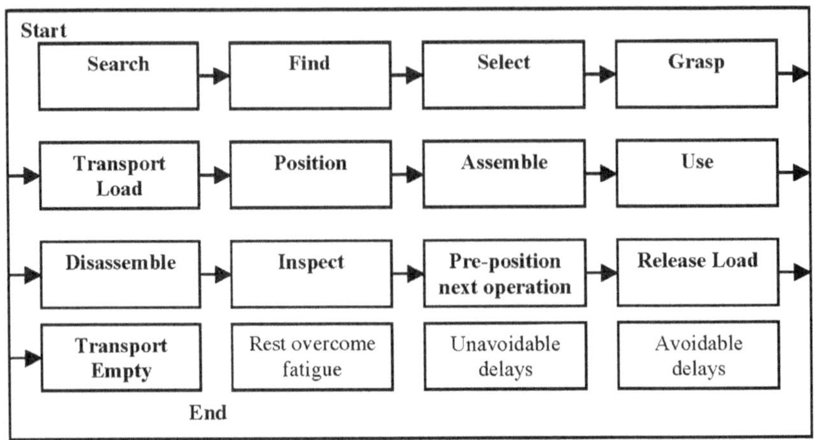

Figure 1-1. Discrete Components of Tasks for Construction Jobsite Operations

Source: Adapted from Matteson and Ivancevich (1977), p. 153.

productivity improvements include eliminating unnecessary work movements, optimizing crew configurations, reconfiguring transportation cycles, optimizing the production rates of equipment, improving the flow of work processes, and improving the layout of the jobsite. These, and other types of productivity improvement techniques, are discussed in this book. Case studies that were performed at construction jobsites are used to illustrate how productivity improvement methods might be used to help increase production rates. Techniques for optimizing the productivity of engineering, project management, and construction management personnel, including time-management studies, are also discussed in this book.

Productivity improvement techniques and methods have been used in the construction industry since the mid-twentieth century, and they were formalized in professional E&C publications starting in the 1970s. In 1972, Clark Oglesby and Henry Parker wrote *Methods Improvement for Construction Managers*, one of the first books to be published on productivity improvement in the construction industry. A few years later, in 1976, Daniel Halpin and Ronald Woodhead authored the book *Design of Construction and Process Operations*. In 1989, Oglesby, Parker, and Howell published *Productivity Improvement in Construction*, which covered many of the important aspects of productivity improvement and included examples of how productivity improvement techniques are implemented in the construction industry. This book was followed by Halpin's *MicroCYCLONE System Manual* and *WebCYCLONE User's Manual*, both originally published in 1990 (Halpin 1990a and 1990b).

In 2004, James Adrian published a new edition of his book *Construction Productivity: Measurement and Improvement*, which provides productivity measurement techniques and formulas used in the construction industry to improve productivity. Since these books were published, many of the techniques and methods for productivity improvement have been refined and updated to include additional areas, such as concern for worker safety, sustainable development methods and materials, hybrid and electric heavy construction equipment, global productivity improvement issues, new computer applications, and global positioning systems (GPS), all of which are discussed in this book.

The Construction Industry Institute (CII) has published the results of several research teams that have produced documents related to construction productivity, including

1. *The Manual of Construction Productivity Measurement and Performance Evaluation* (SD-35).
2. *Work Force View of Construction Labor Productivity* (RR215-11).
3. *Productivity Benchmarking Summary Report* (BMM2011-1).
4. *Craft Productivity Research Program—Phase III* (RS252-1b).
5. *Leveraging Technology to Improve Construction Productivity* (RS240).
6. *Construction Productivity Research Program—Phase III* (RS252-11b).
7. *Construction Productivity Research Program—Phase V* (RS252-1d).
8. *The Construction Productivity Handbook* (IR252-2d).

All of the CII publications provide information on productivity best practices in the construction industry. *The Construction Productivity Handbook* includes the results of a five-year research investigation into productivity practices.

This chapter provides an introduction to productivity improvement programs and definitions for productivity improvement and performance. Production rates in the United States are examined at the national level in terms of how they have been fluctuating during the last half of the 20th century and the first part of the 21st century and how they compare to production rates in the construction industry.

1.1 Definition of Productivity Improvement

Productivity improvement is defined by Dr. D. Scott Sink (1985, p. 2) as "the result of managing and intervening in key transformations or work processes. Productivity improvements will occur if any of the following conditions are made to exist":

1. Output increases, input decreases (output up/input down);
2. Output increases, input remains the same (output up/input consistent);
3. Output increases, input increases but at a lower rate (output up/input up);
4. Output remains constant, input decreases (output constant/input down); and
5. Output decreases, input decreases but at a more rapid rate (output up/input down).

In the same publication, Sink (1985, p. 3) defines productivity management as "a process that entails strategic and action planning and a critical focus on ongoing and effective implementation and the relationship between outputs generated from a system and the inputs provided to create those outputs." Typical inputs include

- Labor (human resources),
- Capital (physical and financial capital assets),
- Energy,
- Materials, and
- Data.

The concept of productivity implies the conversion of inputs, such as labor, equipment, energy, materials, or data, through a transformative process that creates outputs of either goods or services (Sink 1985). The competitive advantage of individual firms is based on how efficiently members of firms are able to perform these transformations. Even slight increases in productivity rates might lead to higher profit margins for firms.

Regardless of whether the perspective is economic, political, psychological, managerial, or engineering, the basic definition for productivity remains the same but the boundaries, the size, the type, and the scope of the system being examined change. Many politicians and economists focus on a **macro systems perspective** of productivity, which includes nations, regions, states, or perhaps industries. This focus complicates productivity issues because there are various organizational perspectives within these different boundaries.

Productivity is normally measured as **output/input** over a specified period. This is the ratio used by the **U.S. Department of Commerce** and the **U.S. Bureau of Labor Statistics** (U.S. Department of Commerce 2012). Productivity ratios are more meaningful in measurements of changes over time. Statistically regressing inputs and time trends reflects the trend rate of growth of productivity; however, this method does not account for short-term productivity changes. Output per labor hour is one indicator of **productivity efficiency**. Output should be related to both labor and capital to determine net savings. But capital per labor hour does not remain constant over time. A measure of **business efficiency** is what firms are trying to achieve when they evaluate their productivity. Performance is influenced by market competition, profitability, government regulations, the reputation of a firm, pride in workmanship, and client demands.

1.2 Global Productivity Rates

For firms to remain competitive in the global marketplace, they have to examine all possible means of improving their competitive advantage, and in the E&C industry, the main competitive advantage for firms is in how efficiently they perform their work. The cost of materials and equipment are relatively equal for most construction firms that operate in the E&C industry; therefore, the only way for firms to compete against other firms is on the basis of unique methods and processes that lead to improvements in productivity rates and allow firms to reduce their bid estimates. If project management or construction management personnel are able to increase productivity rates at jobsites by 1% or 2%, it might increase profits on large construction projects by millions of dollars. In the global arena, where firms from many countries bid on projects and the value of some of the projects exceeds billions of dollars, being able to increase productivity rates might be the determining factor on whether a company is awarded a project and how much profit a firm realizes. In newly emerging economies, productivity rates are affected by factors that might not be as influential in industrialized countries, and these factors are explored in Chapter 11 in order to demonstrate their impact on productivity. Chapter 11 also provides a detailed analysis of **global productivity rates**.

1.3 Productivity and Performance Enhancement

Performance is based on three elements: (1) completing operations as scheduled, (2) maintaining required levels of quality, and (3) achieving project objectives without exceeding budgeted costs. Construction management personnel monitor productivity using project control systems in order to determine when there is a need to closely assess current progress, when mitigation strategies are necessary, and whether a project is meeting its goals. Monitoring budgets, schedules, quality, and safety allows project team members to react more rapidly to mitigate or eliminate potential problems.

In the E&C industry, management personnel work with **performance elements** on a daily basis. In addition to trying to increase productivity by either minimizing costs or maximizing profits, project team members are concerned with reducing the occurrence of accidents by following government-mandated regulations, completing projects on time and within the budget, and providing a high-quality structure.

Productivity improvement programs are managed by **productivity improvement teams**, which provide assistance to project and construction management personnel when they are trying to achieve organizational goals. Productivity improvement team members plan, implement, maintain, and document productivity improvement programs and encourage the use of appropriate leadership techniques and the efficient management of resources.

Productivity improvement programs require a three-stage process that involves determining (1) who will be involved, (2) how programs will be implemented (strategy), and (3) when programs will be implemented by project personnel (solutions). All three of these stages are discussed in the following chapters.

1.4 Different Levels of Productivity Measures

Productivity is measured at all levels of society, and this section discusses productivity measures at the national, industrial, corporate, and personal levels.

National Level

At the national level, productivity is a measure of economic growth and success; therefore, it is monitored by members of government agencies. In the United States between 1944 and 1963, half of the increase in the **gross national product** (GNP) was attributable to productivity gains. Improvements in productivity are partially responsible for increasing real income per capita because they provide a proportionate offset to increasing wage rates and other input costs.

In the United States between 2000 and 2006, the **gross domestic product** (GDP) increased by 33% even though there were excessive budget deficits and the country experienced major terrorist attacks in 2001 that destroyed the World Trade Center in New York City and damaged the Pentagon in Arlington, Virginia. During this period, the economy experienced high levels of productivity, many firms had record profits, tax rates were low, and oil prices fluctuated frequently. Productivity improvements provided a proportionate offset to increasing wage rates, and this helped stabilize inflation. At the national level, improvements in productivity helped to reduce the requirements for additional labor, capital, and natural resources and contributed to domestic firms maintaining a competitive advantage over foreign firms.

Industrial Level

In industries, higher productivity rates lead to a relative decline in the price of goods and services and increase competitiveness nationally and internationally. High productivity leads to decreasing prices, increasing sales, and increasing output. Low productivity rates tend to increase prices, which causes reduced sales and output.

Corporate Level

In corporations, productivity rates influence both profitability and the survival of firms. Increased productivity rates translate into higher profit margins. Lower productivity rates might lead to company reorganizations, dissolutions, or bankruptcies. Corporate-level productivity rates contribute to maintaining national-level productivity goals.

Personal Level

At the personal level, in some circumstances higher productivity rates contribute to personal self-fulfillment and possibly career advancement. When workers are involved in increasing productivity rates, they are contributing to the achievement of corporate goals and also to national productivity goals. Unfortunately, some of the societal benefits produced by higher productivity rates occur because individual workers are more productive than standard expectations, and this might lead to workers becoming disaffected, experiencing increased health issues, or terminating their employment.

1.5 Effects of Productivity Fluctuations

The **American Productivity Center** (1980) in Houston, Texas, developed a model of a **low productivity trap** that explains the consequences of low productivity rates and models the effects of high productivity rates. In this model, low productivity rates cause a spiraling downward effect as prices rise for domestic and exported goods, which in turn leads to reduced sales. Once sales decline, lower utilization rates of plant capacities cause insufficient capital investment in plants, which in turn leads to lower productivity and higher costs per unit. The reverse occurs when productivity increases: the cost per unit decreases, firms become more competitive, and sales increase, which in turn leads to capital investment in the firm and additional gains in productivity rates (American Productivity Center 1980).

From 1990 to 2010, there were wide fluctuations in productivity rates in the United States. Table 1-1 provides the increasing and decreasing nonfarm business output index per quarter over a 13-year period starting in 1996 and ending in 2008. Exercise 1.2 at the end of the chapter requests that readers investigate the nonfarm output for the past five years. Nonfarm output declined during 2001 and 2008, when the United States and global stock markets declined rapidly. Although the nonfarm output recovered in 2001, it did not return to the levels that were prevalent before that year.

Figure 1-2 illustrates the change in annual nonfarm output per hour from 1996 to 2011. Every 10 years since 1953, there has been a decline in nonfarm output. In the early 1970s, the measurable slowdown in productivity resulted in inflation, stagnant income for workers, and increasing international competition. In the 1980s, the federal government implemented new policies to promote increases in worker productivity, and firms adopted new productivity measurement systems and programs for measuring quality control.

A comparison of the nonfarm labor productivity indexes from 1964 to 2011 and the productivity indices for construction during the same period indicates that construction productivity has been steadily declining whereas nonfarm productivity has been constantly increasing.

Table 1-1. Rise and Fall of Nonfarm Business Output Index (1996–2008)

Year	Quarter 1	Quarter 2	Quarter 3	Quarter 4	Annual
1996	3.2	7.6	4.3	5.7	4.5
1997	3.2	7.3	5.8	3.7	5.2
1998	5.6	2.9	5.3	7.9	5.0
1999	3.7	3.7	5.6	8.4	5.2
2000	−0.1	7.5	−0.8	2.2	3.8
2001	−1.1	1.2	−2.9	1.2	0.4
2002	3.5	1.4	3.1	0.1	1.5
2003	1.2	4.3	11.0	1.6	3.1
2004	2.6	5.0	3.8	2.6	4.1
2005	3.7	2.9	4.8	1.2	3.4
2006	6.0	2.9	0.3	1.4	3.2
2007	−0.9	5.8	5.5	−0.7	2.0
2008	0.9	2.8	−1.9	−8.8	2.0

Source: U.S. Department of Commerce (2009).

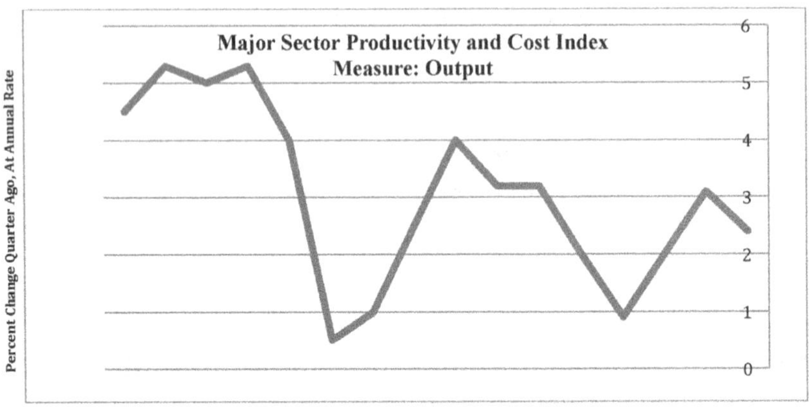

Figure 1-2. Nonfarm Business Output per Hour Measure (1996–2011)

Source: U.S. Department of Commerce (2012).

During the first decade of the twenty-first century, productivity rates increased in the United States. Sectors with rate increases included public utilities (increasing by 5.4%), transportation (4.5%), mining (3.17%), manufacturing (2.6%), and government (1.64%). Unfortunately, construction productivity rates increased by only 0.8%, and because construction is the second-largest economic sector in the United States (the health-care industry is number one) and a more than 14 trillion-dollar industry, improvements in construction industry productivity rates help improve the U.S. economy (U.S. Department of Commerce 2012). From 2008 to 2010, the construction industry, along with the rest of the U.S. economy, experienced a rapid decline attributable to the collapse of the housing market, precipitated by home foreclosures, and the collapse of the mortgage-backed securities market, which led to tight credit markets. During the middle of the second decade of the twenty-first century, the construction industry started to rebound due to pent up demand for facilities that were not built during the recession from 2008 until 2013.

1.6 Organization of the Book

The following chapters in this book discuss productivity improvement techniques that help improve the efficiency of operations in the E&C industry, methods for improving productivity rates, and productivity improvement programs. Chapter 2 provides information on productivity improvement programs, and Chapter 3 discusses the key elements of the construction industry. Chapter 4 discusses human factors and the safety issues that affect productivity rates. Chapter 5 explains how to prepare for conducting productivity improvement studies, and Chapter 6 discusses how to collect and analyze productivity data. Chapter 7 provides four detailed productivity improvement case studies that were conducted on different types of construction projects that demonstrate how to apply the concepts presented in Chapter 6 at construction jobsites.

Chapter 8 introduces engineering and project and construction management productivity improvement techniques. Chapter 9 explains how computer applications might be used for productivity improvement programs, and Chapter 10 provides information on computer simulations and simulation software and a case study that used computer simulation software to analyze productivity rates for heavy construction equipment. This chapter also discusses sustainable alternatives to heavy construction equipment. Chapter 11 discusses global productivity improvement, and it provides global productivity indices, a case study on labor productivity in Nigeria as compared with the United States, and a case study performed in Saudi Arabia. Chapter 12 provides an introduction to sustainability methods and practices that pertain to the E&C industry.

Chapter 13 discusses the different types of sustainable construction materials that are being used to create green structures and it provides information on painting products; steel production and products; concrete production and products; formwork; masonry products; asphalt pavement; fiber-reinforced polymeric composite materials; wood products and wood substitutes; polyvinyl chloride and thermoplastic products; and mineral, mining, and metal products.

Appendix A is a stone-panel curtain-wall installation case study. This case study summarizes the strategies used and the results obtained during a productivity improvement research investigation that evaluated the installation processes for stone-panel curtain walls on a tall office building.

1.7 Summary

This chapter provided an introduction to and definitions for productivity improvement, productivity management, and productivity in terms of inputs and outputs. It also explained the elements of performance and their relationship to productivity. Productivity measures were discussed as they pertain to the national, industrial, corporate, and personal levels. The effects of productivity fluctuations on the economy were explained using examples of U.S. productivity outputs for the nonfarm business sectors and the construction industry. A synopsis of the organization of the book was also included in this chapter.

1.8 Key Terms

American Productivity Center
business efficiency
discrete activities
global productivity rates
gross domestic product
gross national product
low productivity trap
macro systems perspective
output/input
performance elements

productivity
productivity efficiency
productivity improvement
productivity improvement teams
productivity management
productivity rates
U.S. Bureau of Labor Statistics
U.S. Department of Commerce

1.9 Exercises

1.1 On a construction project in the year 2007, masons were earning \$16.00 per hour, and they received a 6% annual increase for the next five years. The masons were able to install 240 bricks per day, or 30 bricks per hour, and the annual increase in productivity was 1% a year. Using this wage information, calculate the following:

a. The labor cost per brick placed for each of the six years,
b. The inflation rate for each of the five years (the inflation rate is the percentage increase in the labor cost per brick for each year from the previous year), and
c. The percentage increase in productivity that would reduce the inflation rates calculated in (b) by half.

1.2 Create a chart of U.S. productivity using data from the U.S. Bureau of Labor Statistics website (http://www.bls.gov). Obtain the data for the last five years by annual periods. Review the latest U.S. Bureau of Labor Statistics report on U.S. productivity that is available online and summarize the key points in the report. What conclusions (or opportunities) may be drawn relative to productivity in the construction industry during the past five years?

1.3 Divide the participants into groups of three or four. Provide each group with 25 assorted Lego pieces. Set up one table on one side of the room and another table on the opposite side of the room, approximately 20 ft (6.1 m) apart. Each group should set their 25 Lego pieces in a pile on the first table. The groups then have to move the Legos from the first table to the second table, but only one person should touch one Lego at a time. Time how long it takes each group to move the 25 Lego pieces from one table to the other, and have each group record their time. As an additional exercise, each group should develop an alternative method for moving the 25 Lego pieces in a shorter period. After the groups have developed a new method for moving the Legos, the groups will use the method they have developed to once again move the Lego pieces from one table to the other, and they will be timed to see if there is any improvement in their time. This exercise should be repeated in successive weeks to determine whether the times continually decline or whether there is a maximum reduction in time that cannot be exceeded no matter what is changed during the activity.

1.4 Explain the differences between productivity improvement programs, productivity improvement studies, productivity, productivity management, and production rates.

1.5 What ratio is used by the U.S. Department of Commerce and the U.S. Bureau of Labor Statistics to measure productivity?
1.6 What are the responsibilities of productivity improvement teams?
1.7 Discuss the influence of productivity at the national, industrial, corporate, and personal levels.
1.8 Explain the difference in the benefits of productivity improvement for public and private firms.
1.9 What general categories of activities are performed during productivity improvement programs?
1.10 What are discrete activities? Provide three examples.

1.10 Optional Project

Table 1-2 provides guidelines for an optional term project that is appropriate for implementation at the completion of Chapters 1–7.

Table 1-2. Optional Productivity Improvement Project

Objective
To select a significant construction operation at a construction jobsite to be studied in detail for a methods improvement study. The term project should incorporate all of the methods and techniques for studying productivity improvement introduced in this textbook. The project is designed to be performed in teams, with the optional team size being five members.
Steps Required to Complete the Term Project
1. Develop procedures for a productivity methods improvement study using the methods and tools presented in this textbook. Briefly document in writing the analysis processes used in the study. 2. Conduct the investigation and collect data over a specified period using the forms and data-collection methods from Chapters 5 and 6, videos, digital photos, samples, questionnaires, and other appropriate procedures. Keep a log of the activities performed to collect field data. Samples of productivity improvement studies are provided in Chapter 6. 3. Perform a thorough analysis of the data collected in Step 2. In addition to the basic analysis, compare the results and draw conclusions from the comparisons. 4. Develop recommendations for method improvements for the construction activity studied and determine the overall percentage that productivity could be improved if the recommendations were implemented. 5. Write a technical report that summarizes the study and that includes the following: a. An introduction b. A synopsis of the construction operations that were studied in the term project c. A brief discussion of the procedures and methods used to collect and analyze the data

continued

Table 1-2. *(continued)*

d. The results
  (1) Work sampling
  (2) Crew balance charts
  (3) Process charts
  (4) Activity sampling
  (5) Field ratings
  (6) 5-minute ratings
e. Analysis of the results
  (1) Productivity analysis
  (2) Video analysis
  (3) Crew balance analysis
  (4) Process flow analysis
f. Conclusions
g. Recommendations
h. References
i. Appendixes

Grading Criteria

Criteria	Percentage of Grade
1. Data development	30
2. Data analysis	30
3. Creativity/innovation	15
4. Documentation/report	15
5. Oral presentation	10
Total	100

References

Adrian, J. (2004). *Construction productivity: Measurement and improvement,* Stipes, Champaign, IL.

American Productivity Center. (1980). "A model for a low productivity trap." *Productivity perspectives,* Houston, TX.

Construction Industry Institute. (2006). *Work Force View of Construction Labor Productivity* (RR215-11), Austin, TX.

Construction Industry Institute. (2011a). *Productivity Benchmarking Summary Report* (BMM2011-1), Austin, TX.

Construction Industry Institute. (2011b). *Craft Productivity Research Program—Phase III* (RS252-1b), Austin, TX.

Construction Industry Institute. (2013a). *Leveraging Technology to Improve Construction Productivity* (RS240), Austin, TX.

Construction Industry Institute. (2013b). *Construction Productivity Research Program—Phase III* (RS252-11b), Austin, TX.

Construction Industry Institute. (2013c). *Construction Productivity Research Program—Phase V* (RS252-1d), Austin, TX.

Construction Industry Institute. (2013d). *The Construction Productivity Handbook* (IR252-2d), Austin, TX.

Halpin, D. (1990a). *MicroCYCLONE System Manual*, Purdue University, West Lafayette, IN <https://engineering.purdue.edu/CEM/People/Personal/Halpin/Sim/MicroCYCLONE/Index.html> (March 2014).

Halpin, D. (1990b). *WebCYCLONE User's Manual*, Purdue University, West Lafayette, IN <https://engineering.purdue.edu/CEM/People/Personal/Halpin/Sim/MicroCYCLONE/Index.html> (March 2014).

Halpin, D., and Woodhead, R. (1976). *Design of construction and process operations*, Wiley, New York.

Matteson, J., and Ivancevich, J., eds. (1977). *Management classics*, Goodyear, Santa Monica, CA.

Oglesby, C., and Parker, H. (1972). *Methods improvement for construction managers*, McGraw Hill, New York.

Oglesby, C., Parker, H., and Howell, G. (1989). *Productivity improvement in construction*, McGraw Hill, New York.

Sink, D. S. (1985). *Productivity management: Planning, measurement, evaluation, control, and improvement*, Wiley, New York.

U.S. Department of Commerce. (2009). *Rise and fall of nonfarm business output index (1996–2009)*, Bureau of Labor Statistics, Washington, DC.

U.S. Department of Commerce. (2012). *Quarterly trends of nonfarm business output index (1996–2012)*, Bureau of Labor Statistics, Washington, DC.

CHAPTER 2

Productivity Improvement Programs

Members of firms in the United States are not always able to rely on an expanding economy to help maintain or increase corporate earnings; therefore, they strive to implement work processes that increase productivity or lower expenses, or they attempt to increase the market share of their firm by capturing a portion of the market share of their competitors. One other method used to increase earnings is raising prices, but this alternative is inflationary; therefore, rather than raising prices, corporate executives prefer to implement other methods that increase output without having to increase input.

The U.S. **Federal Reserve Board** members attempt to encourage moderate economic growth and prevent inflation by periodically adjusting interest rates. Raising interest rates too high restricts economic growth and reduces the corporate earnings of firms, but artificially low interest rates might restrict foreign investment in the United States. Therefore, members of the Federal Reserve Board try to adjust interest rates to a level that prevents inflation but still allows the economy to expand at a moderate rate.

Productivity improvement programs help to achieve the goal of increasing the value added per hour of work, and management personnel are responsible for exploring alternatives that help achieve this goal. Implementing productivity improvement programs creates opportunities for increasing the success of firms and for improving the performance of employees. During the implementation of productivity improvement programs, areas where costs might be reduced are identified and the overall efficiency of the firm is measured by productivity improvement team members. One method for measuring productivity is to develop a graph of the input and the output of a firm that shows how these two variables increase or decrease during comparable periods. Output per labor hour is used as one indicator of productivity. If productivity is measured using different indices, there needs to be a method for measuring inputs. If labor hours are used for measuring productivity, the ratio of output per labor hour might not remain constant over time due to intervening variables that affect productivity.

This chapter discusses some of the techniques used to increase productivity, such as cost reducing innovations and increasing production efficiency. It also discusses some of the benefits of increasing production rates. Other topics covered in this chapter include the importance of being knowledgeable about the industry where the productivity study is being conducted, the evaluation of operations, the role of management personnel in the productivity improvement process, the strategic management approach, the purpose of productivity improvement programs, characteristics of successful productivity improvement approaches, **quality assurance, quality control, lean construction, prefabrication, and modularization.**

2.1 Components of the Productivity Improvement Process

When the profits of a firm are declining, members of management evaluate all of the corporate units to determine the cause of the decline and which individual units are contributing to the decline in profits. One of the major causes of declining profits is decreasing productivity rates caused by human and process inefficiencies. Productivity improvement programs are used by management personnel to analyze both worker and process productivity rates in order to discover why there are inefficiencies and to develop alternative work processes that help increase production rates. Individual work tasks performed during unproductive operations are analyzed to determine which work tasks or work processes are negatively affecting productivity rates.

Once the inefficient work tasks have been identified, productivity improvement work-sampling and process-analysis techniques are implemented to gather data that are used to calculate production rates and to analyze the efficiency of work crews and flow processes. Each individual work task is evaluated to determine the percentage of time that the workers performing the task are actively engaged in productive work. After every work task has been evaluated, the overall productivity rate is computed for the operation. If production rates for the operation are less than 60%, then an additional analysis is performed to gather information that is used to develop alternative work processes that might help to increase productivity.

The results of productivity improvement studies are summarized in productivity reports, and the reports are presented to management, along with recommendations for increasing production rates for each operation. Productivity reports include a synopsis of the analysis techniques used to collect data; a summary of the data collected; the results of the data analysis process; graphs and charts that help illustrate existing and proposed work processes, flow diagrams, and workflow charts; a comparison of the results to existing work processes; specific recommendations for improving productivity rates; and the percentage increase in productivity that might be achieved if the recommendations are implemented for the specific tasks. At this point, it is the responsibility of management personnel to determine whether they are willing or able to implement any of the recommendations in the productivity improvement reports.

If the recommendations are implemented, then productivity improvement team members assist management personnel in training the workers on how to perform the proposed work processes and in directing the workers to change existing workflow patterns. Productivity improvement team members monitor the implementation process to ensure that the workers are correctly applying the new work processes. Productivity improvement team members should also return to the jobsite to conduct periodic inspections to ensure that the workers have not reverted to using the original work processes.

2.2 Reducing Costs and Increasing Production Efficiency

This section discusses some of the methods used for increasing productivity rates, including reducing costs by using innovative technology, increasing production efficiency by implementing productivity programs, increasing knowledge of the industry, evaluating construction operations, and involving management personnel in the productivity improvement process.

Reducing Costs

Methods for implementing productivity improvement techniques include encouraging the application of cost-reducing innovations in technology within an organization and increasing actual efficiency relative to potential efficiency, as reflected in work-measurement data at the technological, production, or organizational level. To increase productivity rates, large firms might invest in research and development, whereas small firms might

- Investigate industry innovations, such as new computer software applications, machinery, equipment, tools, and materials;
- Provide training programs and courses to educate workers on how to implement new or alternative work processes;
- Provide opportunities for employees to participate in industry conferences to increase their knowledge; and
- Provide workers with literature that discusses the latest advances, trends, and innovations in the industry.

Increasing Production Efficiency

Productivity increases reflect internal and external economies of scale in the growth of manufacturing plants, construction firms, and the economy. Increasing specialization of personnel, equipment, and manufacturing plants might lead to improvements in productivity rates. Innovations, investments in equipment and upgrades, the reorganization of production, and technological changes all help to increase productivity rates. Other techniques for improving productivity include increasing productive output, decreasing input rates, reducing energy use, planning efficient inventory methods, and improving the quality of the workforce.

Inventory requirements might be reduced by using more efficient scheduling techniques and synchronization of purchases, production, and shipments. Capital investment might help increase productivity, especially if new products and equipment are more efficient or higher in quality than existing products or equipment because of advances in technology or innovative materials. Other methods for improving productivity include continual training, retraining, and specialized training of workers who operate new equipment and are using new materials or methods.

Learning curves illustrate that productivity is usually lower while workers are first learning to perform a work task and increases as they begin to master their work processes. As a result, unit costs increase when a worker is first learning a task, then decrease rapidly, and then level off. As workers master a work task, their production rates increase, and the time required to complete the task decreases until they reach their optimum production level. If firms are able to retain workers for long periods, then production rates are not affected by the decrease in productivity that occurs when new employees are learning how to perform their work tasks. Unfortunately, in the construction industry, even workers who have been with a firm a long time are required to perform a variety of different work tasks, and this leads to learning curve effects on production rates.

The structure of a firm should be periodically reviewed to determine whether changes should be implemented to increase productivity. Having workers involved in developing work-process alternatives might help to increase efficiency. Also, providing an environment that fosters innovation and encourages suggestions from workers might help to increase productivity rates. Creating a working atmosphere where the employees identify with the objectives of the firm, and are not merely working for competitive pay and benefits, might also help improve productivity rates. Some employees are motivated to perform more efficiently when they are provided with information about the objectives of productivity improvement studies.

One method used to motivate workers to perform at their optimum level of efficiency is providing bonuses to workers when a project is completed on time and within budget. Bonuses might also be used as an incentive for workers to meet intermediate project milestones. Another method to motivate workers used throughout the world is creating production teams that compete to achieve the highest productivity rate. The most productive team is rewarded with gifts or monetary bonuses. In some cultures, the teams are formed based on ethnicity, tribal affiliations, nationality, or regional affiliations. In the United States, teams might be formed based on work experience or by the hometowns of workers.

Knowledge of the Industry

To develop an understanding of an industry, it is useful to identify key individuals and stakeholders who are involved in the industry, along with any existing procedures that should be evaluated and compared with future procedures. Existing procedures that should be reviewed include current work processes, transportation schemes, material ordering processes, material fabrication processes, and methods used to design crew configurations. Existing planning processes and quality-control procedures are evaluated by management personnel to determine whether they should be altered or improved by incorporating new techniques. Identifying the firms within an industry that have high production rates is useful for determining how they achieve those rates. It is also useful to identify firms with low production rates caused by the improper use of technology in order to avoid repeating their mistakes.

Before implementing a productivity improvement program, the productivity improvement team members should identify any processes that are unique to the industry and their related impacts; determine the problems, goals, and interests of management personnel; become familiar with the perspectives of the key players; and observe the roles played by different organizations and supporting organizations, both inside and outside the industry. Chapter 3 discusses some of the key elements of the construction industry.

Evaluating Construction Operations

During the quantitative analysis stage of a productivity improvement program, the productivity improvement team members review the job layout, equipment operations, work processes, transportation schemes, and resource management systems in order to highlight any inefficiencies in existing procedures. Safety procedures are also reviewed to determine whether there is any way to improve or enhance them with the addition of other safety measures. Work processes are evaluated to determine whether alternative work processes are viable, and if they are, suggestions for improvements are developed and implemented to enhance operations. The quantitative analysis also includes performing work-sampling studies and developing recommendations for improvements in work processes.

Involving Management Personnel in the Productivity Improvement Process

Productivity improvement team members should include management personnel in the process of developing productivity improvement alternatives because management-level personnel are responsible for determining whether a recommended alternative is viable. Productivity improvement programs involve risks, especially if they are implemented incorrectly or without proper supervision. Any time work processes are altered, inherent risks arise if workers are not familiar with their work tasks or if their work locations and equipment are moved to other locations. To reduce risks, management personnel should provide detailed instructions to workers and observe the implementation of any new or alternative work process for at least one work cycle.

Productivity improvement team members prepare reports and provide presentations to management personnel that document the productivity improvement process and illustrate potential improvements in quality, performance, safety, productivity, and profitability through the implementation of alternative procedures and work processes. After management personnel have decided which alternative work processes are appropriate, the preferred alternatives are implemented in a controlled environment with support from those involved in the process, and then adjustments are made based on the feedback obtained from the workers. It is also important for members of management to require that the entire evaluation process, and the lessons learned, be documented at the completion of the productivity improvement study.

The support and involvement of management personnel in the development and implementation of productivity improvement programs is important because management personnel control resources and set priorities for their firms. Management personnel are concerned with profitability, long-term planning, and setting performance standards that help their firm remain competitive, and they have to consider these while working within the fiscal-planning cycles of their firm. Decision makers and those responsible for formulating strategies should be involved in implementing productivity improvement procedures and processes, and managers should be trained to advise subordinates when they become disillusioned, disappointed, or frustrated over unattainable expectations.

2.3 Benefits of Increasing Production Rates

Some of the benefits of increasing production rates are discussed in this section to demonstrate why firms should consider implementing productivity improvement programs.

The primary benefits of increasing production rates are increased output and higher profits without increases in inputs such as money, labor, materials, or other resources. In an expanding economy, these types of inputs are available, and it is possible to increase inputs to maximize output. But in a shrinking economy, inputs are restricted, and members of firms have to improve production rates without increasing inputs. Cost-reducing innovations, such as those mentioned in previous sections, might not be financially viable in a restricted economy, and this forces management personnel to evaluate internal work processes to determine whether there are options for increasing production rates. When additional funds, resources, or innovative equipment, tools, and materials are not available, the only remaining input that might be changed is labor. In this scenario, the only means for increasing production rates is improving the productivity of workers.

When worker productivity rates improve, fewer workers may be needed to handle the same number of job responsibilities, and this benefits the firm that employs the workers by reducing its payroll costs. The implementation of new technology might reduce the initial skill-level requirements for workers, and the new, less-skilled workers may be paid lower wages. Higher-paid individuals might be dismissed if lower-paid workers are able to produce at the same rate, and this also lowers payroll costs. On union projects, hiring and dismissing workers on the basis of pay rates may not be feasible because unions set wages and determine who works on projects.

Group motivation influences productivity, and workers as a group often set output restrictions for certain piece-rate jobs without informing management personnel. **Rate busters**, or individuals who produce at rates that are higher than the other members of their work group, are often subject to social sanctions by other group members. Workers might develop output restrictions on easy jobs to avoid management personnel from having higher output expectations. Another pattern is **goldbricking**, which involves workers putting forth the least possible effort on jobs that do not offer bonuses so that it might lead to overtime work and extra pay. If management personnel notice output restrictions or goldbricking, they should intervene and provide incentives for the workers to increase their production rates.

In addition to increasing output and profits, implementing productivity improvement programs might produce the following benefits:

- Improved site layout, which reduces time requirements for transporting materials and workers;
- Less worker interference with the work of others;
- Increased use of equipment, which might reduce labor costs;
- More efficient crew sizes, which reduces labor costs;
- Using less equipment, which reduces equipment costs;
- Timely completion of projects, which prevents having to pay liquidated damages or additional interest on borrowed funds;
- Enhanced awareness of the specific job responsibilities of workers; and
- Improved communication between management and the workers.

2.4 Strategic Management Approach

For productivity improvement programs to contribute to the attainment of organizational goals, they should be supported both intellectually and financially by upper management and be integrated into existing project-planning efforts. Improving the productivity of project-level tasks is part of productivity improvement programs, but it is also part of the corporate-level planning process, both of which require input and feedback from management and the workers.

Successful integration of a productivity improvement program in a firm requires that the program be compatible with the objectives of the **corporate strategic plans** of the firm. Upper-level management personnel develop strategic plans, and then decisions on how to implement them are made at lower levels in the organization. During the implementation of strategic plans, feedback on the effectiveness of the strategic plans is provided to upper management. **Strategic management** involves decision making and implementing strategies that help achieve the objectives of the firm. Productivity improvement programs are part of strategic plans, and their objective is to increase output or production rates (Pierce and Robinson 1985). The strategic planning process includes (1) a **planning development process**, which incorporates various considerations, followed by (2) a **decision-making process**, which supports the objectives of the organization, and (3) an implementation process, which has sufficient controls and feedback to allow for adjustments to changing conditions (Hofer and Schendel 1978).

If members of firms are able to follow strategic plans for several years, it enhances planning processes and creates standards and data that are used to continually improve the planning process. Before management decisions are made, current conditions are assessed and analyzed, and alternatives are proposed for improving existing processes. The planning process should incorporate the values of firms and include **socially responsible considerations**, that is, the concerns of society and impacts on the environment. These considerations should be evaluated to ensure that the decisions made by members of firms do not conflict with its responsibility to society. Having a structured process helps facilitate decision making and the rapid implementation of plans, and providing feedback regarding previous stages is an integral part of the process. Documenting the process helps to integrate the implementation of improvements into future planning efforts.

2.5 Purpose of Productivity Improvement Programs

The purpose of productivity improvement programs is to help solve problems that are related to planning work processes; organizing activities; controlling, scheduling, and financing projects; controlling quality; and managing the workers. The skills and knowledge of individuals within an organization are collectively used to develop productivity improvement programs, and the cooperation of all of the organizational units is secured to ensure the success of programs. Productivity improvement programs require the coordination of various types of resources and the individuals who know how to effectively manage resources. Productivity improvement programs should provide feedback to existing units, include methods for disseminating information, and include processes for implementing corrective measures when necessary. Productivity improvement programs should be sensitive to worker limitations and the nuances of human behavior.

In addition, productivity improvement team members should be cognizant of outside forces that influence productivity rates, such as environmental influences including weather conditions. A productivity improvement program should also generate data that are available for review during the planning stage of other projects. Productivity improvement programs are used to validate existing productivity rates and to explore potential improvements to work processes and work-process flow. Productivity improvement programs provide a method for quantifying productivity, assessing work processes, developing alternative work processes, and developing recommendations for increasing productivity rates.

2.6 Characteristics of Successful Productivity Improvement Approaches

When starting to implement a productivity improvement program, the productivity improvement program coordinator should first outline the intent and scope of the project, identify the resources that will be consumed, and provide a schedule. Next, the coordinator should compile a list of proposed improvement processes and their associated cost estimates. Some of the program characteristics that lead to successful productivity improvement studies are using techniques that focus on assessing current productivity and providing methods for improving productivity that also evaluate safety measures. Productivity improvement studies analyze planned, repetitive work

Table 2-1. Percentage of Time Spent on Nonproductive Work

Nonproductive Activities	Percentage of Time
Waiting for resources	14
Waiting for work assignments	7
Double-handling materials	6
Changes, disputes, and claims	5
Late or inaccurate reports	5
Waste and theft	3
Accidents	3
Punch-list items	3
Rework	2
Substance abuse	2
Total	50

Source: Adapted from Adrian (2004), Figure 10.7.

processes to eliminate nonproductive work. Using 50% as the average percentage of time that construction workers are productive, James Adrian (2004) developed a list of nonproductive work activities and respective percentages; these are listed in Table 2-1. Successful productivity improvement programs evaluate the types of work activities listed in Table 2-1 and develop methods for eliminating or reducing nonproductive activities.

Another element that helps when implementing productivity improvement programs is providing an atmosphere where the workers and the foreman are included in the evaluation process and their input is considered by productivity improvement team members. The workers and foreman are valuable resources that help determine why nonproductive activities continue to occur during work operations. The workers or the foremen might be aware of issues delaying the work but that are not observable.

Another important component of successful programs is the ability to record work operations. Gathering data using human observations is useful, but due to the fast pace of construction operations, it is not always possible to record what every worker is doing at a particular point in time. Having visual recordings allows observers to view operations numerous times, thus increasing the accuracy of the observations. Methods for recording work tasks are discussed in Chapter 5.

A crucial element of productivity programs is the analytical skill of the productivity improvement team members. Advanced analytical skill is necessary to develop alternative work processes and workflow patterns that reduce the time required to complete construction operations.

2.7 Quality Assurance and Quality Control

In addition to increasing productivity at construction jobsites, construction management personnel are also concerned with maintaining high levels of quality. In many instances, if productivity rates increase, quality declines because the workers are focusing more on working faster than on achieving high-quality output. One of the more difficult aspects of managing construction is achieving a balance between higher productivity rates and maintaining quality. According to Adrian (2004, p. 249), quality is the "freedom from errors or deficiencies, conformance or lack of variation, and adding value to a process through the commitment to improvements." At construction jobsites, personnel should be concerned with both quality assurance and quality control (QA/QC).

Quality Assurance

One definition for quality assurance is "all those planned and systematic actions necessary to provide adequate confidence that a product or service will satisfy the given requirements for quality. Quality assurance includes ensuring that the project requirements are developed to meet the needs of all relevant agencies, planning the processes needed to ensure quality of the project, ensuring that equipment and staffing are capable of performing tasks related to project quality, ensuring that contractors are capable of meeting and carrying out quality requirements, and documenting the quality efforts" (Tudor-Saliba/Slattery Joint Venture 1998, p. 8).

Quality Control

One definition for quality control is "the operational techniques and activities that are used to fulfill requirements for quality. Generally, quality control refers to the act of taking measurements, testing, and inspecting a process or product to ensure that it meets specification requirements. It also includes actions by those performing the work to control the quality of the work. Products may be design drawings or specifications, manufactured equipment, or construction items. Quality control also refers to the process of documenting such actions" (Tudor-Saliba/Slattery Joint Venture 1998, Vol. 1, p. 8).

Quality Management Systems

Quality management systems are defined as "that aspect of the overall management function that determines and implements the quality policy. The organizational structure, responsibilities, procedures, processes, and resources for implementing the quality policy" (Tudor-Saliba/Slattery Joint Venture 1998, Vol. 1, p. 8).

Large construction projects normally have **quality-program plans**, **design quality-control plans**, **construction quality-control plans**, **quality-assurance plans**, **vendor quality plans**, and **subcontractor quality plans**. A quality plan is "a written description of intended actions to assure or control quality. Quality plans include the quality assurance plan, the design quality control plan, and the construction quality plan as elements of the overall quality plan" (Tudor-Saliba/Slattery Joint Venture 1998, Vol. 1, p. 8).

To monitor quality on construction projects, members of either the project management team or a quality management team, if there is one on a project, are responsible for planning and implementing scheduled and unscheduled **quality audits** to verify compliance with the requirements of the quality management system. Quality managers are responsible for establishing and maintaining quality management systems and for ensuring that corrective actions are implemented to prevent system shortcomings from occurring again.

Quality managers provide the documentation used to control quality throughout project phases, including design, development, purchasing, construction, system testing, and maintenance. The documents that have to be provided include process instructions, quality control instructions, product specifications, design drawings, and bills of materials. Quality-program plans and supporting procedures and instructions are part of the quality management system, and they are used as the primary basis for document control.

Construction quality-control personnel are required to ensure that detailed inspection instructions and procedures are provided when necessary and to assist personnel in conducting inspections. Quality management personnel test items to determine which do not conform to the required quality standards. They have to properly identify the items, segregate them, report them, determine corrective actions, and then dispose of the nonconforming items according to the procedures outlined in the quality plan. Construction quality personnel are responsible for ensuring that materials are delivered in the proper manner, identified, protected, stored, and issued in accordance with prescribed routines. They are also responsible for conducting regular inspections of equipment and materials to determine whether the materials and equipment are being maintained in a satisfactory condition.

Quality records include project drawings, specifications, change orders, approved shop drawings, inspection reports, laboratory test records, field-test reports, as-built drawings, QA/QC audit reports, and monthly quality reports. All quality-control personnel are responsible for maintaining quality records and for recording quality test results in a timely manner. Testing equipment has to be calibrated, and the calibrations are recorded in calibration logs to ensure that test readings are verified as being accurate.

Some organizations use statistical process control to maintain levels of quality. Statistical methods are used to decrease the incidence of unexpected failure to six standard deviations on the normal distribution, resulting in a probability of 3.4 one-millionth. In manufacturing, quality is maintained by using random-sample testing for a fraction of the output. In engineering, **failure tests** (**stress tests**) are used to test some products until they fail under stress. Products are tested using vibrations, compression, or excessive temperatures and humidity. The results of failure tests are used to improve engineering designs. In addition to failure tests, nondestructive testing methods are also used on structural elements.

2.8 Lean Construction

Another process implemented on some construction projects that affects productivity is lean construction. Lean construction was adapted from **lean production**, a process that "primarily focuses on the reduction of the time from order of any transaction be it assembly, billing, supply, etc., to delivery. This time reduction is achieved by the elimination of waste (the unproductive use of resources) that is captured in 'downtime'" (Howell 1999, p. 6). Lean production focuses on "designing a production system that will deliver a custom product instantly on order but maintain no intermediate inventories. Lean production concepts include:"

- Identify and deliver value to the customer: eliminate anything that does not add value.
- Organize production as a continuous flow.
- Perfect the product and create reliable flow through stopping the line, pulling inventory, and distributing information and decision making.
- Pursue perfection: deliver on order a product meeting customer requirements with nothing in inventory. (Howell 1999, p. 3–4)

Lean construction:

- Has a clear set of objectives for the delivery process.
- Is aimed at maximizing performance for the customer at the project level.
- Designs concurrently products and processes.
- Applies production control throughout the life of the project. (Howell 1999, p. 4)

Howell (1999, p. 7–8) continues, "Lean construction embraces uncertainty in supply and use rates as the first great opportunity and employs production planning to make the release of work to the next crew more predictable, and then working within the crews to understand the causes of variations.... Lean construction results from the application of a new form of production management to construction. Essential features of lean construction include a clear set of objectives for the delivery process aimed at maximizing performance for the customer at the project level, concurrent design of products and processes, and the application of production control throughout the life of the product from design to delivery."

2.9 Prefabrication and Modularization

Prefabrication and modularization have been in use for over a century, but their use escalated during and immediately after World War II, when governments in both Europe and Japan needed to quickly rebuild their countries. Quonset huts are a prime example of how prefabrication and modularization were used during World War II to rapidly build structures to house soldiers, for offices and equipment storage, and for other temporary structures. Some of these structures survived for decades and were repurposed for uses such as married student housing on college campuses, low-income housing projects, and manufacturing facilities.

In the early part of the 21st century, the use of prefabrication and modularization increased dramatically as a result of technological advances. The four main technological advances that led to an increase in the use of prefabrication and modularization are (1) **building information modeling** (BIM) software programs, (2) innovative materials, (3) sophisticated manufacturing facilities, and (4) computer-controlled fabrication equipment (Haas and Fagerlund 2002; McGraw Hill Construction 2011).

Firms are increasingly using prefabrication and modularization on construction projects for a variety of reasons, including being required by owners, time savings, lower cost, compensating for an unskilled workforce or a shortage of workers, and to help decrease site waste. In addition, prework also affects safety at construction jobsites. According to Haas and Fagerlund: "Prework may reduce exposure to weather, heights, hazardous operations and neighboring construction activities. Workers indoors at a fabrication shop are not affected as much by temperature, wind, and precipitation extremes. Since most of the prework is done at grade level, fewer safety harnesses are required and workers are able to focus on the work. Less workers onsite also translates into reduced craft congestion and exposure to ongoing operations" (Haas and Fagerlund 2002, p. 10).

In addition to increases in competitiveness and returns on investments, firms are implementing prefabrication and modularization because of shortages of skilled construction workers; low purchase and installation costs; and reducing the need for on-site resources, including labor and equipment such as scaffolding, staging areas, and material-storage areas. Fabricating construction elements in off-site facilities allow workers to perform their tasks in a climate-controlled environment and promotes consistent quality monitoring; plus, the same workers are able to manufacture the elements each time. Other benefits include reducing jobsite waste by at least 5%, preventing natural habitat destruction, and protecting materials from water damage caused by bad weather (McGraw Hill Construction 2011).

Building information modeling software programs allow firms to provide building product manufacturers with three-dimensional (3D) models that are used to prefabricate building elements in their manufacturing facilities. In addition, modular elements are designed to be integrated into structures using BIM software. Building information modeling software allows designers to test different design scenarios including combinations of build-in-place and modular elements (McGraw Hill Construction 2011). These software programs are discussed in Chapter 9.

Prefabrication and modularization are used most frequently in superstructures; mechanical, electrical, and plumbing systems; and exterior walls. Being able to assemble building elements in a controlled environment helps to improve the quality of the elements. The prefabrication of ductwork helps to keep the ductwork cleaner, which is important when it is going to be used in structures that include clean rooms or medical facilities (McGraw Hill Construction 2011).

One of the major advantages of prefabrication and modularization is the increase in productivity at construction jobsites. Before they decide to include prefabricated or modular elements, members of firms review the design and the jobsite to determine whether using prefabrication or modular elements is possible. The three main factors that influence the use of prefabrication and modularization are "1) job site accessibility, 2) the number of building stories, and 3) the type of building exterior" (McGraw Hill Construction 2011, p. 5).

Once it has been determined that prefabrication and modularization will be used on a project, the schedule is reviewed to evaluate how the integration of prefabrication and modularization will affect the schedule. Because the elements will be prefabricated off site, the workers are able to start building them earlier than if they were being built on site. This normally results in a reduction in the schedule of four weeks or more, depending on the size of the project. In addition to evaluating the schedule, members of the firm will review the budget to determine how the use of prefabrication and modularization will affect cost. In many instances, prefabrication and modularization are able to reduce project budgets by more than 6% (McGraw Hill Construction 2011).

To prefabricate elements, the engineering design has to be completed before the fabrication of the assemblies, and this may require earlier engineering. One engineering concern is whether the fabricated assemblies will be integrated seamlessly into the nonprefabricated assemblies or other prefabricated assemblies (Haas and Fagerlund 2002). Other concerns are related to the size of the prefabricated assemblies. Assemblies may be difficult to transport if they are too large to fit on tractor trailers or if no cranes with the capacity to lift the assemblies on and off trucks are available. According to Haas and Fagerlund (2002, p. 11), "Transportation logistics play a large role in determining prework feasibility. Size and weight limitations, route restrictions, permitting, and the availability of lifting equipment are among the considerations to be made for the coordination of construction."

If all of these considerations lead to a determination that prefabrication or modularization should be used on a project, then the effects on labor productivity should be evaluated for the project. Haas and Fagerlund indicate that (2002, p. 71), "Labor affects both cost and schedule through differences in wages and productivity. Factory conditions associated with prework often provide increased efficiencies due to higher optimization of equipment usage and workplace layout. Shop overhead cranes, welding stations, and fabrication jigs could be utilized for increased production capabilities. Controlled conditions such as indoor work, work at grade level, and level workspaces could also provide potential for increased productivity. Comparisons should be based on all-in wage rates that include bare wages, equipment costs, and productivity factors."

Labor productivity increases with the use of prefabrication and modularization for many different reasons: the work is accomplished at ground level, the workers are able to perform their work in a weather-controlled and well-lighted environment, the skill level of the workers is higher because they engage in repetitive tasks, there are fewer workers in the work space, and there is additional access to automated equipment (Haas and Fagerlund 2002).

Haas and Fagerlund (2002, p. 34) add the following:

> "Careful selection of prework is part of a spectrum and prework complexity determines decision timing. Prework generally requires:

- Earlier decision making
- Integrated involvement of project participants
- Detailed analysis of labor differentials
- Overcoming lack of industry knowledge
- Detailed transportation planning and expediting
- Thorough shop testing and verification
- Careful supply chain management."

2.10 Summary

This chapter provided information on the techniques that are used to increase productivity, such as cost-reducing innovations. The benefits of implementing productivity improvement programs were explained in terms of how they reduce costs or increase production rates. This chapter also covered issues related to implementing productivity programs, the importance of knowing about the industry where the productivity program is being implemented, the evaluation of construction operations, management involvement in the productivity improvement process, the strategic management approach, the characteristics of successful productivity improvement approaches, quality assurance, quality control, lean construction, and prefabrication and modularization.

The next chapter, Chapter 3, discusses the engineering and construction (E&C) industry and its unique environment, along with current trends in productivity rates for the different sectors of the industry. It also includes information on project development, project contract administration models, and project life-cycle considerations that helps to demonstrate why productivity improvement programs are designed differently for firms in the E&C industry.

2.11 Key Terms

building information modeling
construction quality-control plans
corporate strategic plans
decision-making process
design quality-control plans
failure tests
Federal Reserve Board
goldbricking
lean construction
lean production
planning development process
quality assurance
quality-assurance plans
quality audit
quality control
quality management systems
quality-program plans
rate busters
socially responsible considerations
strategic management
stress tests
subcontractor quality plans
vendor quality plans

2.12 Exercises

2.1 How would applying lean-construction techniques help to improve productivity rates during construction projects?

2.2 What is strategic management, and how is it incorporated into firms?

2.3 Explain the difference between quality assurance and quality control.

2.4 The term *rate buster* is applied to workers who produce at levels that are much higher than other workers. How might a firm prevent workers from sabotaging rate busters so that they produce at the same rate as the other workers?

2.5 What are the techniques that might help increase production efficiency?

2.6 What are the three strategic planning processes?

2.7 Of the techniques listed in Section 2.3 for improving productivity rates, which might be implemented without any additional cost to a firm?

2.8 What are the benefits of higher productivity rates?

2.9 How do the actions of the Federal Reserve Board affect economic growth and interest rates?

2.10 Why is it important to understand the competitive nature of an industry before implementing productivity improvement techniques in a firm in the industry?

2.11 What types of initiatives lead to successful productivity improvement programs?

2.12 What should be included in productivity improvement reports that are submitted to management personnel to help communicate the results of a productivity improvement study?

2.13 Which of the cost-reducing innovations are the least expensive to implement, and why are they the least expensive?

2.14 Discuss why it is important to involve management personnel in all of the stages of a productivity improvement study.

2.15 What is the average percentage of time that construction workers are productive according to James Adrian (2004)?

References

Adrian, J. (2004). *Construction productivity measurement and improvement,* Stipes, Champaign, IL.

Haas, C., and Fagerlund, W. (2002). *Preliminary research on fabrication, pre-assembly, modularization, and off-site fabrication in construction,* Construction Industry Institute, Austin, TX.

Hofer, C., and Schendel, D. (1978). *Strategic formulation of analytical concepts,* West, San Francisco.

Howell, G. (1999). "What is lean construction—1999." *Proc., Seventh Annual Conf. of the International Group for Lean Construction,* Univ. of California–Berkeley, Berkeley.

McGraw Hill Construction. (2011). "Prefabrication and modularization." *SmartMarket Report,* New York.

Pierce, J., and Robinson, R. (1985). *Strategic management,* Richard D. Irwin, Homewood, IL.

Tudor-Saliba/Slattery Joint Venture. (1998). *San Francisco Airport extension project—Contract 12YC-120,* Vol. 1, Bay Area Rapid Transit District, San Francisco.

CHAPTER 3

Elements of the Construction Industry

During the past half century, a variety of techniques were implemented to increase productivity in the construction industry. Some methods were successful at improving productivity, and others were implemented because of government legislation requiring the protection of construction workers. In the construction industry, it is difficult to definitively establish measures of worker performance because numerous intervening variables affect the output of workers. The U.S. Bureau of Labor Statistics provides data on productivity rates and worker output, but it does not provide data that are specific to the construction industry. Construction management personnel would benefit if they had access to government statistics directly related to the construction industry. Having access to industry-specific data would provide insight to construction managers who are trying to develop alternative methods for analyzing labor productivity.

It is important for productivity improvement team members to understand the characteristics of the construction industry and to be familiar with (1) the responsibilities of the parties involved in construction projects, (2) how projects are organized, and (3) how internal and external communication flows through companies.

This chapter describes some elements of the construction industry and the contributions of each member of a construction team to the completion of a project. Construction management personnel attempt to identify which work processes benefit from productivity improvement studies; therefore, it is helpful for them to be familiar with how the construction industry operates, how projects are developed, how they are managed, and who the competition is, all of which are covered in this chapter. This chapter also discusses current trends and issues that affect productivity improvement in the construction industry.

3.1 Construction Industry Statistics

The construction industry contributes approximately 6.2% to the $14 trillion economy of the United States, and it employs over five and a half million workers (U.S. Department of Commerce 2012a). Table 3-1 provides information on the growth of the

Table 3-1. U.S. Construction Industry Statistics for 1996, 2005, and 2012

	1996	2005	October 2012
Employment	4.5 million	6.37 million	5.53 million
Expenditures	$588 billion	$1.2 trillion	$872 billion
Percentage of GDP	6%	10%	6.2%
GDP	$9.2 trillion	$12 trillion	$14 trillion

Source: U.S. Department of Commerce (2012a).

Table 3-2. Annual Volume of Completed Construction (Seasonally Adjusted)

Type of Construction	October 2012 (millions of dollars)
Total construction	872,142
Total private construction	592,091
Residential	294,237
Nonresidential	297,853
Public	280,051
State and local	253,895
Federal	26,156

Source: Adapted from U.S. Department of Commerce (2012b).

construction industry between 1996 and 2012, and Table 3-2 shows the dollar volume of completed construction for October 2012.

3.2 The Construction Industry versus the Manufacturing Industry

To demonstrate why productivity improvement programs are implemented differently in the construction industry than in the manufacturing industry, this section discusses some of the unique aspects of the two industries.

Construction Industry

Manufacturing plants are either domestic or foreign, but construction has to be performed wherever a project is required by an owner (although materials might be prefabricated off site or manufactured in foreign countries). Construction projects are completed rapidly, in a matter of months or a few years, and each is built in a different location. Project designs are unique, but similar plans might be adapted for use at different jobsites. Construction jobsites are sometimes located in congested or remote locations, which might restrict access or create transportation challenges.

Unique aspects of construction projects include the numerous safety risks of conducting simultaneous activities where workers might accidentally interfere with other workers. Construction projects are built using sequenced activities, and activity delays commonly occur if preceding activities are not completed on time. Construction projects also involve large or heavy assemblies that create risks when they are lifted into place.

Construction projects are managed using a hierarchy that includes workers, foremen, supervisors, construction managers (optional), engineers who represent owners, and owners. Adversarial contractual relationships sometimes develop on projects for a variety of reasons, including a high degree of owner involvement, disagreements about the original scope of work, claims for additional work, or improper directions.

Construction projects are designed and built to unique specifications, and all of the contract requirements have to be met before a project is considered complete and the retainage is released to contractors. Construction workers move from project to project and from one employer to another in short periods. Many construction operations are performed outside; therefore, workers are exposed to the environment, which might cause them discomfort and delay construction activities.

The main sectors of the construction industry are

- Building,
- Commercial,
- Heavy,
- Highway,
- Industrial,
- Institutional,
- Marine,
- Power, and
- Residential.

The amount of work that is available in each of the construction sectors is affected by

- Government legislation,
- Interest rates,
- Natural disasters,
- Changes in demand,
- Global economic fluctuations,
- Global competition,
- Increasing costs of materials,
- Shortages of materials,
- Shortages of qualified construction management personnel,
- Shortages of qualified construction workers, and
- Cycles of the industries supported by the construction industry.

Manufacturing Industry

In the manufacturing industry, workers and equipment remain in one location, and the materials and assemblies move past the workers on conveyor systems. Employees work in one location for long periods and possibly for their entire career. In manufacturing plants, due to space limitations, assembly processes occur within a limited area. Productivity improvements usually focus on sequencing and manufacturing processes rather than on worker locations or crew configurations. Products are also analyzed to determine whether they might be assembled in a different order or benefit from different parts. Weather conditions are not an issue in manufacturing plants, unless they are not air-conditioned or they are difficult to heat in subzero temperatures.

Because the manufacturing workforce is more stable than the construction workforce, implementing productivity improvement techniques provides longer term effects in manufacturing than in construction. Once manufacturing workers are trained on a new work process, they continue using that process for long periods. Management personnel are more willing to spend additional funds on training programs in manufacturing because the workers will be employed at plants for years or possibly decades. It is easier to evaluate process improvements when workers perform their work tasks while stationary, with little movement between machines or into different work areas.

During the 1990s, an experimental productivity improvement technique called **management by stress** was implemented in the manufacturing plants of one firm. This technique involved a series of alarms, installed above each worker, that went off if a worker became delinquent in performing his or her work task. This technique led to increased worker dissatisfaction and was discontinued shortly after its implementation in the industry. Another method that was implemented in an automobile manufacturing plant, where workers typically performed only one task per vehicle on the assembly line. To improve productivity, specific teams were created to work on the production of individual cars. When a vehicle was completed, the team members signed a photograph of the team and inserted it into the glove box. This technique installed additional pride and a sense of accomplishment in the team members.

3.3 Construction Productivity Improvement

There are numerous techniques for improving productivity in the construction industry. Some of the most common techniques are the following:

- Optimizing the use of heavy construction equipment,
- Optimizing heavy construction fleet configurations,
- Changing crew configurations to eliminate excess workers,
- Improving work processes,
- Improving workflow patterns,
- Retrofitting equipment and tools in innovative ways,
- Improving site layout,
- Improving equipment locations,
- Modifying existing equipment,

- Modifying material delivery schedules, and
- Revising material transportation patterns.

In addition to implementing these techniques, improving individual worker output produces significant gains in productivity rates.

In the construction industry, owners, managers, supervisors, and workers all contribute to the timely delivery of projects and to the profitability of firms. The difficulties associated with the coordination of labor, materials, and equipment; worker training; project meetings; and **procurement** issues are all part of the immediate, short-term priorities that consume the time and attention of construction managers. However, construction managers should also try to focus on the long-term benefits of improving productivity on projects. Having a construction manager focus on productivity improvement and the coordination of improvement activities might positively affect performance, teamwork, and the timely delivery of projects.

One requirement for implementing productivity improvement processes is gathering information about the particular segment of the industry in which a firm is operating, including

- Size,
- Scope of projects,
- Skill level of the workers,
- Organization structure used for projects,
- Contribution the projects make to society,
- Business practices,
- Work environments,
- Strengths and weaknesses,
- Management philosophies,
- Management priorities,
- Union or open shop,
- Traditional work practices, and
- Competitive practices.

Other information that is useful is documentation that describes previous procedures and attempts to improve productivity in the past. The construction industry requires that members of firms be flexible and able to quickly move resources to different locations. Productivity improvement techniques should be integrated into all aspects of the operations of construction firms and be practiced on a continual basis, not merely for short periods.

3.4 Construction Industry Project Team Members

This section describes the personnel who are directly involved in construction projects and some of their responsibilities when they are assigned to work on construction projects. The information in this section provides productivity improvement program team members with a basis for evaluating the roles of project team members during productivity improvement programs.

Owners

Owners are responsible for financing projects, planning projects, providing information to **designers** on project requirements, and operating and maintaining their facilities. During the design and construction phases, owners request modifications to projects that might lead to change orders. Owners may also request that value engineering be integrated into their projects. Value engineering allows designers or contractors to propose alternative designs or construction processes that would save owners money, and it allows designers and contractors to share in a percentage of any cost savings generated by the alternative schemes.

Designers

Designers are selected by owners through negotiations, not competitive bidding. The code of ethics of professional engineering societies prohibits engineers from competing for commissions on the basis of cost because it might compromise the quality of projects. Architects and engineers are responsible for converting design concepts into the plans and specifications used for the construction and operation of projects. In addition to designing projects, designers perform engineering analysis to verify that their designs conform to legal requirements and standards, such as building codes, and to ensure that their designs meet the required factors of safety for each particular type of structure.

Designers should also review their designs to evaluate whether incorporating productivity improvements into designs would help reduce the cost of construction. In addition to creating plans and specifications, designers have to effectively coordinate multiple disciplines, be a liaison to different government agencies, and act as a representative of the owner during construction. Designers work with owners, contractors, vendors, suppliers, lawyers, and employees of government agencies, and it is the responsibility of the owner or the designer to verify conformance to standards, codes, and the contract plans and specifications. Designers might also be responsible for verifying that quality-control procedures are properly implemented and that quality-assurance checks are completed in a timely manner.

Contractors

Contractors interpret contract documents, drawings, and specifications and convert them into physical facilities. Contractors monitor the progress of construction projects and provide leadership and management expertise during construction. Contractors develop bid estimates, create and monitor project schedules, assemble and coordinate the workforce, issue instructions, determine methods, develop and monitor construction operations, conduct jobsite meetings, process change orders, order and expedite materials, monitor project costs, and provide or rent equipment and tools. The main concern of contractors is profitability. If construction costs exceed productivity gains, profits decrease. Contractors work under time constraints, and their work is negatively affected by procurement delays, unskilled workers, shortages of workers, the jobsite environment, site-accessibility issues, equipment mismanagement, poor planning, and many other variables. Improving productivity at jobsites improves profitability. Productivity improvements that contractors are able to implement that might help to increase profits include

- Providing accurate estimates,
- Reducing risks,
- Improving methods and processes,
- Using innovative materials, and
- Using lower cost materials.

Labor Force

The labor force includes foremen and skilled craft workers who work on construction operations. Laborers either work individually or in teams, and they are responsible for transforming plans and specifications into completed facilities. Construction workers use methods that they have developed, or methods designed by engineers and construction management personnel, to perform their work tasks. Construction workers use plans and specifications, along with materials, equipment, and tools, to create projects. Construction workers who do not possess the proper skills and attitude or who are not adequately trained to perform their work tasks have a difficult time performing their work requirements. The output of workers is reduced if materials are not delivered to jobsites in a timely manner. For construction workers to perform their tasks, supervisors or construction managers have to provide them with leadership and training, approve the work as it is being completed, and order any necessary corrections to the work. Productivity at the worker level might be affected by the field support that workers receive from management.

Sureties

The role of **sureties** in construction is to provide bonds to contractors. Bid, performance, and payment bonds are required by law for government projects. Bid bonds provide a guarantee to owners that if the lowest responsible bidder is awarded the contract that contractor will accept the bid. Performance bonds guarantee that the contractor will perform all of the requirements of the project, and material bonds ensure that material suppliers and vendors will be paid if they are not paid by the contractor. Sureties are required to approve change orders because each change order alters the total contract price, and they guarantee only the original-scope-of-work bid price if change orders that increase the scope of work and the bid price have not been approved by the surety.

3.5 Construction Project Development

To initiate construction projects, clients first identify a particular need, and then **feasibility studies** are conducted, either by the owner or an architecture/engineering firm, to provide financial information that is used to determine whether the pro-

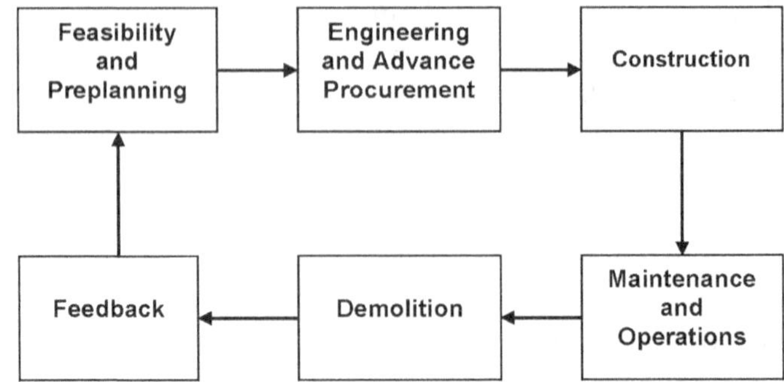

Figure 3-1. Project Development Phases

jected **rate of return** (ROR) meets the **minimum attractive rate of return** (MARR) for the firm. If the projected ROR exceeds the MARR, and the owner decides to proceed with the project, he or she will then use the feasibility study to try to obtain financing for the project and to solicit public or political support for projects.

Figure 3-1 shows the project-development cycle that occurs during the design and construction of projects. The following sections discuss each of the construction-project phases in detail.

Feasibility Studies and the Preplanning Phase

If an owner is having difficultly articulating what he or she desires in a structure, he or she may hire an architect or an engineer to define the initial scope of the project and to help perform a feasibility study. Feasibility studies require owners, or architects and engineers, to work together to evaluate potential project alternatives and project sites. The main thrust of feasibility studies is to determine whether a potential project will be profitable in terms of its ROR, which is the potential amount that an owner would make in profit, including the time value of money. Projections are developed for all of the revenue and income streams, and they are analyzed using either **present worth** (PW) or **equivalent uniform annual worth** (EUAW) analysis. After the engineering economic analysis equations are developed, the ROR is calculated as the interest rate at which the PW of the cost and income streams is equal to zero or the EUAW is equal to zero.

Owners normally compare the estimated ROR to the MARR for their firm. Each firm uses a different MARR, which might be based on the amount of money that an owner is able to obtain if he or she were to invest in other more secure investments, such as **certificates of deposit, money market accounts**, or savings accounts. If several potential projects meet the MARR, then they are each analyzed in terms of the parameters that affect potential projects that are listed in Table 3-3.

To effectively plan potential projects, the following activities are performed:

1. Developing the project scope,
2. Gathering supporting information and data,

Table 3-3. Project Evaluation Parameters

Project Evaluation Parameters
Accessibility to major arterials, such as freeways or feeder roads
Accessibility to mass transit systems
Accessibility to schools or shopping areas
Alignment feasibility studies
Availability of labor
Availability of resources
Availability of utilities
Contract implementation plan
Environmental impacts report/study
Location, including desirability of area
Major facilities sites
Potential business/resident relocations
Potential risks analysis
Project management plan
Real estate acquisition management plan
Relocation plans
Right-of-way availability
Traffic impacts on construction
Value of adjacent property

3. Analyzing the data,
4. Preparing a plan,
5. Disseminating the plan and appropriate information to all affected parties, and
6. Evaluating the results of the planning effort.

During the preliminary planning and feasibility study phase, environmental studies, site evaluations, and economic studies are conducted by engineers. Other documents that might be generated during this phase include initial transportation alignment studies, a facility-sizing report, a description of the project scope, aerial and preliminary surveys, and justification plans for the preferred alternatives.

Engineering Design Phase

The engineering design phase starts after funding is obtained for the initial design effort. All throughout this phase, owners promote their projects in order to secure the support of the local community. The financial plan is refined, as the design evolves during this phase. During the engineering phase, engineers develop the contract, the plans, the specifications, an engineering estimate, and other supporting project and management documentation. Engineers may also develop plans for construction methods and processes and a tentative project schedule.

Once the planning effort is sufficiently far along, the owner may either hire the architects or engineers who performed the feasibility study to develop the project, the contract, the specifications, the plans, and any supporting materials or select another firm to perform these tasks. While a project is being designed, other studies may have to be conducted, such as geotechnical analysis; a traffic study; vibration, noise, utility impacts, and utility relocation studies; additional surveys; and amendments to environmental studies.

Procurement Phase

Some materials required during construction may have a long lead time (that is, they take months or years to be manufactured or fabricated and delivered to the jobsite) and need to be ordered during the engineering design phase. Therefore, procurement starts during the engineering phase, and it continues throughout the construction phase as changes are initiated during construction.

While a project is being designed, the engineers working on the design need to keep the owner informed about any materials that have a long lead time. One example of a **long-lead-time item** is steel that requires months or years to be fabricated in the sizes and shapes requested for each project. Normally, the frame for a structure will be designed in the early design stage, and the other elements will be designed around the structural frame. Any changes made to the design of the steel frames once the steel has been ordered will be more expensive than changes incorporated before the steel is ordered from steel fabricators. Major steel elements, such as columns and beams or prefabricated sections containing columns and beams welded together at steel-fabrication plants, might require a year or more to reach jobsites once they have been ordered by engineers. Other long-lead-time items include major pieces of equipment, such as turbines, and equipment that will be built from unique specifications for a particular project. Turbines for nuclear power plants or hydroelectric power plants usually are ordered years in advance.

The procurement phase includes not only ordering materials but also **expediting** materials to make sure that they arrive at jobsites when they are required for activities. A procurement schedule is developed that indicates when materials are ordered, when they should be delivered, and when they are required for activities.

Construction Phase

The construction phase requires contractors to interpret the contract documents and to construct facilities according to the contract, plans, and specifications. In addition to constructing facilities, contractors perform field modifications and manage construction, including the subcontractors, using project control and quality assurance and quality control (QA/QC) procedures. Construction projects are constructed according to the schedules that are submitted with bid proposals developed by contractors, not owners or their representative. If owners included project schedules in the contract documents, then subcontractors would have claims against owners every time there was a delay on a project.

Testing, Operation, and Maintenance Phase

Before a facility is commissioned for operation, it goes through a series of tests to ensure that all of the systems are operating properly. The testing phase might require a few days for a simple project or a few years, as is the case with nuclear power plants. During the operation and maintenance phase, owners operate and maintain their facility, unless they have made prior arrangements for the contractor who built the structure or another firm to operate it.

Demolition Phase

Once a facility has served its purpose or has been operated for its useful life, it is demolished during the demolition phase, and another facility is constructed in its place. If the facility is not demolished, it may be repurposed and used as a different type of facility. Facilities are demolished using heavy construction equipment, wrecking balls, explosives, or other means. The materials from demolished structures have to be either recycled for use in other facilities or put into landfills. If any of the materials contain hazardous waste, they have to be disposed of in special waste-disposal sites that usually contain liners to prevent the leaching of the hazardous materials into the surrounding soil or water systems.

3.6 Project Contract Administration Models

Several different types of project contract administration models are used on construction projects, and the implementation of productivity improvement programs depends on which type of contractual agreement is being used on a project. Owners select different contractual relationships to provide checks and balances as they try to achieve their desired level of quality and timely completion of projects without exceeding project budgets and the scheduled completion time.

Typical contractual relationships sometimes result in **adversarial relationships** developing between the various parties that are involved in construction contracts. Adversarial relationships develop on projects because each of the major parties is working toward different objectives. Owners want the lowest cost project at the highest quality. Contractors want to complete projects quickly and make the most profit possible. Engineers try to ensure that projects are built according to the plans, specifications, and contract and that the structures will be safe for the occupants.

In addition to avoiding adversarial relationships, there are other reasons owners carefully select which type of contracting procedure to use on a construction project, and some of these include

- **Fast-tracking** to expedite the project construction phase (starting construction before the design is complete),
- Having assistance from members of a construction management firm in maintaining the schedule and monitoring project control processes,
- Obtaining representation for their interests,
- Outsourcing to obtain management expertise that is not available in-house,
- Reducing processing costs associated with design and construction, and
- Reducing the liability of owners.

The following sections discuss the different types of contractual relationships used for construction projects.

Single Prime Contracts

The traditional form of a contractual construction relationship is a **single prime contract**, in which the owner hires an architect or an engineer to design a project and then hires a general contractor to construct the project. This contractual relationship is shown in Figure 3-2. The contractual agreements in force in this contracting arrangement are between the owner and the architect/engineer and the owner and the general contractor. There is no contractual agreement between the architect/engineer and the general contractor. Although in a single prime contract the architect/engineer does not have a contractual relationship with the contractor, he or she may be called upon to administer the contract for an owner and act as an agent for an owner.

Separate Prime Contracts

If an owner has the expertise to manage a project, to perform the duties of a general contractor, and to coordinate **separate prime contracts**, then he or she might use a separate prime contractual agreement, which is the contractual relationship shown in Figure 3-3. When using separate prime contractors, the owner coordinates the work of the separate prime contractors rather than having a general contractor perform this function.

Construction Management Contracts

If an owner needs additional assistance in managing a project, he or she may hire a construction manager or a construction management firm to oversee construction. This contractual relationship is shown in Figure 3-4. A construction manager has a contract with an owner but not with the architect, the engineer, or the general con-

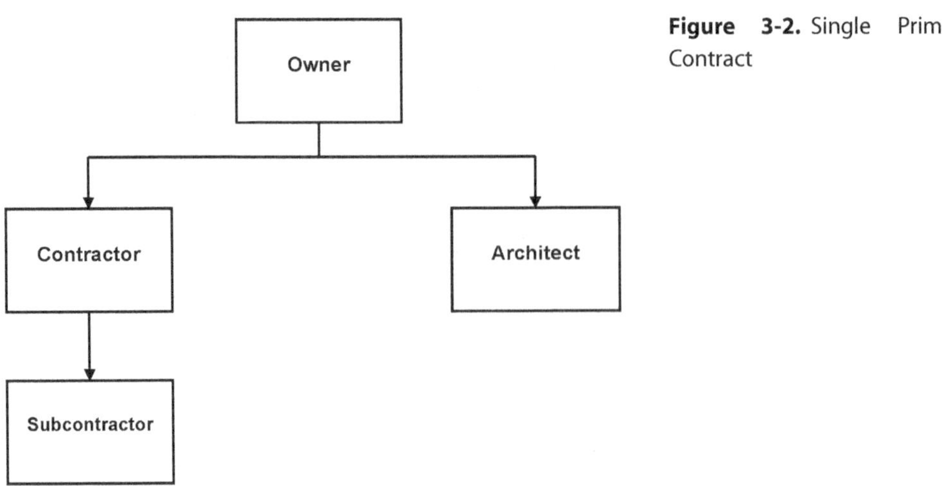

Figure 3-2. Single Prime Contract

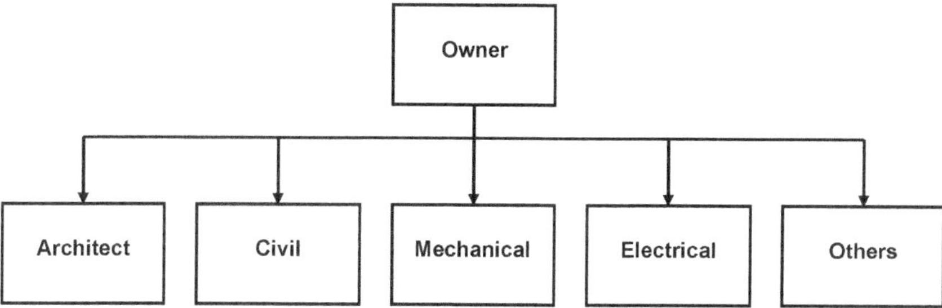

Figure 3-3. Separate Prime Contract

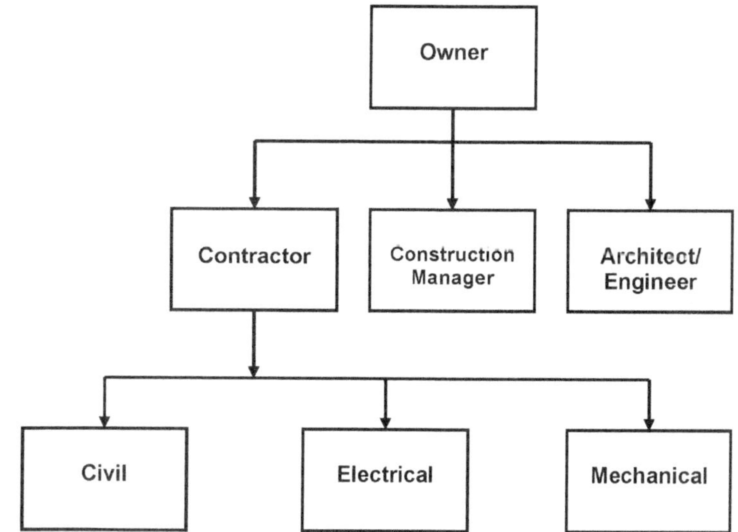

Figure 3-4. Construction Management Contract

tractor, which dilutes his or her authority with these other parties, unless the owner firmly emphasizes the authority of the construction manager. As shown in Figure 3-4, the owner has separate contractual agreements with the architect, the engineer, the general contractor, and the construction manager.

Design-Build Contracts

Another form of contractual relationship that places more responsibility and liability with one firm is the **design-build** contracting method, shown in Figure 3-5. In this contracting agreement, one firm both designs and builds a structure. This contracting method is advantageous because it reduces the cost and time of the project by not issuing contract documents for competitive bids and eliminating the evaluation of multiple bids, and it provides a single point of coordination for projects. The design-build process simplifies the process of transforming the contract documents into a finished structure because a single firm designs and builds the structure. During design-build projects, the engineer of record, who was in charge of the design process, is also available for consultations during construction.

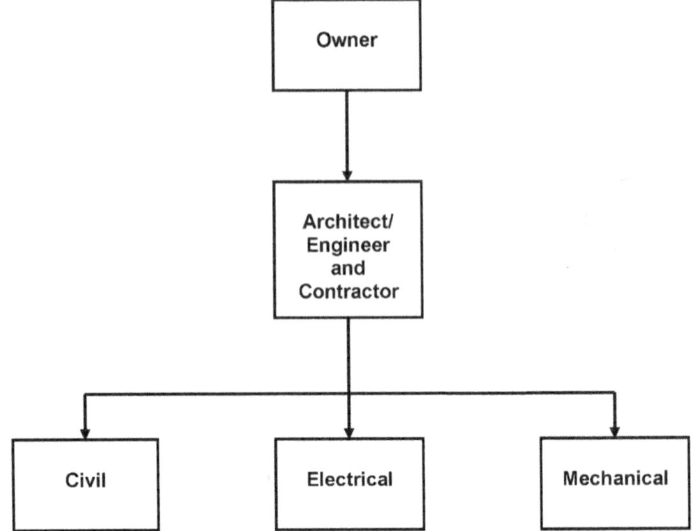

Figure 3-5. Design-Build Contract

The disadvantages of using the design-build approach are that there is no unbiased oversight because a separate contractor does not review the work of the design firm, an independent contractor does not conduct a second value-engineering review, and the owner has access to the expertise of the personnel from only one firm rather than two firms.

In the past, all government construction projects had to be competitively bid, but some states are embracing the design-build approach to construction and allowing it on some projects. For the design-build approach to be used for state government projects, state legislatures have to ratify existing requirements that require competitive bidding on all state construction projects.

3.7 Planning Construction Projects

Construction projects start when an owner decides that he or she needs a new facility or has a reason to remodel or refurbish an existing facility by, for example, modernizing the façade or interior; installing new technology; replacing outdated heating, ventilating, and air conditioning (HVAC) or electrical systems; reorganizing the layout of a structure for a new tenant; or preparing a structure to be sold.

Contractors become involved in construction projects either during the design process, if they are hired to perform a **constructability review**, or when they prepare a bid proposal. A constructability review occurs when a contractor reviews the plans, specifications, and contract to determine whether the design is feasible using existing technology and resources. Contractors also review the methods and processes that would have to be performed in order to build a project as designed by the engineers. One definition for constructability reviews is as follows (Oberlender 1993, p. 146):

> The purpose of this procedure is to provide a framework and process for an organized review of the contract documents by a multidisciplinary team with the objective of reducing conflicting design information to allow rational bidding and to decrease the extent of construction changes. An additional objective is to comment on the monetary efficiency of the design relative to market conditions (i.e., labor, material, and equipment availability) and the design's effect on construction operations. The objective is not a plan check or code compliance review. The procedure should not be construed as providing a warranty of documents free from defects.

When contractors prepare bid proposals that they will submit to owners, they are refining the project-development process by preparing different types of plans. The four types of plans that are prepared are construction, contingency, detailed, and foremen plans, and these are discussed in the following sections.

Construction Plans

Construction plans include an overall outline of how construction operations will be performed and the work performance sequence. This process requires that contractors develop **milestone schedules**, which indicate the sequential timing of activities and show when resources will be delivered to a jobsite. Owners might require contractors to provide **precedence diagrams**, such as **critical path method** (CPM) **schedules**, with their bid proposal. In addition to the required CPM diagrams, many contractors also use **bar charts** to schedule weekly activities or to provide foremen with simplified schedules that he or she is able to understand and implement on a daily basis. Figure 3-6 shows a sample of a CPM schedule, and Figure 3-7 provides a sample construction-planning bar chart.

Contingency Plans

Contractors also prepare **contingency plans**, which are implemented when something changes in the original plan. Contractors may have to implement contingency plans because of delays or acceleration, project scope changes, unforeseen site conditions, labor shortages, work stoppages, accidents, late delivery of critical materials, and other unanticipated conditions.

Detailed Plans

The third level of plans is detailed plans, and these plans are **execution plans** implemented at the task level. Detailed plans should include specific locations where an activity will be performed, the skill level required to perform an activity, and the materials, tools, and equipment required for each activity.

Foremen Plans

The last level in the planning process is **foremen plans**, which are developed to assign work tasks to specific workers and to track the work performance of the workers

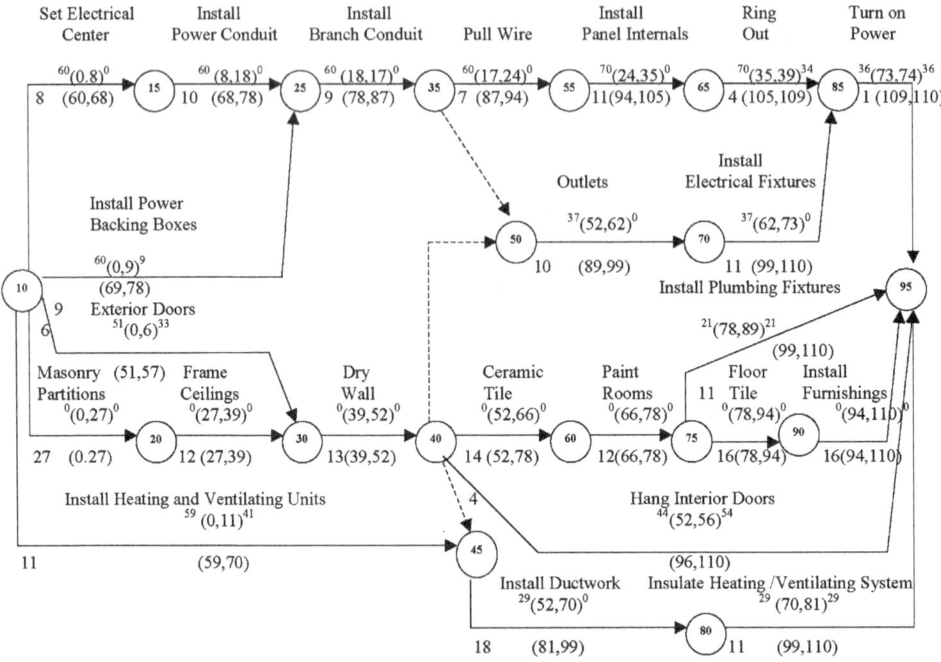

Figure 3-6. Sample Critical Path Method Schedule

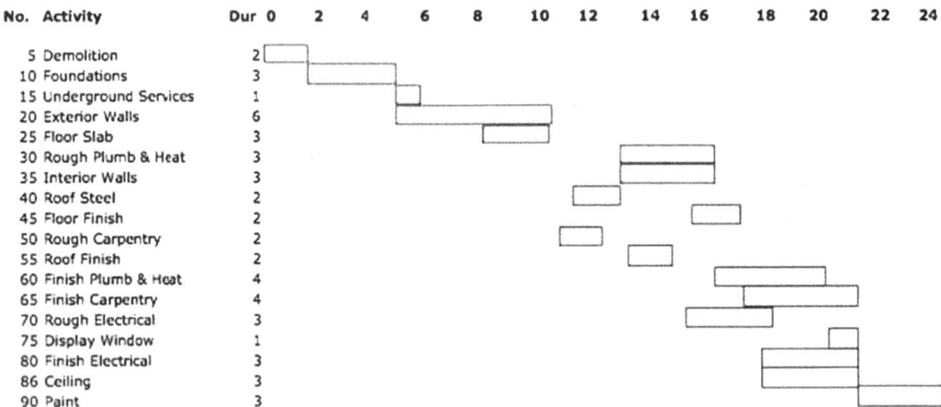

Figure 3-7. Sample Bar Chart

under the direct line of supervision of the foreman. Each foreman uses his or her own technique for tracking the progress of workers. The formal process for tracking worker performance is **time cards**. In this system, each worker charges his or her time to a specific activity, and then the time cards are summarized on a daily and weekly basis and incorporated into labor **cost-accounting systems**. Different methods are available for tracking time cards. Workers are able to manually enter their time and activity numbers, or bar codes are used to designate each activity and the amount of time expended on each activity. There are several project management computer software programs that also assist with this task and they are mentioned in Chapter 9.

Foremen use their experience to assign the appropriate workers to each activity. They might make periodic adjustments to worker assignments in order to expedite the work. If there is a shortage of a particular type of worker, then foremen may consult with the general contractor to determine where workers should be assigned to work. Normally, they will first be assigned to work on critical-path activities. If a construction project is a union project, then there must be a foreman for every seven workers, and this requirement needs to be incorporated into productivity improvement studies.

3.8 U.S. and Foreign Labor Statistics

When U.S. manufacturing output per hour is compared with that of eight other countries over a 10-year period, the output in the United States has been lower than the output in Sweden since the year 1973 (U.S. Department of Labor 2006a). Between 1995 and 2005, the ranking of U.S. output relative to that of seven other countries with available data improved each year, as indicated in Table 3-4. In 1995, four of the countries with available data had higher output rates than the United States, but by 2005, the output in the United States (195.7) was higher than all of the other countries except for Sweden (235.4). Exercise 3.10 provides the reader an opportunity to investigate the foreign labor statistics for manufacturing for the countries listed in Table 3-4 for the past five years.

3.9 Construction Sectors in the United States

The U.S. construction industry comprises several different sectors, and Table 3-5 lists some of the sectors and examples of the types of structures that are constructed in each sector.

Contracting Subcategories

When general contractors construct projects, they might use subcontractors to perform part of the work. Construction contracts state the maximum allowable percentage of subcontracted work. A general contractor who subcontracts all of the work and only manages a project is referred to as a **broker contractor**. Examples of the types of work that are typically subcontracted to specialty firms are listed in Table 3-6.

3.10 Drivers for Productivity Improvement

Because the construction industry is the second-largest industry in the United States, it has a major impact on the U.S. economy, and thus, improving productivity in the

Table 3-4. Foreign Labor Statistics: Manufacturing Output per Hour

	1995	1996	1997	1998	1999	2000	2001	2002	2003	2004	2005	Rate of Increase (percentage)
United States	112.1	116.8	121.7	130.2	136.7	147.7	149.2	165	176.8	186.3	195.7	5.05
France	116	116.7	125.8	132.6	138.7	148.2	150.7	157.4	164.2	170	176.7	3.94
Germany	110.2	113.3	120	120.4	123.4	132	135.4	137	142.4	149	156.9	5.30
Italy	111.1	112.5	113.3	112.5	112.5	116	116.3	114.2	111.3	112.4	112.4	0.00
Netherlands	117.3	120.5	121.2	124.5	129.3	138.5	139.2	143.4	146.4	153.7	160	4.10
UK	103.7	102.8	104.1	105.6	110.9	117.9	121.8	125.1	130.6	137.9	141.4	2.54
Japan	111	116.1	120.7	120.4	124.9	131.7	128.9	133.1	142.3	150.4	154.1	2.46
Canada	112.4	109.7	117	120.7	124.5	129.8	127.4	129.8	130.6	135.9	143.7	5.74
Sweden	125.1	130.2	142	150.7	164.1	176.8	172.6	190.7	204.5	224.6	235.4	4.81

Source: U.S. Department of Commerce (2005); U.S. Department of Labor (2006a).
Note: Index (1992 = 100).

Table 3-5. U.S. Construction Sectors and Examples of Structures in Each Sector

General Building	Manufacturing	Power
Apartments Commercial buildings Educational facilities Government buildings Hospitals Hotels Housing Medical facilities Offices Stores	Automobile assembly plants Electronic assembly plants Textile factories Manufacturing-tooling plants Die-molding plants Plastic-molding plants Rubber-molding plants Food-processing plants	Cogeneration plants Electrical substations Liquefied natural gas plants Nuclear power plants Thermal, hydroelectric, and coal-fired power plants Transmission towers and lines Waste-to-energy plants
Water Supply	**Sewage and Solid Waste Disposal**	**Industrial Processes**
Dams Desalination plants Distribution mains Irrigation canals Potability treatment plants Pumping stations Reservoirs Transmission pipelines	Hazardous waste landfill sites Incinerators Industrial waste facilities Landfills Pump stations Pumping plants Sanitary and storm sewers Transportation facilities for wastes Treatment plants	Chemical plants Nonferrous metal refineries Pharmaceutical plants Pulp and paper mills Steel mills
Petrochemical	**Transportation**	**Hazardous Waste**
Decoking plants Natural gas extraction Offshore oil-drilling facilities Oil-drilling facilities Oil refineries Petrochemical plants Tanker-docking facilities Transmission pipelines	Airports Bridges Canals Highways/roads Locks Marine facilities Piers Railroads Tunnels	Asbestos-abatement facilities Chemical and nuclear waste facilities Hazardous waste process facilities Lead-abatement facilities
Telecommunications		
Cables Data centers Fiber-optic and standard transmission lines Servers Towers and antennae Web hotels		

Table 3-6. Examples of Specialty Contractors

Specialty Contractors	Specialty Contractors
Asbestos abatement	Glazing
Concrete	Ironwork
Demolition/wrecking	Landscaping
Doors and windows	Masonry
Electrical	Mechanical
Excavation	Painting
Fencing	Pipe fitting
Fiber optics	Plumbing/gas
Foundation	Roofing
Framing	Sheet metal

construction industry using productivity improvement studies benefits the U.S. economy. One reason for incorporating productivity improvement methods at construction jobsites is increasing global competition, which requires U.S. construction firms to be more efficient and more energy and environmentally conscious to remain competitive with firms from the rest of the world. Productivity improvement studies might also help firms focus on re-engineering efforts that improve productivity and reduce the amount of resources consumed in the construction industry. When owners demand higher quality projects, it might improve productivity because contractors have to meet the requirements that are established by owners. Management, stakeholders, and stockholders all desire higher profits, and they offer incentives to contractors to increase profits, which requires contractors to explore new methods for improving productivity.

Another driver for improving productivity is that many types of innovative multimedia technologies are now available. These technologies help improve the training of personnel by visually demonstrating information from documentation sources, and they provide a process for documenting methods and processes for litigation.

The potential liability of contractors and increasing arbitration and litigation costs related to environmental and other laws force contractors to improve their productivity so that they will have funds to dedicate to these other areas. New legislation and personal rights advocates have increased the regard contractors have for safety, and this has led to the use of productivity improvement studies to evaluate safety at construction jobsites. Substantial increases in the number of claims for compensation for injuries during construction projects has caused a corresponding increase in insurance costs for firms that are not able to prevent accidents.

The increasing use of **computer-aided design and drafting** (CADD) and building information modeling (BIM) that uses **three-dimensional computer-aided drafting** helps to improve the design process. Contractors now have access to three-dimensional models for projects that include the drawings created by the designers and contractors, models of different potential processes, validation of value-engineering concepts, and as-built drawings. Improvements in computer hardware and software applications that contribute to more accurate project-control systems, which help in the evaluation of productivity, include

- Accounting software,
- Bar code readers,
- Bluetooth technology,
- Cellular networks,
- Computerized scheduling programs,
- Database software,
- Flash memory,
- Global information systems (GIS),
- Global positioning system (GPS) surveying equipment,
- Laptop computers,
- Presentation graphic software,
- Project controls software,
- Scanners,
- Smart phones,
- Spreadsheet software,
- Tablet computers,
- Wireless networks, and
- Word processing software.

Additional information on using computer software programs for productivity improvement programs is discussed in Chapter 9.

Other drivers for productivity improvement are the increasing wages of workers, scarce resources, inadequate access to specific technologies, the increasing cost of material fabrication, difficulties in locating highly qualified skilled workers, and the increasing cost of equipment and gasoline.

3.11 Summary

This chapter provided information on the different aspects of the construction industry, including the phases of construction and the contract administration models used in the industry. It also discussed construction team members, forms of contractual agreements, the different construction sectors, construction project planning, the construction industry as compared with the manufacturing industry, and the project life cycle.

Because the construction industry contributes 6.2% of the U.S. gross domestic product (GDP) and $14 trillion to the economy (U.S. Department of Commerce 2012a), it has a significant impact on the economic health of the country. Members of the construction industry would benefit from improvements in productivity because higher productivity rates would increase profits and also address shortages of qualified labor. Whether the goal is reducing costs, increasing profitability, or providing sustainable and green construction, many different methods for improving productivity rates in the construction industry are available.

3.12 Key Terms

adversarial relationships
bar charts
broker contractor
certificates of deposit
computer-aided design and drafting
constructability review
contingency plans
cost-accounting systems
critical-path method schedules
design-build
equivalent uniform annual worth
execution plans
expediting
fast-tracking
feasibility studies
foremen plans
long-lead-time item
management by stress
milestone schedules
minimum attractive rate of return
money market accounts
precedence diagrams
present worth
procurement
rate of return
separate prime contracts
single prime contract
sureties
three-dimensional computer-aided drafting
time cards

3.13 Exercises

3.1 Provide an example of a current construction contract administration model that is being used by a specific construction firm and discuss how it is being used by the firm.

3.2 Why would it be easier to implement productivity improvement programs in the manufacturing industry than in the construction industry?

3.3 Of the construction industry sectors listed in this chapter, in which one would it be easiest to implement productivity improvement techniques and why?

3.4 Which element of the construction industry has the greatest effect on productivity rates and why?

3.5 Explain the responsibilities of design engineers who work in the construction industry.

3.6 What are the project development phases?

3.7 Which contracting model would be the most efficient to use on a project where the parameters are not well defined when construction starts and why?

3.8 When would owners want to implement separate prime contracts on a project, and why would they do it?

3.9 Of the list of reasons that affect the amount of work in construction, which items are currently affecting the construction industry the most often?

3.10 Using U.S. Department of Labor statistics, look up the foreign labor statistics for the past five years for manufacturing output per hour for the countries listed in Table 3-4.

References

Oberlender, G. (1993). *Project management for engineering and construction,* McGraw Hill, New York.

U.S. Department of Commerce. (2005). *Census 2002 estimates: National occupational employment,* U.S. Census Bureau, Washington, DC. <http://www.census.gov/prod/2003pubs/02statab/labor.pdf> (December 2012).

U.S. Department of Commerce. (2012a). *Manufacturing and construction statistics,* U. S. Census Bureau, Washington, DC. <http://www.census.gov/mcd> (March 2013).

U.S. Department of Commerce. (2012b). *Value of construction put-in-place October 2012,* U.S. Census Bureau, Washington, DC. <http://www.census.gov/construction/c30/c30index.html> (Dec. 2012).

U.S. Department of Labor. (2006a). *Foreign labor statistics,* Washington, DC.

CHAPTER 4

Human Component of Productivity Improvement

The safety of workers has always been a major issue at construction jobsites, where the risk of injury and death is high. Not only are workers negatively affected when someone is injured or killed on site but coping with accidents is also costly to owners and time consuming for construction personnel. Statistics indicate that construction employees account for approximately 4.8% of the total labor force, but they incur 15.6% of all work-related fatalities (U.S. Department of Labor 2011).

In 2008, 969 construction-related fatalities were reported; in 2011, this number dropped to 721 because there were fewer construction projects in the United States. The fatality rate was 9.6 per 100,000 full-time-equivalent workers in 2008 (U.S. Department of Labor 2011). Unfortunately, by 2012 construction deaths were 19.6% of private industry deaths and the number of deaths had increased to 4,383 or 3.2 per 100,000 or 84 per week and 12 per day, which is still a major improvement over 9,880 deaths in 1970. The most common causes of deaths in construction in 2012 were falls, struck by an object, electrocution, and caught-in or between objects (U.S. Department of Labor 2014). Construction safety practices continually need to be reformed because ensuring that work is being conducted in a safe environment is an ongoing process, not an isolated effort. Implementing progressive safety programs, safety checklists, and innovative safety management techniques helps to reduce the number of accidents and fatalities from current accident rates.

Contractors are not the only ones responsible for safety; construction managers and owners are also responsible. Some factors that contribute to the high number of accidents in the construction industry include unclear delineation of responsibility for safety issues, poor management, worker inattention, overcrowded work areas, defective materials, equipment-operator errors, defective tools and equipment, alcohol and drug use, and negligence. Effective safety programs help prevent some accidents, possibly reduce project costs, and increase productivity. Architects, engineers, and construction managers supervise and manage many different types of projects, all of which involve dangerous activities that risk human lives. Construction projects, particularly industrial construction projects, require integrated and intricate operations that involve the coordination of various trades by management personnel.

Figure 4-1. Construction of Liquefied Natural Gas Plant in Bontang, Indonesia

Figure 4-1 shows a liquefied natural gas plant exchanger that was built to replace an exchanger that was damaged in an accident. This figure helps to illustrate the complications and expenses attributable to accidents, especially in industrial construction projects. The exchanger in Figure 4-1 was being built after the original exchanger blew up in an accident that claimed 30 lives. While the exchanger was offline and being replaced, the owner lost $1 million per day because he could not sell the liquefied natural gas product normally produced by the exchanger.

The phases of projects and the activities required to build projects are carefully planned so that workers are able to complete them on time and within budget. In addition, the safety measures planned and implemented on projects have to comply with the requirements of state registration laws, codes of professional conduct and ethics, building codes, and safety laws; therefore, safety measures should be carefully planned and coordinated by management personnel. Engineers, construction managers, safety managers, and members of safety committees should be cognizant of their duties and responsibilities in order to comply with safety requirements. Construction management personnel are responsible for trying to prevent accidents, mitigating existing or potential safety violations, appropriately training personnel to function during inclement weather, motivating workers, and preparing them to perform their work tasks under adverse working conditions. If working conditions are not carefully evaluated during the project-planning stage, projects may not be completed on time.

This chapter examines the human component of construction and provides suggestions for addressing the physical limitations of workers, inclement weather conditions, the monotony of repetitive tasks, worker fatigue, safety issues, and worker **motivation**, all areas that affect productivity. Providing a safe and well-organized jobsite environment helps contribute to the successful completion of construction projects. Additionally, understanding and accommodating the physical limitations of workers helps create working environments that reduce the incidence of accidents and contributes to increasing productivity rates.

4.1 Factors Affecting Worker Performance

Construction workers are often required to perform demanding physical tasks and repetitive activities, operate dangerous machines and equipment, work in unstable conditions above or below ground level, and work during extreme weather conditions, all of which lower productivity rates. Productivity rates are also negatively affected by the mental and physical characteristics of workers and by boredom. The following sections discuss how these and other factors affect construction workers.

Physical and Mental Requirements for Construction Workers

The **physical requirements** for performing construction tasks are different for each construction trade. Some positions, such as finishing concrete or operating light machinery, require only a normal level of physical exertion, yet other tasks, such as setting steel or pipe fitting, require lifting and moving heavy materials. For some occupations, such as firefighting and law enforcement, workers must meet minimum strength and agility criteria. Other occupations have minimum weight requirements. In the construction industry, some companies have requirements that set minimum lifting abilities and agility levels for workers. When construction firms enforce physical requirements, they are trying to reduce the number of worker compensation claims filed for injuries resulting from workers attempting to perform tasks beyond their inherent physical capabilities.

The mental aspect of construction work entails maintaining high levels of concentration over long periods. Varying the tasks of workers throughout the day helps to mitigate mental concentration problems. Workers have difficulty remaining alert and staying attentive when they are performing repetitive tasks that lead to boredom. When workers become bored and distracted while performing tasks, they are not as alert to danger as they are when they are focused on performing their work tasks.

Adjustments should be made to **work-break cycles** because this helps improve worker performance and optimize production rates. The amount of rest time allotted to workers should be varied until an optimum production rate is achieved between breaks. To help alleviate mental fatigue, 20% to 25% of a work period should be devoted to rest. Also, changing activities after 30 to 50 min and using a **work-rest cycle** of 10 min and 2 min helps workers remain alert. Other techniques that might help to optimize production rates are monitoring heart rates and oxygen levels during and after a task is completed by workers. It is possible to monitor oxygen levels using a bicycle **ergometer**, other types of monitors, or smart phone applications. Heart rates should not increase beyond 140 to 160 beats per minute for long periods (this range varies depending on the normal heart rate of individuals). Workers might also develop **stress fatigue** while they are performing physical labor. Stress fatigue is caused by several factors, including

- A cramped working environment,
- Excessive noise levels,
- Continual vibrations,
- Glare,
- Excessive heat or cold, and
- Arguments among workers.

Construction workers should not be expected to continuously perform tasks that require excessive strength or heavy lifting because this could lead to overexertion. If workers are required to overexert themselves, their productivity rate declines with each successive hour. One method for preventing declines in productivity attributable to overexertion is to rotate work tasks during the course of each day. If workers start the day lifting heavy materials, they could be reassigned to other tasks that require less exertion after their first break. If it is not possible to reassign workers to less strenuous tasks, then additional breaks should be provided during the day. The extra breaks could be used for discussing work or safety issues, providing instructions, explaining how activities should be proceeding, or discussing how activities should be completed by work crews. Official breaks should not be used for these purposes, or the workers will resent the intrusion into their personal time.

Effects of Extreme Temperatures on Workers

Extreme temperatures negatively affect the production rates of construction workers. Prolonged exposure to temperatures that are over 80°F (26.7°C) or below 40°F (4.4°C) could cause workers to perform their assigned tasks at slower rates. At high temperatures, workers may start to dehydrate if they are not drinking sufficient levels of fluids that contain salt or if they are not taking high levels of buffered salt supplements.

Heat Stroke

Construction workers might develop **heat stroke** when they are working for prolonged periods in the hot sun. Between 2003 and 2009, there were 230 heat-related deaths among U.S. workers, 81 (35%) in the construction industry. In addition, during this same period, 15,370 heat-related injuries and illnesses occurred, with 4,110 (27%) of them in the construction industry (U.S. Department of Labor 2014).

Heat stroke causes workers to become disoriented and to move slower, their heart rate increases (**tachycardia**), they might exhibit flushed cheeks and irrational behavior, and they become lethargic. In severe cases of heat stroke, workers exhibit hot, dry skin and body temperatures over 105°F (40.6°C). If any of these symptoms are present, a worker should not be allowed to perform his or her assigned tasks because, not only will performance be substandard, but it could result in safety hazards. Immediate treatment is required and involves having workers lie down in a cool location, soaking their clothing in water, and applying cool water and ice packs to their skin (Cho et al. 2008). If workers become overheated, management personnel have several options, including

- Shutting down job activities,
- Providing shelter,
- Providing air conditioning, and
- Increasing water and salt intake.

Cold Weather Issues

During cold weather, workers might wear several layers of clothing along with **gloves**, hats, earmuffs, and ski masks, and this could impair their sensory perception. Workers might not be able to hear instructions or heavy equipment approaching the area where they are working. While wearing gloves, workers might not be able to properly grip their tools or materials. In cold temperatures, the clothing of workers could be pierced by tools or materials when an object hits them and they may not be aware of it. Skin exposed to temperatures below 0°F (–17.8°C) for prolonged periods might be damaged by frostbite, which causes serious complications and, in some instances, leads to the loss of extremities, such as fingers or toes. At very low temperatures, such as –50°F (–45.6°C), frostbite usually occurs within minutes or even seconds. During cold weather, management personnel should realize that the blood flow rate of workers decreases, which leads to muscle contractions, tissue damage, and possibly unconscious workers. In cold temperatures, workers require frequent breaks inside heated structures to warm up. Steel tools and materials should not be touched with exposed skin in freezing temperatures because this could damage the skin.

Working in cold temperatures burns more calories than performing the same functions in moderate temperatures; therefore, workers should consume additional calories to compensate for caloric deficiencies. Workers could become fatigued more rapidly when they are burning calories beyond normal levels. For a 160-lb (72.57-kg) male, approximately 5 ft 8 in. (1.727 m) tall, the normal caloric burn rate is 1 kcal/min. In order to sustain all of the basic metabolic functions, under normal working conditions a worker needs to consume enough calories to allow for the burning of 1 kcal/min.

4.2 Occupational Safety and Health Administration

The **Occupational Safety and Health Administration** (OSHA) is the government agency responsible for publishing safety requirements and investigating safety violations at construction jobsites. It is a federal agency, but there are also state occupational safety and health agencies. The federal agency has the authority to assess fines for violations of safety requirements and for worker injuries or deaths. Workers have the right to request inspections by OSHA if they are being required to work in unsafe environments. The Occupational Safety and Health Administration inspectors visit construction jobsites to determine whether jobsite practices comply with established laws, standards, and regulations, and they have the authority to impose fines if standards are not being met.

The Occupational Safety and Health Administration was established by federal law, and it has jurisdiction over all industries, except for mining and the railroad industry. The **Occupational Safety and Health Act** of 1970 (PL 91-596, 84 Stat. 1590) states that every employer involved in commerce is required to provide a workplace that is free from health and safety hazards: "An act to assure safe and healthful working conditions for working men and women; by authorizing enforcement of the standards developed under the Act; by assisting and encouraging the States in their efforts to assure safe and healthful working conditions; by providing for research, information, education, and training in the field of occupational safety and health; and for other purposes."

Employers are required to comply with all federal and state safety laws, requirements, and OSHA standards and to keep records of occupational illnesses and injuries. In the construction industry, the OSHA safety inspections require either a voluntary commitment for an inspection or a search warrant. When an inspection is being conducted, the employer and a representative of the employees have the right to accompany the inspector. When an OSHA violation is discovered, a citation is issued that describes the violation, the penalty assessed for the violation, and the time frame in which the violation must be corrected by the employer. All OSHA citations have to be posted close to where the violations occurred at jobsites. Employers have 15 days to contest a violation and have a review commission hold a formal hearing. The Occupational Safety and Health Administration citations are enforced through the **U.S. Court of Appeals**.

Employees are allowed to request inspections by filing a complaint with the U.S. Department of Labor. If an employee files a complaint, he or she cannot be discriminated against or discharged for doing so. The **U.S. Supreme Court** has ruled that employees have the right to refuse to perform any task that they think might result in serious injury or death. If there is imminent danger to the safety or health of employees, then an inspector notifies the **U.S. Secretary of Labor**, and if the danger is not eliminated, then the Secretary of Labor may seek a temporary injunction to close down the part of the jobsite where the danger exists.

The Occupational Safety and Health Administration (2014, p. 1) requires that companies report

- Occupational deaths,
- Injuries,
- Loss of consciousness,
- Restriction of work or motion,
- Transfer to another job,
- Work-related illnesses, and
- Medical treatments other than first aid.

Accident incidence rates are calculated using the following formula and reported to OSHA:

accident incidence rate = [(number of cases or days lost per year) × (number of employees × 2,000 work hours per year per employee)]/total employee hours per year (4-1)

An example of the accident incidence rate using this formula is the following:

[(50 cases or days lost per year) × (100 employees × 2,000 work hours per year per employee)]/200,000 hours per year = 50

Recordable cases are classified as follows (Clough et al. 2005, p. 428):

1. *Total recordable cases*: The sum of all recordable occupational injuries and illnesses, including deaths, lost-workday cases, and cases without lost workdays.
2. *Deaths.*
3. *Total lost-workday cases*: The sum of cases involving days away from work or days of restricted activity.
4. *Nonfatal cases without lost workdays*: The sum of cases that are recordable injuries or illnesses that do not result in death or lost workdays, either days away from work or days of restricted activity.
5. *Total lost workdays*: The sum of days away from work and days of restricted work activity.

Occupational Safety and Health Administration regulations include over 200 sections and 1,000 subsections that pertain to construction. The sections that apply to construction are grouped into 26 subparts. According to Dr. James Hinze, in his book *Construction Safety* (2006, p. 105), the sections that most commonly apply to construction are the following:

1. Housekeeping (Section 25, Subpart C);
2. First aid, medical services, and sanitation (Sections 50–51, Subpart D);
3. Personal protection (**hard hats**, eye and face masks, safety nets, safety lines, etc.) (Sections 100, 102, 104, 105, Subpart E);
4. Traffic control (Sections 200–202, Subpart G);
5. Hand tools (Sections 300–305, Subpart I);
6. Ladders and scaffolds (Sections 451–452, Subpart L);
7. Cranes, derricks, etc. (Section 550, Subpart N);
8. Earthmoving equipment (Sections 600, 602–604, Subpart O);
9. Excavation and trenching (Sections 651–652, Subpart P);
10. Concrete and concrete forms (Sections 701–703, Subpart Q);
11. Structural steel (Sections 751–752, Subpart R);
12. Cofferdams (Section 802, Subpart S); and
13. Explosives (Sections 900–911, Subpart U).

The complete set of OSHA regulations is available on the website http://www.OSHA.gov.

4.3 Workers' Compensation Insurance

Workers' compensation insurance is required for workers, and employers pay for it on a monthly basis. This insurance reimburses employees for lost wages if they are injured while working at jobsites. Workers' compensation insurance is required in all 50 states, as set forth by **workers' compensation acts**. Workers' compensation is paid by employers into a government fund; the amount paid is calculated by multiplying the wages earned by the workers in the craft by a multiplier specific to that craft. The fund covers employees when they sustain a personal injury or contract a disease and there is a direct causal relationship between the work environment and the injury or disease. Workers' compensation benefits are not awarded because an employer caused an injury or disease; claims are based on proving a connection between the work performed and the injury sustained or the disease contracted by the employee. Fraudulent workers' compensation claims and lawsuits are a major reason for the increase in concern over safety issues.

Employees who accept workers' compensation waive their right to sue their employer for causing the injury or disease. At the time workers' compensation benefits commence—shortly after an injury occurs—employees may not yet know the long-term consequences of their injuries, and if they accept workers' compensation benefits, they will not have the option to sue their employer at a later date, when they are more aware of the severity of their injuries. Workers' compensation benefits include lost wages, economic loss, past and future medical expenses, and in the event of a death, benefits to dependents. Workers' compensation claims do not cover punitive damages or compensation for pain and suffering, and that is why some claimants file lawsuits rather than accepting what they are entitled to through their insurance policy. Employees whose injuries are self-inflicted or the result of their own **willful misconduct** are not eligible for workers' compensation benefits.

4.4 Safety Issues Related to Productivity Rates

Construction accidents negatively affect production rates due to the amount of time lost when accidents occur and when they are being investigated to determine their cause. Firms are required by law to monitor lost-time accidents, which are accidents that result in a worker or workers not being able to perform their normal work function. Construction contracts define responsibility and authority for safety on a project (some more concisely than others). Defective or unclear clauses in contracts might lead to confusion about who is responsible for safety during construction. Contracts should clearly specify who is responsible for safety and include hazard-prevention requirements.

Losses Resulting from Accidents

The types of losses that result from construction accidents or work-related diseases include direct and indirect costs. **Direct costs**, also referred to as insurable costs, are "medical costs, premiums for compensation benefits, liability, and property losses" (Hinze and Applegate 1991, p. 538). **Indirect costs** are those associated with the items listed in Table 4-1. Additional types of losses include loss of life, bodily injuries, loss of earnings, loss of skills or services, and loss of a family member.

Table 4-1. Indirect Costs of Construction Accidents and Work-Related Diseases

Indirect Costs of Accidents and Diseases
Accident investigations
Damage to property, tools, etc.
Frequent lawsuits
Hiring of new employees
Jury awards
Lost supervisory time attributable to assisting with accident investigations
Lost time of the injured employee
Lost time of other workers (assisting or curiosity)
Workers' compensation claims

Source: Adapted from Hinze and Applegate (1991), p. 538.

The **Construction Industry Institute** funded a research investigation that determined the ratios of indirect costs to direct costs for two categories of injuries from the data collected from a variety of different firms. The resulting percentage of costs per jobsite for medical case injuries, for which workers required medical attention before they were able to return to work, were as follows (Hinze 2006, p. 57):

- Injured worker costs—40%
- Material and equipment damage—22%
- Crew costs—13%
- Impact costs—13%
- Supervisory and administrative costs—12%

The jobsite cost breakdown of restricted activity or lost-workday cases, where workers could not return to work the day after being injured or were assigned different tasks to perform, were as follows (Hinze 2006, p. 58):

- Injured worker costs—50%
- Impact costs—16%
- Crew costs—13%
- Material and equipment damage—11%
- Supervisory and administrative costs—9%
- Replacement worker—1%

Maintaining accurate accident and injury records helps to monitor loss control. Being able to analyze previous accident records might lead to accident prevention through altered or improved work techniques, tools, equipment, or work processes. The type of information that is useful during a loss-control analysis includes data on accidents measured by frequency, severity rates, cost, the days of lost time, and the number of lost-time injuries.

Occupational Health Hazards

In addition to safety concerns, health issues are gaining in importance in the construction industry. The most common health hazards include heat, dust, **aspiration** (inhaling into the lungs) of mineral particles or **asbestos**, noise, **dermatitis** (skin rashes), burns, and toxic chemicals. In addition to injuries caused by construction accidents, workers could be exposed to other occupational health hazards, such as airborne dust, **silica dust** (which causes silicosis), fumes from solvents, heat exhaustion, excessive noise, toxic substances, and radiation.

Asbestos is a natural substance that is mined and used in the manufacture of insulation board, Sheetrock™, wallboard, ceiling tile, floor tiles, and other construction materials. As asbestos breaks down or is disturbed by drilling or other means of penetration, it releases a fine dust that is toxic to humans, but the effects of asbestos exposure may not be apparent for decades. Asbestos poisoning manifests itself as silicosis, which is a lung disease that can be fatal; therefore, asbestos was banned in Great Britain in 1985 by the Health and Safety Code of Practice and in the United States in 1970 by the Environmental Protection Agency. Other countries still use asbestos in the manufacture of wallboard and other construction materials. Construction workers should be aware that they might be exposed to asbestos dust while installing or demolishing wallboard that was originally installed before 1970 in the United States or that was manufactured in a country that has not banned the use of asbestos.

Occupational diseases are also a serious problem in construction. Direct costs, claims, and the loss of workers all increase considerably when the health of workers is compromised by some type of hazard. Several measures might be used to mitigate identified hazards, such as removing materials containing asbestos and substituting less toxic materials. Enforcing the use of protective equipment is also important. Jobsite personnel should be aware of any new health issues and implement methods proven to reduce health hazards.

Safety Equipment

Hard hats, **safety goggles**, **steel-toed boots**, gloves, **face masks**, and **respirators** are among the many safety devices that should be used at all times to protect workers. Workers might also wear **double-layer clothing** to protect themselves from puncture wounds and other abrasion injuries. Steel-toed boots have steel inserts in the toe region of the boots to protect feet from falling objects. The construction worker in Figure 4-2 is wearing steel-toed safety boots.

Workers who operate equipment and the condition of the equipment should be carefully monitored by construction management team members. Old equipment should be repaired and inspected periodically for defects, workers should be properly trained to operate modern equipment, and safeguards should always be incorporated into the design of machinery with no provisions for disengaging safety features. According to MacCollum (1990, p. 19), "Many of the machines used in today's production activities do not have the safeguards needed to provide adequate hazard prevention. Unfortunately, U.S. employers do not require that the equipment they

Figure 4-2. Concrete Placement with Worker Wearing Steel-Toed Boots

buy come equipped with all needed safeguards." Equipment safeguards should be specified in contract provisions.

Contractors are responsible for providing appropriate safety equipment and for ensuring that workers properly use their safety equipment. In addition to preventing accidents from occurring, contractors try to minimize the losses associated with accidents. Two of the methods that help to minimize losses are including insurance costs in budgets and integrating safety measures as a continuous process from inception to the completion of projects.

Activities should be planned so that workers will not be required to hold heavy objects or materials above shoulder level for long periods because this might cause back or arm injuries, cause workers to drop their loads, or accelerate worker fatigue. Workers should be provided with the proper support mechanisms for the materials being lifted or held aloft. Each task should be evaluated to determine the following:

- Whether workers are able to see or feel the point of application,
- Whether workers will have to lift and hold objects,
- Whether materials have to be carried to new locations,
- Where tasks are positioned,
- Whether force has to be applied to complete tasks,
- Whether a normal posture is maintained, and
- How far a worker has to reach to complete a task.

Conducting an analysis of these items helps to increase the productivity of workers by preventing workers from injuring themselves or experiencing discomfort while they are performing their work tasks.

Worker Fatigue Cycles

The workers at construction jobsites exhibit **fatigue cycles** that include lower production rates at the beginning of the day, a production peak during the middle of the morning, and declining rates later in the morning. This cycle is repeated in the afternoon, with overall lower production rates than are achieved during the morning, especially late in the workday. The lower production rates at the beginning of the day could be attributable to workers getting a late start; work not being assigned to workers prior to the start of the work; and workers not having drawings, work plans, tools, or equipment. Workers may also purposely slow down work to stretch it out over longer periods than necessary so that they do not have to start new tasks or so that they are able to earn overtime pay (Oglesby et al. 1989).

When heavy construction equipment or other types of machinery are used, the work pace is set by machine cycle times. The cycle times should not be so rapid as to compromise the safety of workers, even if workers are able to keep pace with the machines.

In addition to daily worker-fatigue cycles, there are weekly production-rate fluctuations. Typically the production level averages are as follows (Hinze 2006):

- 70% Mondays
- 85% Tuesdays
- 70% Wednesdays
- 60% Thursdays
- 58% Fridays

Productivity rates are also affected by whether workers have their peak energy in the morning, in the afternoon, or at night. Twenty-five percent of the population has its peak energy in the morning, 25% peak at night, and 50% are able to perform at their peak at any time. Tasks should be assigned to workers who are at their peak, and tasks that require high levels of mental or physical attention should be performed during the peak period of a worker (Oglesby et al. 1989).

Rotating shifts could severely lower production rates because workers are not as alert when they are coming off a different shift to start a new shift. Workers need several days to adjust to new sleep routines. In addition, they require 1.5 days of rest per week to sustain normal output levels, and that is why working overtime does not always lead to proportional gains in productivity. Overall productivity falls during the first week of overtime work, and after five weeks, productivity for 50 h of work is comparable to that of 40 h of work without overtime.

Members of management are also concerned about fatigue because it might lead to

- Exhaustion,
- Heat dissipation,
- A lack of oxygen,
- Exceeding inherent capabilities, and
- Decreased pulse rates.

4.5 Preventing Construction Accidents

An analysis of the contributing factors in any type of hazard situation is important for determining how to minimize or prevent risks. Accidents might occur repeatedly when the same mistakes are made, yet identifying the principal causes of accidents and ways to avoid them might be possible by gathering relevant data and conducting investigations. Knowing the sources of accidents is useful in building prevention programs, improving construction practices, designing structures and new equipment, and implementing different management methods. Being able to analyze the data collected at construction jobsites is important in determining the factors that frequently cause accidents.

One benefit of accident reduction is cost savings from preventing injuries and property damage. Improving safety measures helps reduce costs because accidents result in higher insurance premiums and increased overhead costs. If work is not interrupted because of accidents, there are no lost-time workdays, and the project has a better chance of being completed on time. Large companies and some government agencies have a lower incidence of accidents among their employees because they generally have more effective safety programs and are more efficient at enforcing existing programs. If management-level executives of large firms are supportive of safety programs, it may result in better performance by supervisors and workers. Support from management is demonstrated by the close monitoring of safety performance, effective communication and project coordination, the nature of the contract, and budgets.

Although implementing safety management programs is costly, it is worthwhile. In some firms, if costs begin to rise, corporate executives have to decide where to cut costs, and unfortunately, safety is an item that could be sacrificed in difficult economic times. The relationship between safety performance and economic pressure is not clear because it is hard to establish whether reductions in safety programs lead to a higher frequency of injuries.

In addition to being required to follow federal, state, and local laws related to jobsite safety and health, contractors also have to adhere to safety requirements that are set by owners. The following are examples of the types of safety requirements that owners might impose at construction jobsites (Hinze 2006, p. 105):

- Hard hats, eye protection, hearing protection, and safety shoes should be worn by all workers;
- First-aid supplies should be available for all work areas;
- All jobsite personnel should know workers who have had first-aid training;

- No visitors should be permitted on the premises without authorization;
- All visitors should comply with applicable safety regulations;
- The jobsite should be kept clean and free of debris;
- Temporary chain-link fencing or an approved equal should be furnished by the contractor and installed around construction areas;
- Persons under the influence of alcohol or drugs should not be permitted on the premises;
- Thorough equipment safety checks should be conducted regularly;
- No smoking should be permitted on the premises; and
- Everyone should comply with posted speed limits on the site.

Construction foremen, project superintendents, and construction managers play an important role in defining safety performance on their projects. Orientation programs, the work environment, the pressure of deadlines, and relationships with coworkers are decisive factors, but conforming to OSHA regulations and local building codes also helps to reduce accidents. Management personnel should set an example when they are working at construction jobsites by observing all safety requirements. Figure 4-3 shows a project manager and his assistant at work at a construction jobsite where a water tank is being constructed by their firm. Both of these representatives of management are observing safety regulations by wearing hard hats and steel-toed boots.

Managers have to address hazards that are particular to their segment of the construction industry. In building construction, for example, one of the main safety focus areas is workers falling off structures and worker injuries caused by falling objects. Twenty-five percent of the injury costs in building construction result from falls. Figure 4-4 shows construction workers working on the repair of an oil tank on top of a temporary steel structure. All of the workers shown in Figure 4-4 should be wearing safety harnesses.

Figure 4-3. Tank Construction Project with Management Personnel Wearing Hard Hats

Figure 4-4. Workers Not Wearing Safety Harnesses at Oil-Tank Reconstruction in St. Croix

The types of questions that should be answered to help prevent falls include the following:

- Are safety harnesses needed?
- What types of scaffolding or ladders are appropriate for the job?
- Should perimeter guards be used around floor edges?
- How will floor openings be protected?
- How should materials and workers be transported safely to different levels?
- How should the public be protected both inside and outside jobsites?
- How will supervisory and maintenance inspections to manage loss control be provided?

When these questions are addressed before the start of construction, production rates increase because workers do not have to be concerned with these issues while performing their tasks. Workers know that safety requirements are being met, and this allows them to spend their time concentrating on their work tasks rather than worrying about their safety. Similar lists of safety concerns should be developed for each segment of the construction industry before the assignment of work tasks, and each of the issues on the lists should be investigated to determine the best method for addressing safety concerns.

Figure 4-5 shows workers being properly held aloft in a safety cage rather than hanging over the edge of the corrugated metal structure. Some other safety measures that might be implemented to reduce the number of accidents related to construction falls are locating potential fall hazards, fall protection harnesses, safety nets, and the use of retractable lines (Occupational Safety and Health 2014, p. 57).

A variety of measures are used in construction to prevent accidents and applied when necessary at construction jobsites. Appropriate signs and signals, fences around

Figure 4-5. Workers Performing Their Work in Safety Cage

Figure 4-6. Intricate Rebar-Tying Operation for Concrete Slab

the perimeter of the jobsite, organization of the work, identification of tools and hazardous substances, appropriate use of construction procedures, appropriate job coordination, and use of protective clothing are examples of hazard-reduction techniques. Construction projects involve intricate operations that require management personnel to coordinate members of different trades to help prevent accidents from occurring during construction. Figure 4-6 shows workers tying numerous types of rebar for a concrete slab. Construction operations, such as the one in Figure 4-6, require management to plan the execution of activities with precision to keep members of different trades from interfering with the operations performed by other workers.

One of the most hazardous jobs in the construction industry is excavation (OSHA 2011).

Excavating and trenching are regulated by OSHA standards in the same manner as other construction activities. Specific standards for excavating are provided in OSHA Standard 1926.651 and excavation operations are discussed in OSHA Directive TED 01-00-015 [TED 1-0.15A] (1999, January 20). *Excavations: Hazard Recognition in Trenching and Shoring* (OSHA 2014).

In 2011, the OSHA (2011) reported that approximately 200 people had died in trench cave-ins since 2003. The **National Institute for Occupational Safety and Health** (NIOSH) indicated that approximately 35 workers die each year in accidents related to excavating activities (National Institute for Occupational Safety and Health 2011). The number of fatalities varies each year, according to the **Standard Industrial Classification Code** (SICC), but most excavation-related deaths occur during the construction of sewer lines.

The major causes of accidents related to excavations are

- Using accepted engineering practice rather than adapting pratices to current conditions,
- Soil type and conditions,
- Recent excavations,
- Weather conditions (freezing and thawing),
- Surcharged loads,
- Vibration and other types of movement, and
- Incompetent workers. (Suruda et al. 1988, p. 29)

Occupational Safety and Health Administration construction standards for excavations were revised in 1990 and contain general requirements that have to be followed during excavation at all construction jobsites. Examples of OSHA construction standards include the following (OSHA 2014):

- Surface encumbrances: must be removed or supported if they present a hazard.
- Proper identification of utilities: localize and identify before digging lines such as gas, water, telephone, and electricity.
- Access and egress: must be provided by ladders, ramps, or steps on trenches deeper than 4 ft (1.22 m).
- Exposure to traffic: use reflectorized warning vests when excavation is taking place near traffic.
- Overhead loads: employees should be aware of the loading and unloading of equipment.
- Emergency response: provide the equipment that is required for emergencies such as lifelines, stretchers, first aid, and harnesses.
- Hazardous atmosphere: excavations dug near chemical plants, sewer lines, and landfills could cause atmospheres to be formed by liquids, gases, and vapors that seep through the soil.

Other safety procedures that should be implemented during excavations include the following (Suruda et al. 1988, p. 37):

- *Sloping*: The sides of excavations should be adequately sloped. It is the simplest method of cave-in control, by preventing the roll back of loose material. The angle of the slope depends on soil type, loads, and distress.
- *Shoring*: It has been proven that most of the deaths involve trenches in which the walls have not been shored. These structures press against sidewalls and prevent cave-ins. Shoring is expensive for the extra materials involved, which is one reason why many times this method is not used.
- *Shielding*: Protects employees from cave-ins by providing a sheltered space where employees are able to work. It is a device shaped like a box, consisting of two steel plates separated by structural members, open at the bottom and the top. The steel plates face the sidewalls of the trench. Work is performed between these two plates, and as it progresses, the device is dragged along the bottom of the trench by means of other equipment.

4.6 Causes of Construction Worker Injuries

For a long time, the most common causes of injuries at construction jobsites were overexerting oneself, falling, and being struck by and against objects (Culver et al. 1993). The distribution of the causes of injuries is provided in Table 4-2.

The causes of accidents listed in Table 4-2 result in strains, sprains, cuts, and lacerations. A distribution of types of construction injuries is provided in Table 4-3. Strains and sprains represent the most frequently occurring injury, followed by cuts, lacerations, and punctures. Some of the methods for reducing these types of injuries include proper signalization, proper equipment, braces, awareness, and improved communication (Yates and Terrero 1997).

Because lifting is the most common activity performed at construction jobsites, when construction accidents occur, the back is the body part that is injured most frequently. Activities that cause back injuries vary widely, with the most common ones being constructing concrete formwork, placing rebar, and moving concrete tubes. The complexity of back injuries makes it difficult to propose specific preventive

Table 4-2. Distribution of Causes of Construction Jobsite Injuries

Cause of Injuries	Percentage of Accidents
Overexertion	24
Struck by an object	22
Fell from an elevation	14
Other	10
Struck against an object	8
Reacted physiologically	7
Fell from the same level	7
Rubbed or abraded	4
Caught in between two objects	4

Source: Adapted from Yates and Terrero (1997), p. 25, Figure 1.

Table 4-3. Distribution of Types of Construction Jobsite Injuries

Injury	Percentage of Total Injuries
Sprains and strains	37.9
Cuts, lacerations, and puncture wounds	17.9
Fractures	14.0
Other	9.5
Contusions, crushed body parts, and bruises	8.4
Nonclassifiable	7.4
Dislocations	2.1
Multiple injuries	2.1

Source: Adapted from Yates and Terrero (1997), p. 26, Figure 2.

Figure 4-7. Oil-Tank Destruction by Hurricane in St. Croix

measures. Back injuries may be reduced by improving health and hazard surveillance systems, redesigning jobs and tools, training workers to properly lift materials, increasing automation, using lifting devices, and screening workers to determine whether they have any existing health problems (Bernold and Guler 1993, p. 610).

4.7 Construction Failures

Construction workers are injured or killed, not only in construction accidents, but also when there are **construction failures**. Construction failures occur when a structure or a part of a structure that is being built either collapses or is distressed to such a degree that it is not able to safely serve its intended purpose, as was the case with the oil tank that collapsed during a hurricane in St. Croix, shown in Figure 4-7 (Janney 1986).

The most frequent causes of structural failures were determined using data from industry professionals and the federal OSHA databases of construction accidents and failures for a 10-year period. The results of this study indicated that the most frequent causes of construction failures were (1) formwork failures and collapses, (2) inadequate temporary bracing, and (3) overloading or impact during construction (Yates and Lockley 2002). Other causes of construction failures are listed in Table 4-4.

From the OSHA investigation data, 29 case studies were extracted and reviewed to provide examples of the most prevalent construction failures, the materials involved in the failures, the injuries and fatalities that resulted from the failures, the OSHA penalties assessed, the OSHA penalties paid, the gravity of the failure, and the most

Table 4-4. Causes of Construction Failures

Causes of Construction Failures
Construction sequences not consistent with design considerations
Damage to prefabricated elements during their handling and erection
Decisions by individuals with insufficient knowledge or education
Designs not reflecting the actual construction loads and field conditions
Failure to have a qualified person in charge
Failure to use the materials specified
Financial pressure to complete a project early
Improper definition of responsibility
Inadequacy of systems during erection
Inadequate original design unknown to the contractor
Inadequate training and education
Incompetent supervisors
Incomplete connections, e.g., installing a few bolts and intending to complete the bolting process later
Incorrect crane operations
Insufficient or improper checking of shop drawings
Lack of common sense, including intoxicated or drug-impaired workers
Lack of safety equipment
Lack of proper inspections
Nature, gross design error, terrorism, or contractor negligence
No consideration for soil conditions
Not thinking fast enough
Poor communication between the designer and the constructor
Poor communication, failure to follow design plans, failure to follow recommended industry practices, and carelessness
Supporting members damaged by other prime contractors as they are installing their work (e.g., ductwork or plumbing)
Unauthorized modifications to the construction specified in the contract document
Unknown or erroneous geotechnical information
Unreasonable schedules
Working too fast

Source: Adapted from Yates and Lockley (2002), p. 11.

probable cause of the failure. Table 4-5 provides a summary of the case study investigation.

The investigation report also included a list of methods for reducing construction failures that are provided in Table 4-6.

Recommendations for Conducting Forensic Failure Investigations

The research investigation conducted by Yates and Lockley (2002, pp. 15–16) also developed recommended steps for conducting a forensic failure investigation.

1. Select a principal investigator who may act independently in the investigation, or as the manager of a staff of engineers and technicians.
2. Assemble (if needed) a staff of in-house engineers and technicians and/or outside specialists.
3. Investigate the scene of the failure as quickly as possible.
4. Conduct an overall visual examination of the failure site.
5. Generate as many hypotheses of causes of failure as possible.
6. Record visual information using photographs, video cameras, sketches, or drawings.
7. Collect samples for field and laboratory testing.
8. Conduct field tests or arrange for laboratory testing.
9. Conduct eyewitness interviews.
10. Review documents relating to the design and construction of the facility.
11. Review the structural design and perform an analysis of the structure.
12. Analyze the data collected and draw conclusions.
13. Prepare and submit a written report to the sponsor of the investigation.

Recommended Format for Reporting the Findings and Conclusions of a Construction Failure Investigation

The following information provided by Yates and Lockley (2002, p. 16) should be collected to properly document a failure investigation:

- Name of the case study.
- Location (city, state).
- Date of failure (month, day, year).
- Time of day.
- Type of project.
- Materials or equipment involved.
- Injuries or fatalities.
- Weather.
- Ownership of the facility (private, local, state, or federal government).
- Name of person, or persons, conducting the investigation.
- Description of the failure. An abstract should be prepared that answers the following two questions: (1) What happened? Describe the construction activities leading up to the failure and the failure itself including damage. (2) Why did it

Table 4-5. Results of 29 Case Studies on Construction Failures

Case Number	Type of Failure	Materials Involved	Injuries (I), Fatalities (F)	Penalties Proposed	Penalties Paid	Gravity of Failure	Most Probable Cause of the Failure
1	Building	Structural steel	1 I, 1 F	$23,560	Unknown	10	Inadequate temporary bracing
2	Building	Concrete	0 I, 1 F	$2,000	Unknown	4	Inadequate temporary bracing
3	Building	Steel and concrete	1 I, 0 F	Unknown	Unknown	3	Overloading—construction
4	Building	Structural steel	0 I, 1 F	$7,000	Unknown	8	Inadequate temporary bracing
5	Building	Concrete and soils	3 I, 1 F	$1,440	Waived	4	Inadequate temporary bracing
6	Building	Structural steel	1 I, 1 F	$1,074	Unknown	4	Inadequate temporary bracing
7	Bridge	Concrete formwork	7 I, 1 F	$6,000	Unknown	8	Formwork failure
8	Bridge	Structural steel	1 I, 2 F	$24,000	Unknown	10	Inadequate temporary bracing
9	Tower	Structural steel	0 I, 3 F	$16,920	Unknown	10	Inadequate temporary bracing
10	Truss	Wood	7 I, 0 F	$510	Unknown	2	Inadequate temporary bracing
11	Truss	Wood	5 I, 0 F	$2,000	Unknown	4	Inadequate temporary bracing
12	Truss	Wood	3 I, 0 F	$200	$200	1	Inadequate temporary bracing
13	Truss	Wood	None reported	N/A[a]	N/A[a]	3[b]	Inadequate temporary bracing
14	Steel	Joists and deck	0 I, 2 F	$960	Unknown	3	Inadequate temporary bracing
15	Steel	Structural steel	1 I, 1 F	$13,500	$9,500	10	Inadequate temporary bracing
16	Steel	Structural steel	0 I, 1 F	$1,600	$800	4	Inadequate temporary bracing
17	Concrete	Concrete	1 I, 1 F	$7,500	$2,800	8	Overloading—construction
18	Concrete	Concrete	3 I, 0 F	$1,000	$1,000	4	Inadequate temporary bracing
19	Connection	Wood	1 I, 1 F	$4,375	Waived	7	Inadequate temporary bracing
20	Connection	Wood	None reported	N/A[a]	N/A[a]	1[b]	Inadequate temporary bracing
21	Formwork	Concrete and forms	2 I, 0 F	$200	Unknown	1	Inadequate temporary bracing
22	Formwork	Concrete and forms	4 I, 0 F	$380	Unknown	2	Formwork failure
23	Formwork	Concrete and forms	0 I, 1 F	Unknown	Unknown	3	Formwork failure
24	Formwork	Wood	3 I, 0 F	$740	Unknown	3	Formwork failure
25	Formwork	Scaffolding	4 I, 0 F	$635	$635	3	Formwork failure
26	Formwork	Concrete and forms	5 I, 0 F	$200	Waived	1	Formwork failure

27	False work	Concrete and forms	5 I, 0 F	$3,000	$1,500	4	Formwork failure
28	False work	Concrete and forms	8 I, 1 F	$350	$250	2	Formwork failure
29	Foundation	Concrete and soils	None reported	N/A[a]	N/A[a]	2[b]	Inadequate temporary bracing

Gravity of Failure[c]

Gravity	Penalty	Assessed
Number	From (dollars)	To (dollars)
01	0	200
02	201	500
03	501	1,000
04	1,001	2,000
05	2,001	3,000
06	3,001	4,000
07	4,001	5,000
08	5,001	7,500
09	7,501	10,000
10	10,001	and up

Source: Adapted from Lockley (1997).

[a] Inspection conducted by a consulting forensic engineer. No penalty was assessed.

[b] Value established subjectively.

[c] Gravity is a number from 01 to 10 representing the gravity or severity of the violation. Values are regarded as follows: 08–10, high; 04–07, moderate; and 01–03, low. Values are from penalties assessed by OSHA or state occupational safety agencies.

Table 4-6. Methods for Reducing the Incidence of Construction Failures

Methods
Clearly defining the responsibilities of the engineer, the fabricator, and the contractor
Conducting constructability reviews during the design stage
Educating and training the construction team
Educating the contractor's superintendents about temporary bracing and stability
Establishing a comprehensive quality-assurance/quality-control plan
Furnishing full-time inspection by an independent construction professional
Furnishing site-specific safety plans that are approved by the engineer
Having a professional engineer design and supervise construction of temporary structures
Having a professional engineer with construction experience review all construction activities, including selection and positioning of cranes, concrete pumps, and truck movements, before the performance of the work
Having construction personnel certified for temporary structure design and construction
Having design engineers review all shop drawings, shoring, and formwork design
Having structural engineers review temporary bracing and shoring designs, details, and construction and submit forms to public agencies indicating such review
Having the engineer of record design and detail the critical connections
Having the structural engineer inspect construction on a full-time basis
Hiring contractors and subcontractors based on a prequalification system for quality, safety, and liability
Including dead loads during construction
Including structural redundancy in the design to avoid progressive collapse
Providing in-depth inspection by local authorities rather than a cursory walk-through
Providing realistic construction schedules
Providing stronger contractor supervision to avoid shortcuts by workers

Source: Adapted from Yates and Lockley (2002), p. 9.

happen? Present the most probable cause of the failure, secondary causes, and the relevant factors affecting the cause.

- Lessons learned. Prepare a practical summary of lessons that may be used to prevent a similar failure.

Recommended Checklist for Investigating and Documenting Structural Failures

The following checklist by Yates and Lockley (2002, p. 16) was developed for use by construction failure investigators and it contains comprehensive steps and procedures to follow during an investigation. The following checklist provides guidelines for helping to plan, prepare, and execute a successful investigation:

1. Select a principal investigator to conduct and manage the investigation.
2. Provide in-house personnel to assist in the investigation.
3. Retain outside specialists as consultants.
4. Assemble an inspection kit to probe, measure samples, and record the wreckage.

5. Make an overall site examination to evaluate the scope and nature of the failure.
6. Generate as many hypotheses of causes of failure as can be developed based on past records of similar failures.
7. Establish a coordinate system for defining the location of fallen debris.
8. Record the position and orientation of the debris after the collapse.
9. Protect the evidence so that it can be properly documented.
10. Completely document the site of the failure using photographs, videotapes, sketches, or drawings.
11. Use a recording device to capture thoughts and impressions as they occur.
12. Collect a written or taped record of eyewitness accounts to the failure.
13. Collect and tag samples for field and laboratory testing.
14. Complete field tests and document the results.
15. Arrange for all necessary laboratory testing.
16. Review the contract documents, inspection reports, project schedules, project correspondence, weather records, and any other related documents that reflect the history and life of the structure.
17. Examine all violations of safety procedures that may have caused the failure.
18. Review the structural design and either perform or arrange for an independent structural analysis to determine what behavior would have been anticipated for the structure.
19. Examine and record deviations from the design.
20. Determine whether there was any unconventional use of materials or unusual construction erection techniques.
21. Determine whether there was a curtailment, even partial, of on-site construction by the design professionals.
22. Review National Weather Service records to log the weather during the period being investigated.
23. Make a final analysis of the data and arrive at any conclusions.
24. Prepare a written report summarizing the findings and recommendations.

This checklist is provided to aid investigators in planning construction failure investigations, collecting and recording field and test data, generating a failure theory, and analyzing and drawing conclusions related to failures.

4.8 Drug and Alcohol Use at Construction Jobsites

Another area that contributes to accidents and slows production rates is the use of drugs or alcohol by construction workers while they are working at jobsites. Most construction jobsites require workers to provide samples for drug testing on a random basis in order to screen for drug use. Engineers and management personnel may also be subjected to drug tests. Drug and alcohol use has been linked to a high proportion of construction accidents, and their use also contributes to reductions in the productivity rates of workers. Random and for-cause testing could be effective, and it has been shown that after implementing testing, accident rates are reduced or minimized (Timura 1990). Testing is not a total solution, but it helps reduce accidents. Some measures that help to prevent drug use include pre-employment and job-applicant testing, random testing, and periodic testing (Altayeb 1992, p. 781).

Consumption of **illicit** (illegal) **substances** while working at jobsites is an important and frequently discussed issue because it may alter proper performance and may result in lost productivity (physical and mental), poor decision making, absenteeism, turnover, and involvement in accidents, among other consequences (Altayeb 1992). A study published by the **U.S. Department of Health and Human Services** in 1990 indicated that drug-abusing employees are 3.6 times more likely to involve themselves or another person in a workplace accident (Timura 1990).

4.9 Communicating Safety Requirements

According to Smith and Roth (1991, p. 362), "Construction managers, because of their unique contractual position, cannot avoid some degree of responsibility and, therefore, liability for safety and the project safety program." Managerial style influences the implementation of safety programs. The safety manager or entity responsible for safety on a project should be able to communicate effectively and execute all measures related to safety. Safety has to be integrated at all levels of management, and project managers and site superintendents should be able to comply with safety requirements. If responsibility for safety is not well defined, organization and control may suffer, and consequently, the application of safety measures could be compromised during construction.

Construction-worker safety is usually the responsibility of project managers or construction contractors, but safety performance could also be dictated by the design of the structure, even though OSHA does not usually place any responsibility on designers (Hinze and Wiegand 1992). In a study conducted among major U.S. design firms, one-third of the respondents stated that "specific consideration [was] given to the safety conditions for construction workers" (Hinze and Wiegand 1992, p. 679). The goal of a design firm is the creation of a final product, and the design may not emphasize worker safety. It is difficult, but not impossible, to accommodate designs that consider safety at the construction stage.

Safety is more manageable at the project level, and a safety program is easier to implement if it is specific and concise and if supervisors play a decisive role in the program. Safety measures should be established during company meetings and at jobsite meetings, held either weekly or biweekly, and there should also be special safety meetings with supervisors, project managers, construction foremen, and others. Other forms of safety control include effective communication between workers and supervisors or coordinators and regular inspections to help determine whether work is being performed properly and in a safe manner.

Providing periodic inspections is one method for detecting and eliminating safety hazards. Another technique for emphasizing the importance of safety at construction jobsites is providing workers with regular **toolbox meetings** or **tailgate meetings** with the project foreman. These meetings could be used to discuss

- Recent accidents and how they might have been prevented,
- Losses resulting from current activities,
- Safe operations,
- Techniques for preventing accidents, and
- Hazards involved in working with equipment.

In addition to monitoring work practices to verify that safety regulations are being followed, it is important to ensure that workers are physically qualified and have the proper skills to perform their work tasks. Managers should provide training and proper supervision. If the workers do not receive instructions on how to correctly perform their work tasks, it may increase the number of accidents that result from the improper use of work methods.

4.10 Heavy Construction Equipment Safety

Many of the accidents that occur at construction jobsites are attributable to heavy construction equipment failures. Production rates decline rapidly when construction equipment fails, and costs escalate while workers are idle during unscheduled equipment repairs or damage mitigation. Several operator safeguards that help prevent heavy construction equipment accidents include

- Rollover protection,
- Noise mufflers,
- Operator protection from heat and dust,
- Equipment adaptability,
- Reduced exposure of the operators to moving parts, and
- Operation of equipment according to the manufacturer's recommendations.

Construction equipment should be fitted with available safeguards, and equipment safety checks should be conducted to detect hazards, including periodic inspections and maintenance, frequent replacement of parts, training for operators on the proper use of equipment, and examinations for employees on jobsite and equipment operating rules. Construction equipment may be newer, such as the loader shown in Figure 4-8, or it could be older, like the grader shown in Figure 4-9, which does not have all the modern safety features available on newer equipment.

4.11 Safety Education and Site Safety Programs and Plans

To keep workers productive and to prevent lost time attributable to accidents, workers should be educated on how to safely perform their work tasks. There are a variety of methods for distributing safety information. The more common practices used at construction jobsites include

Figure 4-8. Modern Heavy Construction Equipment Loader

Figure 4-9. Old Heavy Construction Equipment Motor Grader

- Leaflets,
- Safety posters,
- Safety films,
- Safety bulletin boards,
- Letters from management personnel,
- Payroll inserts,
- Safety incentives and awards,
- In-house safety publications, and
- Accident facts on construction operations.

Jobsite safety programs should accomplish the following purposes (Oglesby et al. 1989; Adrian 2004).

1. *Proper identification of authority*: This assigns responsibility for safety and health to the appropriate entity: the owner, the construction manager, or whomever it states in the contract.
2. *Contract assignment of safety coordinator*: One or more people, depending on how large a project is, should be assigned to coordinate, inspect, and ensure that safety regulations are being followed at jobsites.
3. *Description and identification of hazardous operations*: A plan should describe all hazardous operations that are going to occur throughout the project. These include testing, shoring, and so forth. A record should be kept of inspectors and the test results.
4. *Adequate equipment for emergency procedures and general rules*: A record of inspections and test results should be kept, along with a list of the measures that should be implemented in case of an emergency. Procedures and rules should be established and followed at all times. These include storage of materials and removal of toxic materials. An appropriate emergency plan, to be implemented when an accident occurs, should exist.
5. *Other*: Safety instructions should be provided to new employees as required by OSHA.

Safety programs should be adequately managed by defining objectives, planning, and assigning responsibilities. The following characteristics should be considered when writing a safety plan or program (Oglesby et al. 1989; Adrian 2004):

- A safety plan should be focused on the areas where losses are typically experienced, and its goal should be to implement safe work practices and eliminate workplace hazards.
- Employees should be familiar with the rules and procedures that should be followed because a safety program should not be implemented if employees do not know the safety requirements.
- The safety level of projects should be measured, and existing safety requirements should be followed at all times. Safety procedures should keep pace with changing project activities.
- Project safety is accomplished through distribution and implementation of authority, responsibility and adequate descriptions of what is expected of all participants in the project, with each individual knowing his or her own job.
- A safety program should establish clear and attainable objectives, and it should state how these objectives should be met by assigning specific objectives to individuals.
- Motivation and recognition are key factors in a successful safety program. Employees should be motivated and encouraged to promote and practice safety. Employees could also help in identifying critical factors in order to establish safety rules and procedures.

4.12 Effects of Tool Use on Performance

The improper use of tools and the use of incorrect tools reduces production rates and might also lead to injuries. Not using the proper tools for a particular task may cause workers to exert additional pressure on the tools or to work in an awkward or unnatural position. The characteristics of tools that should be evaluated before assigning them for use are weight, size, shape, temperature during use, vibration, sharp edges, and protrusions. Tools should be matched to the task and worker abilities to increase production rates.

4.13 Construction Accident Case Studies

The accidents that occur at construction jobsites might be prevented by the adoption of safety measures. Information related to the outcome of accidents is available in legal documents, such as the ***New York Jury Verdict Reporter***, which was used to locate case studies for this section. The case studies that follow provide information on the different construction trades, the characteristics of particular accidents, the causes of the accidents, and the decisions that resulted from related lawsuits. Additional information is provided on how the accidents might have been prevented with safety measures. The cases selected are representative of typical construction jobsite situations, and these types of accidents could be prevented if safety measures are enforced by jobsite personnel.

Case Study 1: Asbestos Handler Crushed—Wrongful Death

On December 7, 1989, an asbestos handler was killed by a semitrailer that fell on him while he was removing asbestos at a renovation project. According to his spouse, the decedent was asked by his employer to take bags of asbestos and place them in the semitrailer, which was freestanding. He and a coworker were loading the bags in the trailer when they felt it shift and become unstable. The decedent crawled under the trailer to see what was wrong, and the nose of the trailer tipped, caught him underneath, and crushed him to death.

The wife of the decedent claimed that the trailer manufacturers and those responsible for providing the trailer were negligent in that antiloading devices, which would have stabilized the trailer and prevented it from tipping, were not in place. A member of the transportation company said that the worker had to know how to use the antiloading devices on the trailer and that warnings about securing the front end of the trailer had been provided to the worker. All the defendants stated that the decedent was contributory negligent for going beneath the trailer (Aberdeen's Concrete Construction 1992, p. 15; *New York Jury Verdict Reporter* 1991b). The case was settled on April 1, 1993. This accident might have been prevented if, once the structure became unstable, the employee had checked the manufacturer's warnings before going underneath it. The contractor or the party responsible for safety should have instructed workers on how to properly use the equipment.

Case Study 2: Construction Worker Hit by Motor Vehicle—Traffic Not Properly Channeled around Site

A sewer maintenance worker was supervising the cleaning of storm drains and manholes when he was struck by a vehicle that had broken through a metal barricade. The impact of the vehicle fractured his femur. The project supervisor had placed the barricade and cones too close to the worksite, which prevented them from properly channeling traffic away from the site. Similar accidents had occurred for the same reason, with no injuries. The defendant claimed the accident was the result of negligence by the operator of the vehicle, who was driving at an excessive speed. Proof was provided that the supervisor had instructed the workers on how to set up the barricades.

On January 15, 1992, the contractor was found to be 80% liable because his workers had set up the barricades, and the vehicle operator was 20% negligent for driving at an excessive speed around the work area. The accident occurred on November 5, 1988, and the worker did not return to work for three and a half years (*New York Jury Verdict Reporter* 1992a). This is a typical case of workers either not following instructions or being careless. If the project supervisor had explained how to properly set up the barricades, the workers might not have set them up incorrectly. In addition to the barricades, signs should have been strategically posted to indicate the work area. The vehicle operator should also have known to reduce his speed when approaching a construction site if signs and barricades were present.

Case Study 3: Carpenter Injured—Failure to Provide Lifting Device

On February 24, 1986, a carpenter was directed to install an I-beam header for an 8-ft (2.44-m) picture window in the wall of a five-story building. The beam weighed 250 to 300 lbs (113.4 to 136.1 kg). The carpenter, who was the lead person on the site at the time, had requested a mechanical lifting device from his employer, but the employer denied his request and directed that the work proceed with two additional laborers whose native language was not English. While lifting the beam, the carpenter determined that it did not fit into position, and as they attempted to lower it, the beam fell onto the shoulder of the carpenter. The carpenter stated that the two laborers did not understand his command to shift the beam so that the men would be holding the weight equally.

The injured carpenter stated that his employer was negligent for failing to furnish a mechanical lifting device and for providing him with non-English-speaking assistants. The employer stated that a mechanical lifting device was unnecessary (*New York Jury Verdict Reporter* 1992b). The carpenter sustained a herniated disc (back injury) and developed further complications. This case was settled on June 6, 1992.

The employer might have not been aware of how difficult it was to perform the work required without the use of a mechanical device. He should have been more familiar with the construction methods to be used, and at the request of his employee, he should have carefully analyzed the situation to determine whether or not other methods could be used to safely lift the beam. The lead carpenter should have verified the procedures with the workers before executing any action. The carpenter should have provided the owner written notice that the work would not be performed unless the proper equipment was provided to the workers.

Case Study 4: Mason Falls from Defective Ramp

A mason fell approximately 3 ft (0.91 m) off a ramp at a construction site. The mason said that the contractor had violated Labor Law 240 because the ramp was defective and not nailed together properly. Among the injuries suffered by the mason were multiple leg and ankle fractures (*New York Jury Verdict Reporter* 1993). The court held in a final verdict on July 1, 1991, that the contractor was liable. The accident could have been prevented if the ramp and other structures were inspected after they had been installed to verify whether they had been secured properly. If a worker knows a defect exists, he or she should notify the appropriate authority.

Case Study 5: Plank Used as Brace in Doorway Situated 5 ft above Floor—Causes Cerebral Contusion

On September 17, 1987, an electrician was walking through a construction jobsite, tracing wires and reading blueprints, when he struck his head on a 2 × 4 plank that was braced in a doorway. The plank was 5 ft (1.52 m) above the ground, and according to the electrician, it was difficult to see because it was unpainted and unmarked. The general contractor disputed the position of the plank, stated that it was easy to see, and said that the electrician should have paid more attention to his surroundings. An architect testified that the brace was unsafe, and because the site was busy, the plank should have been painted or an orange flag should have been placed on it (*New York Jury Verdict Reporter* 1991a). The court determined that the general contractor was 70% liable for what occurred, the owner was 30% liable, and the subcontractor was not liable at all. Proper signage is mandatory at all construction jobsites. The plank should have been properly marked by the general contractor.

These cases are samples of the types of accidents occurring on construction jobsites that might have been prevented by emphasizing safety measures to employees. The injuries sustained by the workers in cases 1, 2, 4, and 5 could have been prevented with routine safety inspections. Safety inspections should have consisted of inspecting equipment and assisting devices after their installation and also checking signals, fencing, barricade placement, and other safety measures.

Improper worker instruction also contributes to the number of accidents that occur at construction jobsites. One of the most common causes of accidents is carelessness or indifference by workers, and another is workers not following proper construction procedures. All of the cases demonstrate how important communication is between workers and supervisors. In some situations, the implementation of safety measures is not enough to prevent accidents. Many of the contributory causes of accidents, such as traffic, pedestrians, and other factors, are external to a project. Workers and supervisors should be aware of these types of factors even though they are difficult to control.

Providing assisting equipment is important in accident prevention as this type of equipment helps laborers perform difficult tasks that have to be accomplished alone or with limited help from coworkers. Assisting equipment simplifies tasks, increases productivity, and may reduce the incidence of injuries by helping laborers perform strenuous activities, such as lifting the I-beam in case 3.

4.14 Managing Workers and Work Processes

For construction workers to properly perform their assigned work activities, they should have adequate information about those activities. Management personnel should have access to all of the available documentation required to effectively communicate work assignments. The documentation required to develop work assignments is listed in Table 4-7.

These documents are used to determine spatial relationships, construction sequencing, site accessibility, and fabrication sequences. When analyzing productivity, the documentation in Table 4-7 should be used to improve construction systems, to simplify the design or combine elements, to make it easier to understand information, to improve construction sequencing, and to improve the use of equipment or tools and communication.

Some of the ways productivity might be improved through human factors are listed in Table 4-8. Other items that affect productivity are listed in Table 4-9.

Some techniques that are useful for communicating the requirements of work tasks to workers include

- Providing physical layout plans;
- Providing detailed work plans;

Table 4-7. Documentation Used to Develop Work Assignments

Documentation
Bills of materials
Block and task drawings
Contracts, plans, and specifications
Equipment availability information
Estimate
Material delivery schedules
Physical models or building information models
Project schedule
Reference materials
Shop drawings
Supplier information
Trade information
Union contracts, if applicable
Work quantities
Worker availability information

Table 4-8. Methods for Improving Productivity through Human Factors

Methods
Adjusting work-break cycles
Implementing assembly-line production processes
Awarding bonuses
Effective crew size
Eliminating overcrowded work areas
Improving working conditions
Increasing crew sizes
Using modularization
Using prefabrication
Reducing labor requirements through the redistribution of work assignments
Reducing worker fatigue
Accelerating the schedule
Using advanced technology
Scheduling around the weather

Table 4-9. Factors That Affect Productivity

Factors
Availability of resources
Crew configurations
Dust
Equipment operations
Extreme winds
Hazardous waste mitigation issues
Material-handling processes
Noise levels
Reduction or elimination of site constraints
Seasonal weather variations
Site accessibility
Site layout
Tool availability
Transportation times for materials
Warehousing techniques
Worker experience levels
Worker interferences
Work processes

- Providing distances, elevations, locations of work stations, and equipment;
- Including detailed flowcharts, plans, isometrics, and cross sections; and
- Indicating the number of workers, craft-type categories, positions of workers, and movement of people and equipment.

4.15 Motivating Workers

Productivity rates are directly influenced by how motivated workers are when they are performing their work tasks. It is the responsibility of project management personnel to motivate workers to perform faster and more efficiently. Motivation is defined in many ways, including

- A drive to satisfy a need,
- An incentive to do something,
- Stimulation to do something one does not want to do, and
- The direction of an individual's behavior that one chooses when several alternatives are available.

An early definition of a motivation theory was provided by Greek philosophers when they described **hedonism** as "people behaving in a fashion that will maximize their pleasure" ("Hedonism" 2006). When trying to motivate workers, managers have to consider three elements that affect motivation levels: (1) the individual being motivated, (2) the job, and (3) the work situation. When managers try to motivate workers, they are really trying to optimize the talents of the workers.

Managers use **formal incentives, informal incentives, formal disincentives**, and **informal disincentives** to motivate workers. Formal incentives include money, bonuses, promotions, awards, special privileges, and choice of work schedule. Informal incentives include praise, encouragement, acceptance by groups, minimal supervision, and respect of peers or management. Formal disincentives include disciplinary action, demotions, layoffs, and the withholding of privileges. Informal disincentives include rejection by others, criticism, assignment to undesirable jobs, scrutiny from the supervisor, and lack of cooperation from coworkers.

It is not always the current situation that motivates or fails to motivate workers. The culture of the work environment also influences motivation. Workers are influenced by the perceived consistency and equity of the reward system; the perceived workplace structure, including rules, procedures, and red tape; and the perceived emphasis on rewards or punishment. Not all workers are motivated by the same things. Some workers are motivated by

- Achievement,
- Advancement,
- Interpersonal relationships at work,
- Job security,
- Recognition,
- Responsibility,
- Salary,
- Work affiliations, and
- Working conditions.

In many work situations, the anticipation of rewards is important, and this means that rewards should be tied to the behaviors desired by organizations. People choose to perform differently according to what they perceive to be a fair exchange. Workers compare their payoffs to those that others receive, and if a state of inequity exists, they may slow down their production rates to compensate. Workers consider rewards to be relative; it is not only how much they receive but also how much they receive compared to others.

One method for motivating workers is to set goals and link incentives to the goals. Employees normally set goals for themselves while they are working, and their work behavior could be affected by influencing their goals. Allowing employees to participate in the goal-setting process helps to motivate them, as does setting more specific goals. Incentives should be related to goal attainment, and the goals that are set should be difficult but attainable.

Positive reinforcement and **negative reinforcement** affect the production rates of workers. Positive reinforcement strengthens desired behaviors, and negative reinforcement reduces potentially damaging behaviors. The least effective type of reinforcement is **partial reinforcement**, or occasional, inconsistent reinforcement of behavior by management personnel. Partial reinforcement causes confusion among workers because the correct behavior is inconsistently reinforced and the workers are not sure which behavior is the desired behavior.

4.16 Monitoring and Controlling the Human Element

Productivity rates are affected by a variety of human factors that management might or might not be able to control, including

- Acclimation,
- Blood flow rates,
- Breathing patterns,
- Cold,
- Demands of tasks,
- Dizziness,
- Environmental factors,
- Ergonomics,
- Fatigue,
- Heat,
- Instability,
- Nausea,
- Noise levels,
- Overexertion,
- Sweating,
- Tool availability, and
- The abilities of workers.

Management personnel should be aware that all of these items cause variations in production rates. Managers are responsible for carefully monitoring workers to determine whether they are healthy enough to continue with their assigned work tasks. The human element of productivity improvement is unpredictable and difficult to control at construction jobsites. The most effective way to address the human factor is through closely monitoring workers during weather extremes and under unusual working conditions.

4.17 Summary

This chapter provided information on the human element of productivity and on factors that affect worker performance, including fatigue and extreme temperatures. The Occupational Safety and Health Administration was discussed in relation to how it monitors construction jobsite safety. The use of workers' compensation insurance in the construction industry was also discussed in this chapter.

This chapter reviewed the effects of safety issues on productivity rates, the causes of worker injuries, construction failures, and methods for reducing construction failures. Safety programs were discussed, and five case studies that illustrate the legal perspective of construction accidents were presented in this chapter. The last part of the chapter discussed managing workers and worker motivation.

4.18 Key Terms

accident incidence rates
asbestos
aspiration
construction failures
Construction Industry Institute
dermatitis
direct costs
double-layer clothing
ergometer
extreme temperatures
face masks
fatigue cycles
formal disincentives
formal incentives
gloves
hard hats
heat stroke
hedonism
illicit substances
indirect costs
informal disincentives
informal incentives
jobsite safety programs
motivation
National Institute for Occupational Safety and Health
negative reinforcement
New York Jury Verdict Reporter
occupational diseases
Occupational Safety and Health Act
Occupational Safety and Health Administration
partial reinforcement
physical requirements
positive reinforcement
respirators
rotating shifts
safety goggles
silica dust
Standard Industrial Classification Code
steel-toed boots
stress fatigue
tachycardia
tailgate meetings

toolbox meetings
U.S. Court of Appeals
U.S. Department of Health and Human Services
U.S. Department of Labor
U.S. Secretary of Labor
U.S. Supreme Court
willful misconduct
work-break cycles
work-rest cycle
workers' compensation acts
workers' compensation insurance

4.19 Exercises

4.1 Explain how a worker should be treated for heat stroke.

4.2 What physiological changes occur when a worker is subjected to temperatures below freezing for prolonged periods?

4.3 Which of the motivation techniques discussed in this chapter would be the most effective to use with construction workers and why?

4.4 What would be the four most effective methods for distributing safety information at construction jobsites and why?

4.5 What techniques could be used to prevent construction workers from becoming fatigued while they are lifting heavy construction materials for long periods?

4.6 Of the occupational health hazards listed in this chapter, which are the three most prevalent ones at construction jobsites and why?

4.7 Evaluate the safety risks of setting steel using the criteria listed in Section 4.4 in the subsection on Safety Equipment.

4.8 What are three techniques that might be used to persuade construction workers to tie off (hook their safety harness to the structure) while they are working on the upper floors of structures?

4.9 Which of the operator safeguards for heavy construction equipment operators is the least expensive to implement and why?

4.10 Why does productivity decline when construction workers are assigned to rotating shifts?

4.11 What are the three most frequent causes of construction failures according to the longitudinal study that was conducted using OSHA data?

4.12 List six of the personal safety devices that should be used by construction workers.

4.13 In the Construction Industry Institute study on construction safety, what were the cost categories for medical case injuries?

4.14 Why is it dangerous to work around asbestos when either building or remodeling structures?

4.15 What are eight of the safety prevention measures that might be used to prevent accidents?

4.16 What are the major causes of accidents related to excavations?

4.17 What are some of the safety measures that might be used to prevent cave-ins during excavations?

4.18 Of the causes of construction jobsite injuries listed in Table 4-2, which are the easiest to prevent, and how might these types of injuries be prevented at construction jobsites?

4.19 Contrast construction failures with construction safety issues.

4.20 Was one contributing factor prevalent in all of the legal case studies described in Section 4.13?

References

Aberdeen's Concrete Construction. (1992). "*Ways to reduce injuries and workers' compensation rates.*" (37), 167–168.

Adrian, J. (2004). *Construction productivity: Measurement and improvement*, Stipes, Champaign, IL.

Altayeb, S. (1992). "Efficacy of drug testing programs implemented by contractors." *J. Constr. Eng. Manage.*, 118(4), 780–789.

Bernold, L., and Guler, N. (1993). "Analysis of back injuries in construction." *J. Constr. Eng. Manage.*, 119(3), 607–621.

Cho, A., Bodilly, L., Blair, S., and Parsons, J. (2008). "Arizona workers are old pros at sweating in the southwest." *Engineering News Record*, Aug. 11.

Clough, R., Sears, G., Sears, K. (2005). *A practical guide to contract management*, John Wiley and Sons, New York.

Culver, C., Marshall, M., and Connolly, C. (1993). "Analysis of construction accidents: The workers' compensation database." *Professional Safety*, no. 38, 22–27.

"Hedonism." (2006). *Webster's New Universal Unabridged Dictionary*, Simon and Schuster, New York.

Hinze, J. (2006). *Construction safety*, Prentice Hall, Upper Saddle River, NJ.

Hinze, J., and Applegate, L. (1991). "Costs of construction injuries." *J. Constr. Eng. Manage.*, 119(3), 537–550.

Hinze, J., and Wiegand, F. (1992). "Role of designers in construction worker safety." *J. Constr. Eng. Manage.*, 118(4), 677–684.

Janney, J. R., ed. (1986). *Guide to investigation of structural failures*, American Society of Civil Engineers, New York.

Lockley, E. (1997). "*Construction failure investigation techniques.*" Ph.D. dissertation, Polytechnic School of Engineering of New York University, New York.

MacCollum, D. (1990). "Time for change in construction safety." *Professional Safety*, no. 35, 17–20.

National Institute for Occupational Safety and Health. (2011). *Preventing worker deaths from trench cave-ins*, Department of Health and Human Services, Washington DC. <http://www.cdc.gov/niosh/docs/wp-solutions/2011-208/pdfs/2011-208.pdf> (April 2014).

New York Jury Verdict Reporter. (1991a). 9(19), 7.

New York Jury Verdict Reporter. (1991b). 9(22), 2.

New York Jury Verdict Reporter. (1992a). 9(11), 2.

New York Jury Verdict Reporter. (1992b). 10(16), 1.

New York Jury Verdict Reporter. (1993). 10(41), 15–16.

Occupational Safety and Health Administration (OSHA). (2011). *OSHA publishes new educational materials on working safely during trenching operations,* U.S. Department of Labor, Washington, DC. <https://www.osha.gov/pls/oshaweb/owadisp.show_document?p_table=NEWS_RELEASES&p_id=20796> (March 2014).

Occupational Safety and Health Administration (OSHA). (2014). *OSHA Technical Manual TED 01-00-015 (TED 1-0. 15A Chapter 2 – Excavations: Hazard Recognition in Trenching and Shoring,* Washington, DC. <https://www.osha.gov/dts/osta/otm/otm_toc.html> (March 2014).

Oglesby, C., Parker, H., and Howell, G. (1989). *Productivity improvement in construction,* McGraw Hill, New York.

Smith, G., and Roth, R. (1991). "Safety programs and the construction manager." *Journal of Construction Engineering and Management,* 117(2), 360–371.

Suruda, A., Smith, G., and Baker, S. (1988). "Deaths from trench cave-in in the construction industry." *Journal of Occupational Medicine,* 30(7), 552–555.

Timura, M. (1990). "Construction workers should be part of effort to improve workplace safety." *Occupational Health and Safety,* 59(10), 74.

U.S. Department of Labor. (2011). *Census of fatal occupational injuries,* U.S. Bureau of Labor Statistics, Washington, DC.

U.S. Department of Labor. (2014). *U.S. Labor Department aims to protect workers from heat related injuries,* <https://www.osha.gov/SLTC/heatillness/index.html> (March 2014).

Yates, J., and Lockley, E. (2002). "Documenting and analyzing construction failures." *Journal of Construction Engineering and Management,* 128(1), 8–17.

Yates, J., and Terrero, N. (1997). "Construction industry safety measures." *Cost Engineering Journal,* 39(2), 23–32.

CHAPTER 5

Preparing to Conduct Productivity Improvement Studies

Construction industry productivity is influenced by a variety of variables, such as government regulations, the state of the economy, the unique practices of each firm, management styles, and the attitudes of workers. Construction jobsite operations are also affected by outside influences, including voluntary or mandatory quality-control programs. Productivity rates decline when there are frequent quality and safety inspections and also when there are frequent design or schedule changes during construction. In addition, if the workers at construction jobsites are not properly trained to perform their assigned work tasks, productivity might decline. The location of a project (whether remote or urban), material shortages, and scheduling delays also negatively affect productivity. Local, national, or global material shortages and increases in the cost of materials causes lower production rates attributable to changes in the scope of work, rework, or schedule delays to adapt to delays in receiving materials or substituting different materials. If the retention rate for workers is low, productivity declines because of the additional time required for new workers to master their work tasks. While developing a plan for improving productivity, productivity improvement team members should examine these influences to determine whether they are negatively affecting current production rates.

This chapter provides information on how to prepare for conducting productivity improvement studies. It discusses the responsibilities of productivity improvement team members, ways to plan for productivity improvement studies, techniques for preparing productivity study interview guides, how to define the scope of productivity improvement studies, potential work improvement areas, the **scientific analysis method**, methods for investigating productivity improvements, **learning curves**, the effects of overtime on productivity rates, and work-sampling techniques.

5.1 Productivity Improvement Team Member Responsibilities

Either engineering or construction management personnel are responsible for investigating declines in productivity and for identifying whether declines are caused by deficiencies in management, the workers, the owner, or technical aspects of the project. These personnel rely on productivity improvement team members to assist them in their evaluation. Productivity improvement team members help identify work task problems and potential solutions, revise existing work processes, develop new work processes, and work with management personnel to prioritize work issues and increase production rates. They are responsible for collecting data at construction jobsites, analyzing the data, and proposing alternative work processes. They also test potential productivity improvement processes by implementing pilot programs that provide feedback that is used to refine proposed programs. If a pilot program is successful, then it is integrated into jobsite work processes.

Productivity improvement team members should review management and project priorities before they begin to analyze work processes and evaluate production rates, as this helps them to become familiar with the objectives of the firm. Once they have reviewed the objectives of the firm, they develop the scope of the productivity improvement study. Next they focus on evaluating the level of effort expended by workers. Then they identify how work processes will be analyzed, determine data requirements and the data already available, and design a knowledge-acquisition system for gathering data during the study.

Productivity improvement team members perform a variety of tasks, including explaining to management personnel how productivity improvement programs are implemented at construction jobsites. They also provide management with a schedule of the activities included in the study and meet with the workers involved in the study to explain how the study will be conducted and how the workers will be involved in the productivity study. Productivity improvement team members design the work-sampling techniques for collecting data for the studies and the strategies that should be used to implement the recommendations that result from the productivity improvement studies.

Productivity improvement team members outline the steps required to implement the productivity improvement program. They then review existing site procedures and processes and develop a work plan that addresses how the study will analyze existing procedures and processes. During this phase of the study, they also determine what personnel and equipment will be required for collecting data during the study. Other responsibilities of the productivity improvement team members are

- Planning site visits,
- Scheduling and conducting interviews with jobsite personnel,
- Reviewing jobsite activities and work requirements,
- Evaluating whether there are more efficient ways to use equipment or perform work processes,
- Evaluating the potential use of new materials or technology,
- Developing alternative crew configurations, and
- Determining whether material movement sequences might be revised to reduce the time required to transport materials,

All of these activities help the team properly evaluate project parameters and design productivity improvement programs.

Productivity improvement team members develop training programs to demonstrate to workers how to increase their production rates, use their tools properly, and read the diagrams of proposed work process improvements. These programs also cover the use of work environment photographs, proposed work-process changes, and simulation exercises. Productivity improvement team members explain to workers how they should perform the proposed alternative work processes and how the alternative work processes will help to increase production rates. When team members attempt to introduce alternative work processes or improvements to existing methods, the workers may not immediately adopt them; therefore, the team members might have to provide additional instructions or periodically monitor whether the workers are implementing the new or revised methods.

5.2 Productivity Improvement Study Interview Guides

Before a productivity improvement study is initiated, the productivity improvement team members prepare the interview guides they will use to solicit information from the owner, construction management personnel, foremen, and the workers. The following sections discuss the importance of using interviews in productivity improvement studies and the types of information that should be obtained during the interviews.

Owner and Construction Management Personnel Interviews

The productivity improvement team members interview the owner and construction management personnel to gather information that will be helpful during the development of the productivity improvement study. The following types of questions are useful when collecting information during these interviews:

- Did previous projects funded by the owner finish on time and within budget?
- What types of project control systems are being used on the project?
- How familiar are the owner and construction management personnel with productivity improvement programs and the processes required for analyzing productivity rates?
- What is the time frame for conducting the productivity improvement study?
- Would any lessons learned from previous projects be helpful for increasing the productivity rates on the current project?
- Would the owner be able to provide additional resources if required to help increase productivity rates, and if so, what type of resources?
- Does the contract contain a liquidated damage clause that requires the project to be completed by a certain date, and will penalties be assessed for each day that the project exceeds the stipulated date?
- Has the owner ever used productivity improvement techniques on any of his or her other projects, and if so, what were the techniques?

- Is documentation available that would be helpful to review before evaluating productivity on the project, and if so, what types?
- Are there any constraints on the project, such as scheduling, budget, or personnel constraints?
- Is the owner willing to try new and innovative techniques on the project, and if not, is he or she willing to try them on only a select few project activities?
- Is the owner willing to participate directly in the productivity improvement study, and if so, in what capacity?
- Are there any obstacles to implementing new or revised work processes?
- Are there limitations on the funds available for the productivity improvement study, and if so, what are they?
- Is the project a union or a nonunion project?
- Is the owner familiar with three-dimensional building information modeling (BIM) software programs and their capabilities, and have BIM programs been used on any of his or her previous projects?
- Does the owner have access to BIM or simulation software programs that could be used on the project to model existing and potential work processes?
- Does the owner want to include computerized simulation modeling software programs in the productivity improvement study, and if so, how much is he or she willing to pay for the software programs?
- Are there different types of heavy construction equipment available for use on the project for excavation, trenching, lifting, placing, or other operations than what is currently being used on the project?
- Does the owner have any personnel available who could directly assist with the productivity improvement study, and if yes, how many and what are their qualifications and backgrounds?
- Is there a firm deadline for the completion of the productivity improvement study, and if yes, when is it?
- Does the owner have a specific goal or a set percentage for productivity improvement on the project, and if so, what is the goal?
- Is it possible for the contractor to be directly involved in the productivity improvement study, and if so, in what capacity?
- Is the owner interested in follow-up studies to verify the effectiveness of the work improvements implemented during the productivity improvement program?
- What are the desired outcomes of the productivity improvement program?

Once the interviews have been conducted, the answers obtained from the interviews should be summarized in a report that is reviewed by all of the productivity improvement team members.

After the interviews are completed, the productivity improvement team members should provide the owner with information on how the study will be conducted, what tasks will be performed by each of the team members, how the data will be analyzed, and how the potential benefits of the proposed alternatives will be determined in the study. Team members should explain to the owner that if productivity improvement programs are conducted properly, the potential benefits include increased productivity rates, more efficient work processes, a reduction in the time required to complete activities, jobsite layout improvements, crew size reductions, improved safety, and improved worker communication.

At the conclusion of the study, the productivity improvement team members should provide the owner with a report that includes the results of the study and that recommends alternative work processes and information on how profits, safety, and productivity will be enhanced if the recommendations in the study are implemented on the project.

Foremen and Worker Interviews

Productivity improvement team members should also interview the foremen and the workers and observe their work activities. A variety of questions are used to collect information from foremen and workers because their input is important for establishing the parameters of the productivity improvement study and analyzing work tasks during the study. A sample foreman interview guide is included in Chapter 7 in Table 7-3. It is useful to obtain information related to tools; equipment; personnel; management practices, including planning issues and priorities; information related to worker performance; feedback systems; methods used for problem solving; decision-making processes; and jobsite configurations.

The questions in the interview guide should require specific answers rather than a yes or no response in order to gain additional insight into work processes. It is not sufficient to know only whether certain work tasks are performed by the workers. The foreman and worker interview responses are incorporated into the productivity improvement analysis; they are used to determine whether work processes should be modified; and they provide insight on worker performance, worker motivations, work processes, and jobsite organization.

5.3 Defining the Scope of Productivity Improvement Studies

Productivity improvement studies require (1) a **planning stage**, (2) a **development stage**, and (3) an **implementation stage**. During the planning stage, different approaches are selected for analyzing productivity issues. The development stage is used to refine the approaches for reviewing and analyzing research materials. Strategies for implementing different approaches in order to define how to accomplish each work task are developed during this stage. During the implementation stage, work-process activities are analyzed using the strategies developed during the planning and development stages. The procedures for implementing productivity improvement programs are similar to the ones used for implementing project plans. According to Kerzner (2005), project plans include

1. Determining how much funding is available for projects,
2. Determining when a project has to be completed and whether there is any flexibility in deadlines,

3. Defining the limits or boundaries of a project,
4. Investigating different methods for achieving project objectives,
5. Locating potential viable alternatives that could be used to achieve project objectives,
6. Evaluating the alternatives to determine the best method for achieving project objectives,
7. Explaining the alternatives and their implementation to project team members,
8. Implementing the alternative selected and monitoring its performance outcome,
9. Documenting the outcome of implementing the alternatives.

5.4 Potential Work Improvement Areas

When the scope of a study is being defined, it helps to divide work processes into smaller tasks to facilitate the monitoring of performance. When work improvement techniques are implemented, innovative changes can be applied in several areas, including the following (Taylor 1967):

- *Hand and Body Motions*—Worker movements, and the sequence in which they are performed, along with how they are performed, could be altered to make it easier to perform tasks or to reduce the time required to perform tasks.
- *Work Stations* (including tools, work locations, how the work is laid out, and equipment)—The area immediately surrounding a worker, or the tools and equipment that are required to perform a task, could be rearranged or replaced to increase productivity.
- *Work Processes and Work Sequencing*—Work processes, how materials are delivered to work stations, the number of work stations, or the composition of the work all could be rearranged or altered to increase productivity.
- *Product Design Revisions and Material Delivery*—The products and construction elements could be redesigned or altered at the fabricator to facilitate their shipping and installation at construction jobsites.
- *Construction Materials*—The materials being used during construction could be modified or fabricated differently in order to increase production rates for assembling materials.

According to Taylor, there are five classifications used to distinguish between the areas where productivity improvement techniques are utilized on projects and they are (Taylor 1967):

1. *Class One: Hand and Body Motion Alterations*—**Methods Time Measurement** [MTM] procedures could be used to analyze individual body movements or motions and the time required for each movement or motion. [Methods time measurement is covered in detail in Chapter 6].
2. *Class Two: Tools, Equipment, and Workplace Alterations*—The foreman, work crew members, or a tool specialist are involved in replacing or altering the tools or equipment being used by the work crew.

3. *Class Three: Work Process Alterations*—Changing the sequence of operations or adding or deleting operations in any manner that requires the assistance of foremen, engineers, or management personnel.
4. *Class Four: Product Design Alterations*—Changes in the size, shape, form, appearance, tolerances, or finishes of products, including any changes to the plans. This requires assistance from the workers, foremen, and inspectors.
5. *Class Five: Raw Material Alterations*—When project personnel or purchasing department personnel alter the type of materials; the quality of materials; the quantity of materials; the chemical composition of materials; or the form, shape, or the appearance of materials.

An alteration in one of these classes may require a corresponding change in one or several of the other classes.

Making changes to the specifications for raw materials and to the design of the final products may require modification of the plans. Some changes require modifying only the layout of the jobsite and the pattern of hand and body movements, but changes to items in Class 4 or 5 normally affect more than one item. In the higher classes, more items have to be altered before being able to implement changes, and more workers are affected by these changes.

5.5 Scientific Analysis Method

To conduct a productivity improvement study, specific steps from scientific analysis methods are used to evaluate the study parameters and implement new work-process methods. The analysis procedures used in scientific analysis methods require seven steps (U.S. Department of Defense 1963, Chapter 3, p. 6):

1. Determining the problem and its associated objectives.
2. Determining the procedures to be used for analyzing tasks and grouping tasks into appropriate units.
3. Developing lists of both desirable and undesirable features of the alternatives.
4. Determining how to describe new or modified methods.
5. Determining how to check the desirability of new or improved methods.
6. Providing the educational materials that are required to train workers or operators on how to perform new or modified procedures.
7. Providing methods for determining the amount of time required for a job and a method for controlling the quality of work.

The scientific method for solving problems consists of the following steps (U.S. Department of Defense 1963, Chapter 5, p. 4):

1. *Aim*—Set an objective for the job to be improved and select improvement criteria. This involves a rough determination of the amount of change warranted and the consequences of installing various types of changes.

2. *Analysis*—Divide current methods into smaller parts that can be understood and evaluated. This step is performed by applying an analysis technique to each type of task.
3. *Criticism*—Evaluate each part of the job and divide each part into good or bad categories by comparing them to a characteristic list of desirable or undesirable qualities for each type of work.
4. *Redesign of the Job*—Discard undesirable parts of the job, simplify necessary parts, add new and easier parts, and combine parts to create a new job. Sketches of jigs, fixtures, and new tooling techniques are useful during this stage.
5. *Test*—Evaluate the new method in the same manner as the old method, using the same characteristic list. Be sure that the improvements meet the criteria set in Step 1.
6. *Trial*—Test a sample installation using one operator and a mock-up of the task.
7. *Application*—Install the new method, train operators, write **standard operating procedures**, and revisit the process periodically so that there is not a reversion to old methods.

For problems with intermediate degrees of definition, various multistep strategies might be appropriate. One seven-step strategy for analyzing these types of problems is the following (Wankat 1983; Woods et al. 1984; Woods 1987):

1. Problem definition
2. Exploration
3. Planning
4. Implementation
5. Verifying results
6. Generalizing to other activities
7. Documentation

A seven-step process is also used to analyze semidefined problems. The first step is having productivity improvement team members define the problem and then perform a productivity improvement study. In the second step, potential constraints and the criteria used to identify potential solutions are defined. To identify known and unknown issues, a layout of the work processes and an abstract representation of the fundamental relationships between them should be drawn (Wankat 1983).

The third step involves exploring the problem dimensions, including the type of problem, its pattern (routine or nonroutine), the parts of the pattern that are routine, potential solutions, the control systems that should be used, the data available, and the data that should be acquired during the study (Wankat 1983).

The fourth step is defining the formal logic, determining the steps required to develop a problem flowchart that illustrates the interrelationship of work processes, and determining which equations should be used to calculate productivity rates. The fifth step is applying the equations developed in the fourth step to generate results. The sixth step is validating the results obtained in the first five steps by verifying the calculations and comparing the results to external criteria. The seventh step is analyzing the results and developing recommendations for future improvements to work processes (Wankat 1983).

Some considerations that should be reviewed during problem-solving processes are worker motivation, the skill level of workers, the experience level of workers, similar work processes, the learning abilities of the workers, communication processes, worker knowledge of work processes, the perceptions of the workers, coordination techniques, and leadership styles (Wankat 1983).

5.6 Planning Productivity Improvement Studies

When planning to conduct a productivity improvement study, productivity improvement team members should review previous productivity investigations performed on the project to determine whether any recommendations resulting from the studies were successfully applied to the project. All of the techniques available for increasing productivity are evaluated to assess their applicability to the current project, and the team members determine what has to be obtained from management personnel in order to implement new work processes. A system is devised for recording the results of the study to provide detailed documentation for future projects.

Data on specific areas that influence the progression of work processes is collected. Areas for investigation include internal influences, such as the factors that affect resources, planning, worker behavior, operational methods, and work strategies. To prepare for a productivity improvement study, productivity improvement teams inventory current practices and procedures, review administrative controls, determine what assets are available for tasks, and identify the strategies currently being used to accomplish tasks. Team members also determine whether technical problems might hinder the implementation of any changes and whether there are worker shortages or morale problems.

Before productivity improvement studies are conducted, productivity improvement team members develop plans that help to ensure the proper timing of work to avoid the performance of outdoor activities in inclement weather. The plans also delineate viable layout locations for site work, ensure that tools and equipment are available, schedule procurement activities, schedule a risk analysis and time for developing contingency plans, incorporate funds for administrative salaries, and provide a forum for designing and implementing the reporting systems that will be used on the project.

Before conducting a productivity improvement study, productivity improvement team members also decide whether they will use **informal assessments**, **formal assessments**, or both for collecting data. During informal assessments, work processes are observed for a short period and are assessed on the basis of the observations. Typical informal assessments do not provide sufficient observation time to allow for accurate conclusions. Conducting additional observations improves the accuracy of the conclusions and helps to determine whether

- The materials and tools being used are readily available and suitable for the task,
- The work procedures and their sequencing are being conducted efficiently, and
- The tasks are assigned to members of crews in a manner that best utilizes the abilities of the workers and that maximizes the amount of time they are working productively (Oglesby et al. 1989, p. 135).

Formal assessments are initiated when there are schedule slippages or cost overruns. An example of a formal assessment procedure is exception reporting, which is used to highlight delays and cost overruns. Project team members are responsible for determining the reasons delays occur and for suggesting mitigation strategies that, if implemented, would provide a means for expediting activities and achieving scheduled milestones (Oglesby et al. 1989). Exception reporting, which only reports activities that are not meeting predefined criteria, is a classic example of using a systematic process for evaluating worker performance on a macro level.

5.7 Learning Curves

Learning curves are used to summarize productivity improvement through repetition. They are a mathematical modeling process used for predicting or measuring improvements in the productivity of workers over time while they are performing repetitive work tasks. Learning curves are used to model productivity during repetitive work and to capture the cumulative efforts of experience in performing the same work tasks multiple times. When learning curves are developed, they should incorporate deviations from typical routines and unusual circumstances that negatively affect job progress.

Productivity improves when (1) the individual workers are trained before they start performing their work tasks, (2) workers become more proficient at performing their work tasks the longer they perform them, (3) workers observe how other workers are performing their work tasks, and (4) workers receive positive reinforcement from management. Learning-curve models are used by engineers or managers to determine how quickly workers are able to master their work tasks. When learning curves are plotted arithmetically, as shown in Figure 5-1, the expression for a hyperbolic curve is (Adrian 2004, p. 233)

$$Y = AX^{-n} \tag{5-1}$$

where Y = work hours required to produce the Xth unit of work; A = measured value of the cost or work hours required to produce the first unit of work; X = number of work units produced; $n = (\log A - \log Y)/\log X$ = derived exponent that describes the work output variation.

When plotted using logarithmic coordinates, as shown in Figure 5-2, the curves become straight lines; n, the variation, then equals the slope of the line; and the following relationship is used (Oglesby et al. 1989, p. 138; Adrian 2004, p. 233):

$$r/s = (j/i)^n \tag{5-2}$$

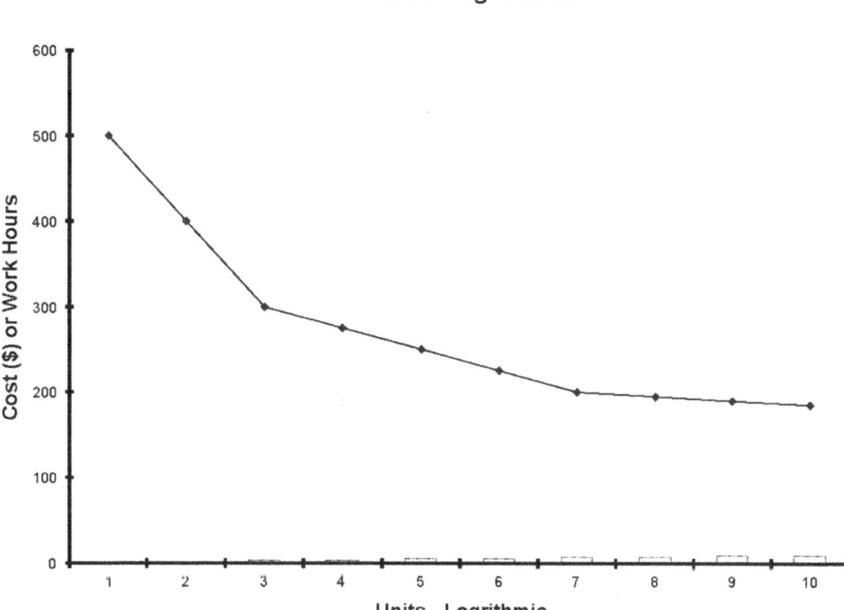

Figure 5-1. Arithmetic Plot of Learning Curve

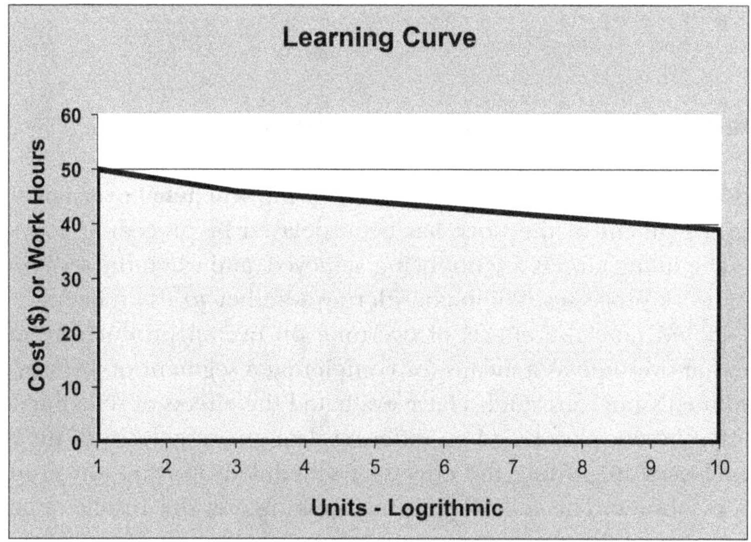

Figure 5-2. Logarithmic Plot of Learning Curve

where i = unit sequence number of the earlier unit; j = unit sequence number of the later unit; r = variable for unit i (measured in dollars or work hours); s = variable for the unit j; $n = (\log r - \log s)/(\log i - \log j)$; and the percent of the curve = 2^n.

Eqs. (5-1) and (5-2) apply to situations where there is an independent relationship. The construction industry has many activities with dependent and independent characteristics; therefore, a system with multiple independent inputs may be expressed as follows (Adrian 2004, p. 235):

$$Y = X_1 + X_2 = a(X_1) + b(X_2) \tag{5-3}$$

An interdependent system may be expressed as follows:

$$Y = b_1 b_2 X \tag{5-4}$$

And a system with multiple interdependent inputs may be expressed as follows (Adrian 2004, p. 236):

$$Y = dX^{\alpha}{}_{1}X^{\beta}{}_{2} \tag{5-5}$$

where d, $\propto$, and β = constants related to production functions.

The smoothness of the learning curves could be disrupted by slower work progress during holidays, breaks between similar projects, or training periods. Workers may also have to increase their work efforts because of unforeseen site conditions, equipment or material delays, strikes, or for other reasons. Learning curves indicate that as workers master the intricacies of their work tasks, the time required to complete work tasks decreases. When workers are beginning to learn new work tasks, the time required to complete work tasks is longer than the time required after they have mastered their work tasks, and there is a rapid decline in time requirements while workers are learning how to perform their work tasks (Oglesby et al. 1989; Adrian 2004).

5.8 Effects of Overtime on Productivity Rates

Construction management personnel consider using **scheduled overtime** to expedite work when a segment of the work has been delayed by preceding activities, when projected scheduling targets are not being achieved, and when the owner would like to expedite work processes. When considering whether to use scheduled overtime, managers should note the effects of overtime on overall productivity rather than merely viewing overtime as a means for completing a segment of work by the designated deadline. Numerous studies have evaluated the effects of scheduled overtime, and although these studies relied on different data sets, a majority of them reported similar conclusions regarding the effects of scheduled overtime on productivity at construction jobsites. The following paragraphs discuss the results obtained from several of these studies.

In one of the initial studies on the effects of scheduled overtime on productivity, the National Electrical Contractors Association (NECA 1969) determined that productivity decreased with each successive week of overtime in a linear progression up to the end of the fourth week. Table 5-1 lists the productivity decline that occurred during the NECA study.

Another study on the effects of productivity on scheduled overtime, published by the Business Round Table (1974), showed that for a 50-h work week, productivity

Table 5-1. NECA Overtime Study Results

Work Crew Configuration	Productivity Rate at End of 4 Weeks (Percentage)
6 9-h days	87.5
6 10-h days	79.8
7 8-h days	75.0
7 9-h days	69.0
6 12-h days	63.5
7 12-h days	47.5

Source: Adapted from NECA (1969), Figure 11.

Table 5-2. Loss of Productivity during Concrete Operations with Overtime

Days	Daily Hours	Total Weekly Hours	Inefficiency (Percentage) 7 Days	Inefficiency (Percentage) 21 Days
5	9	45	2–4	6–8
6	9	54	3–5	7–11
7	9	63	10–12	21–23
5	10	50	5–7	11–13
6	10	60	6–8	13–17
7	10	70	13–15	26–29
5	11	55	8–11	16–18
6	11	66	11–13	20–22
7	11	77	19–21	36–41

Source: Adapted from Adrian (2004), Table 8.2.

declined each successive week for eight weeks before leveling off at approximately 71% of normal productivity. The same study showed that for a 60-h work week, productivity declined during the first six weeks and leveled off during the seventh week at 64% of normal productivity. For both scenarios, productivity rates spiked during the second week before continuing to decline (Business Round Table 1974).

Studies published by the American Subcontractors Association, the Associated General Contractors of America, and Associated Specialty Contractors Inc. (1979) and by the American Association of Cost Engineers (1973) indicated similar results to those cited in the 1974 Business Round Table study.

A study of workers involved in concrete work that was conducted in Chicago in 1982 also demonstrated losses in productivity during overtime (Adrian 2004). Table 5-2 provides the results obtained from the study grouped by hours worked rather than, as in the original study, by number of days.

In another study, titled "The Effects of Scheduled Overtime and Shift Schedule on Construction Craft Productivity" (Construction Industry Institute 1988, p. 33), the following conclusions were reached about scheduled overtime:

- Previous studies by the Bureau of Labor Standards, the Business Round Table, and others were not consistent predictors of productivity loss during overtime schedules for construction projects in this study.
- Even individual crews working an overtime schedule on the same project do not demonstrate consistent productivity trends.
- Productivity does not necessarily decrease with an overtime schedule.
- Absenteeism and accidents do not necessarily increase under overtime conditions.

Another Construction Industry Institute publication, which summarizes the productivity research of Dr. Randolph Thomas (1990, p. 50), concludes that the data suggest

> that the six day week is about seven percent (absolute) more efficient than the seven day week. The ten hour workday compared to a nine hour workday results in a loss of efficiency of about four percent (absolute). The twelve hour workday increases the inefficiency by another seven to eight percent (absolute).

The scheduled overtime studies cited in this section indicate that it is difficult to predict the effects of scheduled overtime on productivity at construction jobsites. All of the studies suggest that productivity declines when workers are required to work scheduled overtime for long periods, but there is no consensus on how much productivity declines.

5.9 Methods for Investigating Productivity Improvement

Productivity improvement programs use various data collection techniques. In some situations, using only one or two data collection techniques provides adequate data, but for detailed productivity improvement studies, multiple techniques provide more accurate data. Some of the techniques for gathering data for productivity improvement studies are

- Computer modeling,
- Computer simulation,
- Digital photographs,
- Direct observation,
- Interviews,
- Methods Time Measurement (MTM),
- Questionnaires,
- Time-lapse photography,
- Work-process videotapes, and
- Work sampling,

All of these methods are described in detail later in this chapter and in the following chapters.

5.10 Work Sampling

Work sampling is used to estimate the amount of time that workers are actively engaged in productive work activities. Rather than observing workers for the entire time they are working on their tasks, random observation samples are recorded and analyzed to estimate the percentage of time that workers are productive or unproductive during a particular work period. The data collected through work sampling are extrapolated to cover the entire work period in order to estimate the overall percentage of time that workers are actively engaged in work.

The accuracy of the observations recorded using work sampling is dependent upon the number of times that the workers are observed while they are performing their normal work assignments. If workers were observed only a few times during an 8-h day, the data could not be used to reliably predict the percentage of time that the workers were productive during the work period. Additional samples would have to be recorded to increase the accuracy of the predictions.

The number of observations required for a work-sampling study is dependent upon the accuracy desired for the study. If a study were used only to determine a pattern of behavior or deviations from a desired behavior, then a minimal amount of samples would be sufficient to provide adequate data to satisfy the study objectives. If a study requires a high level of data accuracy to develop performance standards, rather than general objectives, then numerous samples would have to be recorded and analyzed during the study (Oglesby et al. 1989).

Work sampling is used to estimate how time and resources are used on activities when other data are not available from records or existing documents. Work-sampling techniques are useful for analyzing jobs that are nonrepetitive and highly variable. Work samples provide information about existing situations that is useful when setting specific goals or objectives. Work sampling is also useful for determining work assignments and the appropriate distribution of work, especially when there is a shortage of workers. The data are also helpful to have when allocating worker bonuses, developing equipment use rates, and determining which work areas would benefit from productivity improvement studies that assess how time is being expended or wasted during activities. Work sampling provides data on cyclic variations and their effect on worker performance (Oglesby et al. 1989).

Work-sampling techniques are useful when funds are limited for productivity improvement studies because conducting work-sampling studies requires minimal time and effort compared with the other activity-sampling techniques discussed later in this chapter and in Chapter 6. Only one person is required to perform a work-sampling study during short periods over several days. The randomness of the observations does not affect the study because it is intentionally designed for collecting random data. If the data collection process is interrupted, it can be continued later without negatively affecting the results of the study. Work sampling is not an efficient method for studying work activities that occur over a large work area or for evaluating the performance of one operator or worker. Work sampling is more useful for recording observations on whether members of stationary work crews are working effectively.

When recording work-sampling observations, observers need to ensure that the workers are not aware of them and do not know that a productivity improvement study is being conducted. Otherwise, the workers could intentionally skew the results by performing their work faster or slower than normal. Work-sampling techniques are used to collect data on the performance of workers under typical working conditions. Then the data are analyzed by productivity team members to determine whether the workers are performing at optimum levels, whether materials are being used efficiently, whether the jobsite layout could be improved, whether the workers have adequate tools and equipment, and whether the equipment is being used to its fullest capacity. The data collected during work-sampling studies should be collected in a consistent manner, and the activities being observed should not be dependent on other activities that precede or succeed them because waiting for other activities to be started or completed could alter the results of the data-collection process.

Work-Sampling Statistics

The goal of work sampling is to record a sufficient number of samples in order to accurately evaluate work processes. To collect reliable data, the observers should be familiar with statistical methods for ensuring the validity of the data. Work sampling for productivity improvement involves observations and the classifying of work tasks to evaluate work processes. Work sampling is performed by direct observation, through interviews, or through questionnaires. As the number of direct observations increases, the accuracy of the analysis improves, but the desire for accuracy has to be balanced against time and the cost associated with collecting additional data.

The degree of certainty about the results obtained from only samples of data, rather than analyzing entire data sets, are expressed using three statistical methods: (1) **confidence limit**, (2) the **limit of error**, and (3) the proportion of the sample having (or failing to have) the characteristic observed during a study. According to Oglesby et al. (1989, p. 149), "The first two terms are statistical terms and the last term is a physical condition that is mentioned because it affects the relationship between sample size and the values of the other two factors."

Confidence Limits

Confidence limits are a measure of the dependability of results; therefore,

> a confidence limit of ninety-five percent indicates that an answer could be relied upon ninety-five percent of the time or, conversely, that the answer could be wrong five percent of the time, such as one in twenty determinations. The confidence limit is often chosen according to the purpose of the sample. When human subjects are involved, the tolerable limit for acceptance may be one in one million, i.e., a confidence limit of 99.9999 percent.

> When sampling concrete blocks, to be wrong one time in twenty times, a confidence limit of ninety-five percent might be acceptable. The higher the confidence limit obtained, the lower the chance of being wrong, and the larger the number of observations or samples that have to be made. If certainty, such as a hundred percent confidence limit, is required then every item must be inspected. (Oglesby et al. 1989, p. 149)

Limit of Error

According to Oglesby et al. (1989, p. 149), the limit of error "expresses the accuracy of estimated results. If checking for cracked blocks, it is the percentage variation on either side of the value obtained by sampling within which the results of the true value is expected to fall, given a prescribed confidence limit. It could be stipulated that the estimate of damaged blocks, based on sampling, will fall within plus or minus ten percent of the total number of blocks that were actually damaged and that this result could be depended upon ninety-five times out of one hundred."

Category Proportion

One definition of **category proportion** is provided by Oglesby et al. (1989, p. 149):

> The proportion of the sample that is expected to have a given characteristic. It is the portion of the sample that is expected to have the characteristic being measured and it affects (mathematically) the size of the sample required to meet the pre-established confidence limit and limit of error. For a block example, perhaps one out of ten cracked blocks in each lot would be acceptable, so the category proportions of cracked versus uncracked blocks are set at ten and ninety percent, respectively. Similarly, if workers or equipment were productive during thirty-seven percent of the work period, the category proportions of productivity versus lost time would be 0.37 versus 0.63 respectively.

The national average for productivity in construction is 50%; therefore, category proportion charts are rarely used because a category proportion of 50% is only minimally less accurate (by a few percentage points) than category proportions that are higher (Oglesby et al. 1989). When a category proportion of 50% is used, the numbers of samples required for the limits of error of 1%, 3%, 5%, 7%, and 10% are 9,600, 1,067, 384, 196, and 96, respectively. Because the cost of obtaining 9,600, or even 1,067 samples, could be prohibitive, a limit of error of 5% is normally used (Oglesby et al. 1989). The following formula provides a method for more accurately calculating the number of observations required for each confidence limit and category proportion (Oglesby et al. 1989, p. 151):

$$\text{number of observations required} = N = \{K^2[P(1-P)]\}/S^2 \qquad (5\text{-}6)$$

where K = number of standard deviations, which vary with the desired confidence level; P = category proportion; and S = limit of error.

5.11 Recording Jobsite Observations

Sufficient amounts of data are required to adequately study work processes, and graphical representations of data facilitate the understanding of the results of a study for nontechnical personnel. Sketches that include detailed measurements of site layouts, key cross sections, and elevations could be used to demonstrate study parameters. Detailed flowcharts of operations are used to demonstrate the flow of work processes, the movement of workers and materials, and the positions of workers and equipment. To develop detailed flowcharts or sketches of construction operations, project records are reviewed and then construction operations are recorded using a variety of techniques. Chapter 7 describes three case studies and references a fourth case study in Appendix A that demonstrates the use of a variety of methods for recording jobsite observations.

Project Records

The value of examining project records, such as schedules, cost-control data, estimates, or budgets, is directly proportional to the quality of the historical data records. Historical records help provide insight into past problems or information on where to focus future studies. Cost records and schedules are compared with budgets for various activities to determine whether there are large deviations that need further investigation. However, historical information may be inaccurate; therefore, a productivity improvement team should validate the accuracy of the information before it is incorporated into a new study.

Within organizations, information that is useful during productivity improvement studies comes from a variety of sources, and examples of information sources are listed in Table 5-3. Safety records and reports should be reviewed for insight into the work environment. The reduction of costs associated with accidents could be a by-product of productivity improvement studies. Quality-control records should be reviewed to determine the quality of the product and quality-control procedures. Records on the costs and delays associated with rejected work or materials should be reviewed because delays affect the morale of workers.

Photography

Digital photography, smart phones or tablet computers with digital cameras, and other photographic equipment provide an opportunity for recording jobsite activities. Examples of the types of situations that are easier to explain using photography include overloaded trucks, overcrowded work areas, improper storage of tools and materials, excessive crew sizes, a lack of workers, improper use of equipment, use of the wrong equipment, oversized or undersized equipment, the location of equipment, the location of workers, poor jobsite supervision, unsafe working conditions, safety violations, and defective materials. Smart phones and tablet computers are useful for instantly forwarding information to field or home offices and to the personnel who are performing the productivity improvement study.

Table 5-3. Sources of Information for Productivity Improvement Studies

Sources of Information	Sources of Information
Bid package reviews	Labor reports
Change order log	Material cost reports
Correspondence	Milestone reviews
Cost estimates	Network analysis
Daily construction reports	Procurement reports
Daily contractor meetings	Productivity reports
Design change notices	Progress curves
Design milestone reports	Project manager's log
Design reviews	Quality-control reports
Drawing logs	Request for information logs
Equipment usage reports	Schedules
Exception reports	Strike notices
Expediting reports	Submittal logs
Time cards	Superintendent's log
Time sheets	Test reports
Field labor reports	Variance analysis reports
Inventory reports	Visual observations
Job logs	Weather logs
Labor cost reports	

Source: Adapted from El-Nagar and Yates (1997).

Digital Video Recording

Digital video recording technology is used to gather data for productivity improvement studies. If digital video recordings are used to gather data, it helps if the recorders have the ability to slow down individual frames. Video recorders capture fast-moving processes, are able to record several hours of video, and provide high-quality video that is saved using resident or cloud storage. Recording processes by cycles or specific tasks makes it easier to analyze worker activities. The date and time of each recording should be recorded in the camera and in a written record.

Digital video recordings consume a large amount of hard drive memory if they are transferred to personal computers. Several hours of digital video could require hundreds of gigabytes of computer memory. Large video files are stored on CD-ROMS or **passport flash drives**, which have 500 gigabytes to terabytes of memory, or remote servers. Digital video camera batteries last only a few hours so alternating current (AC) power may be required if long periods are being recorded for a study.

Time-Lapse Photography

Because storing digital videos requires large amounts of memory, **time-lapse photography** may be a more viable alternative for recording jobsite activities. Time-lapse cameras allow activities to be recorded for long periods because the camera does not record continuously. Time-lapse cameras record at set intervals of seconds, minutes, or hours, and when the recording is played, it shows the events as occurring in fast motion rather than at a normal speed. Time-lapse recordings cannot be used to analyze time-dependent processes, but they are useful for analyzing crew sizes, worker locations, equipment usage, material-transport processes, and for determining the location of activities.

Advantages of Digitally Recording Construction Activities

One advantage of using digital videos or time-lapse photography for productivity improvement studies is the ability to view operations without on-site interruptions and distractions. Being able to compress time, reverse, stop, or fast-forward the video of activities provides more insight into work processes than merely observing an activity one time. Another advantage is the ability to watch digital videos multiple times, which provides an opportunity to study all of the crafts involved in activities, as well as multiple opportunities to look at interferences, shortages of workers, and unbalanced crews. Digital video recordings provide a historical record that is compared with work activities after work-improvement techniques have been implemented at jobsites. They may also be reviewed while dealing with legal claims and disputes. Other methods for recording work processes at construction jobsites are covered in Chapter 6.

5.12 Summary

This chapter provided information on how to prepare to conduct productivity improvement studies and outlined the responsibilities of productivity improvement team members. Techniques for preparing interview guides and suggestions for questions to be used during interviews were included, along with information on defining the scope of productivity improvement studies. Potential work improvement areas were highlighted, and the scientific analysis method was explained in terms of how it applies to productivity improvement programs. Methods were introduced on how to plan productivity improvement studies and to investigate productivity. The effects of learning curves were explained to illustrate how productivity is affected while workers are learning new work tasks. The effects of scheduled overtime on productivity rates were discussed in terms of whether overtime improves overall productivity. Work-sampling techniques were introduced, along with a discussion of confidence limits, limit of error, digital recording techniques for collecting data, and the applicability of project records to productivity studies. The next chapter will provide detailed information on specific data-collection methods and formats.

5.13 Key Terms

category proportion
confidence limit
Construction Industry Institute
development stage
formal assessments
implementation stage

informal assessments
learning curves
limit of error
Methods Time Measurement
passport flash drives
planning stage
scheduled overtime
scientific analysis method
standard operating procedures
time-lapse photography

5.14 Exercises

5.1 Is it beneficial to use scheduled overtime to increase productivity rates? Why or why not?

5.2 Explain what limit of error means in terms of sampling products that are being manufactured or built at construction jobsites.

5.3 Outline the responsibilities of productivity improvement team members.

5.4 What is included in the training programs developed by productivity improvement team members for construction workers?

5.5 What do owners need to know about implementing a productivity improvement program?

5.6 What are the benefits to owners of implementing productivity improvement programs?

5.7 What types of activities are conducted during the development stage of a productivity improvement program?

5.8 When work improvement techniques are implemented, what are the five areas where changes might be applied to help improve productivity rates?

5.9 What are the steps for implementing the scientific analysis method?

5.10 What types of information used in productivity improvement studies might be gathered by interviewing foremen or workers?

5.11 During a productivity improvement study for a concrete placing operation, what internal sources of information would be the most useful?

5.12 When should informal assessments be used in conducting productivity improvement studies?

5.13 Why is it important to understand how workers and their work output are affected over time as they are learning to perform new tasks?

5.14 How is work sampling used in productivity improvement studies?

5.15 What are some of the advantages of using work sampling in productivity improvement studies?

5.16 What are confidence limits, and how are they used during productivity improvement studies?

5.17 Why is it important to have many different sources of information, such as those listed in Table 5-3, when conducting productivity improvement programs?

5.18 Why would it be more advantageous to use time-lapse photography than digital video recordings of jobsite activities for analyzing productivity at construction jobsites?

5.19 What are the five classes summarized by Frederick Taylor where productivity improvement techniques are used on construction projects?

5.20 In addition to owners, who should productivity improvement team members interview at construction jobsites while conducting their studies?

References

Adrian, J. (2004). *Construction productivity measurement and improvement*, Stipes, Champaign, IL.

American Association of Cost Engineers. (1973). "Effects of scheduled overtime on construction projects." Morgantown, WV: Prepared by the Construction Users Anti-Inflation Roundtable, *American Association of Cost Engineers Journal*, 15(5).

American Subcontractors Association, Associated General Contractors of America, and Associated Specialty Contractors Inc. (1979). "Owner's guide on overtime, construction costs, and productivity." Report No. 2141, Washington, DC.

Business Round Table. (1974). *Effects of scheduled overtime on construction projects, in coming to grips with some major problems in the construction industry*, New York.

Construction Industry Institute. (1988). "The effects of scheduled overtime and shift schedule on construction craft productivity." Source Document 43, Austin, TX.

El-Nagar, H., and Yates, J. (1997). "Construction documentation used as indicators of delays." *Journal of the American Association of Cost Engineers International*, 39(8), 31–37.

Kerzner, H. (2005). *Project management*, Wiley, New York.

National Electrical Contractors Association (NECA). (1969). *Overtime and productivity in electrical construction*, Washington, DC.

Oglesby, C., Parker, H., and Howell, G. (1989). *Productivity improvement in construction*, McGraw Hill, New York.

Taylor, F. (1967). *Principles of scientific management*, Norton, New York.

Thomas, R. (1990). "Effects of scheduled overtime on labor productivity: A literature review and analysis." Source Document 60, Construction Industry Institute, Austin, TX.

U.S. Department of Defense. (1963). *Methods study: Department of Defense joint course*, Washington, DC.

Wankat, P. (1983). "Analysis of student mistakes and improvement of problem solving on McCabe-Thiele binary distillation tests." *American Institute of Chemical Engineering Symposium*, Series 79, 228(33), 231–236.

Woods, D. (1987). "How might I teach problem solving?—Developing critical thinking and problem-solving abilities." *New Directions for Teaching and Learning No. 30*, Jossey-Bass, San Francisco.

Woods, D., Crowe, C., Taylor, P., and Wood, P. (1984). "The MPS program." *Proc., American Society of Engineering Education Annual Conference*, Washington, DC.

CHAPTER 6

Productivity Improvement Data Analysis Techniques

The evaluation of existing construction operations during productivity improvement studies requires two phases. In the first phase, the productivity improvement team members collect data. Then, during the second phase, the data are analyzed and used to determine whether there are any alternative processes that would help increase productivity.

During the data collection phase, it is important for productivity improvement team members to gather sufficient amounts of data samples. These samples are then analyzed to provide a detailed synopsis of how tasks are being performed by workers. The information that results from the data analysis phase provides a basis for determining whether there is a potential for increasing productivity rates for each activity through the introduction of alternative work processes. After recommendations for changing or altering work processes have been developed, explanations are provided for why each new alternative work process is being recommended for adoption.

Productivity improvement studies analyze work processes, work-crew size deficiencies or inefficiencies, the availability of the correct type and adequate quantities of materials, the use of equipment, work-process flow, the transportation schemes for material movement, site layout configurations, and the monitoring of work processes by management personnel and supervisors.

This chapter discusses techniques for collecting, organizing, and analyzing the data used to identify potential productivity improvement techniques and develop implementation plans. It also includes examples of the types of forms that are useful for recording data, such as **work sampling**, **field ratings**, **5-minute ratings**, **work-process flow diagrams**, and **crew balance charts**. The Occupational Safety and Health Administration (OSHA) is briefly mentioned in this chapter and discussed in more detail in Chapter 4. This chapter provides guidelines for productivity improvement meetings, data presentations, **work-process analysis**, the recording of work-process operations and activities, **work-count analysis**, and **work-measurement analysis**. Systems engineering is also introduced to demonstrate the similarity between it and productivity improvement studies.

6.1 Occupational Safety and Health Administration

The data analyzed during a productivity improvement study should include information on violations of regulations that pertain to the work environment and the protection of workers. The Occupational Safety and Health Administration and state-level occupational safety agencies are responsible for investigating whether safety and health issues are being avoided or properly addressed at construction jobsites. Specific information about federal OSHA regulations is listed on the website (http://www.osha.gov). Some information about the OSHA and other safety-related issues is discussed in Chapter 4.

6.2 Guidelines for Productivity Improvement Meetings

During the implementation of productivity improvement programs, members of productivity improvement teams conduct meetings at jobsites to communicate the program objectives and to explain the program implementation plan. The meetings should include foremen, craftsmen, laborers, engineers, managers, and independent specialists, all of whom should provide input into the productivity improvement process. At the conclusion of the productivity improvement study, meetings are again used to explain how the productivity improvement analysis was performed and to present recommendations for improvements to work processes.

Productivity improvement meetings should be augmented with photos, digital video recordings, or time-lapse video recordings of the work processes that are being discussed and with supporting information that helps to explain work processes. During productivity improvement meetings, productivity improvement team members should explain what is affecting productivity rates and why production is declining for each activity under review. They should focus on the proposed work-improvement techniques and their applicability to the work processes under review during the study.

6.3 Guidelines for Data Presentations

When productivity improvement team members are presenting the results of their study, they should distribute an outline of their presentation, including its important elements, its sequence, and relevant data, charts, graphs, and supporting materials. In addition, a written summary of the conclusions of the study and a list of action items should be distributed to the individuals attending the presentation. Photographs, digital video recordings, and time-lapse photography of work processes should be used during the presentation to illustrate the work tasks and work-process flow concepts discussed during the meeting.

If references are used during oral presentations, they should be properly cited by providing the title of the reference, its author, the date of publication, the publisher, and the location of the publisher (city and state). If articles are being referenced, then the citation should also include the volume and issue number for the journal and the page numbers of the article. If information from interviews is being presented, the name of the person who was interviewed should be included, along with the date of the interview, a title that describes the subject of the interview, the job title of the person interviewed, and the name of his or her employer. Examples of citations for a book, an article, and an interview are as follows:

Book: Murphy, R. (2005). *International project management.* Thomson Higher Education, Mason, OH.

Article: Scott, S. (1993). "Dealing with delay claims: A survey." *International Journal of Project Management,* 11(3), 143–153.

Interview: Jones, M. (2009). "Project legal issues" (Telephone interview). Project Manager, XYZ Construction, Colorado Springs, CO, May 9.

6.4 Work-Process Analysis Procedures

During a work-process analysis, every detail of an operation is evaluated including the methods being used; the suitability of tools, equipment, and materials; and the appropriateness of work locations. First, existing methods are analyzed to identify whether all of the operations are necessary, and then alternative work-process methods are reviewed to determine whether any of the alternative work processes would be applicable to the operation.

To perform a work-process analysis, a plan is developed that provides specific steps for evaluating work processes. The plan should include a list of who is responsible for performing each stage of the analysis and specific steps for how each stage will be conducted and evaluated by productivity improvement team members. The plan should explain the purpose of and reason for the study, set a time frame for completion, and indicate how it will be performed, where it will be performed, and how each process will be recorded, evaluated, and incorporated into the study.

Data are collected for productivity improvement studies using field ratings, 5-min ratings, and work sampling. Three techniques used to develop a work-process analysis are the following:

1. *Work-process flow diagrams*: Sketches of how equipment, materials, and workers traverse jobsites.
2. *Work-process charts*: Charts used to monitor and record how materials are being transported and used by workers.
3. *Crew balance charts*: Histograms that are used to record how each member of a work crew is expending time on each activity.

These techniques are more effective when they are used to evaluate repetitive tasks.

The following sections explain these techniques plus other analysis tools that are available for use when collecting data for productivity improvement studies.

Field Ratings

Field ratings group work activities into two classifications at the time the observations are recorded by observers: working or not working. If a worker is classified as *working*, this indicates that he or she is actively engaged in performing a useful activity. A classification of *not working* indicates that a worker is idle and not contributing to the achievement of the work objective. For a worker to be designated as actively engaged in work, he or she should be carrying, holding, or supporting materials or participating in physical activities such as those listed in Table 6-1. Examples of activities categorized as *not working* are listed in Table 6-2.

At the beginning of the field-rating data-collection process, the observer records the date; the title of the job; the name of the specific activity being observed, such as welding, building formwork, placing concrete, installing electrical conduit, or

Table 6-1. Examples of Activities Categorized as *Working*

Effective Work Activities
Discussing work
Filling in time cards
Giving instructions to workers
Holding a tag line
Holding materials for other workers as they are being installed
Installing materials
Listening to instructions
Measuring and cutting materials
Operating a machine or a piece of equipment
Reading blueprints
Supporting a ladder
Transporting materials
Writing orders

Source: Adapted from Oglesby et al. (1989), p. 175.

Table 6-2. Examples of Activities Categorized as *Not Working*

Nonwork Activities
Conversing but not actively working
Observing self-operating machines
Riding equipment rather than operating it
Standing or sitting without performing any function
Waiting for someone else to finish their work, including waiting for wheelbarrows to be loaded or for a hoist to move materials
Walking but not moving materials

Source: Adapted from Oglesby et al. (1989), p. 175.

excavating soil; the weather conditions; and any remarks pertinent to the study location, jobsite conditions, or the workers.

Observers should position themselves in a location where they are able to observe as many workers as possible but that is also not intrusive so that the workers are not consciously aware of their presence. Once the observers are situated, they should record the time of day of their first observation. The data recorded at each interval include the time, the number of individuals observed, and the number of people working. This recording process is duplicated at either random or predetermined intervals during the designated time interval. Figure 6-1 is a sample blank field-rating form and a completed field-rating form.

Field Rating - Blank							Date:	
Crew Analysis:							Page:	_of_
Job:								
Weather:								
Remarks								
Observation No.	**1**	**2**	**3**	**4**	**5**	**6**	**7**	**8**
Time of Day								
No. Individuals on Site								
No. Individuals Observed								
No. Individuals Working								
% Working								
Add Foreman/Personal Time (10%)								
Field Rating Index								
Type of Operation:								
Examples of Single Observation								

Field Rating - Completed							Date:	2/07/09
Crew Analysis:							Page	1 of 1
Job: High Rise Structure								
Weather: Calm 68 degrees								
Remarks								
Observation No.	**1**	**2**	**3**	**4**	**5**	**6**	**7**	**8**
Time of Day	8am	9	10	11	12	1pm	2	3
No. Individuals on Site	132	125	142	150	80	80	80	133
No. Individuals Observed	122	100	130	135	80	80	80	104
No. Individuals Working	59	65	70	80	70	70	70	69
% Working	48%	65%	54%	59%	88%	88%	88%	70%
Add Foreman/Personal Time (10%)	10%	10%	10%	10%	10%	10%	10%	10%
Field Rating Index	**58%**	**75%**	**64%**	**69%**	**98%**	**98%**	**98%**	**80%**
Type of Operation:								
Examples of Single Observation								

Figure 6-1. Blank Field-Rating Form and Completed Field-Rating Form

Source: Courtesy of Dr. Fred Rahbar.

Table 6-3. Sample Field-Rating Report

Field-Rating Report Number F1234	Date: 11/22/2012
Number of individuals at the jobsite	132
Number of individuals observed	122
Number of individuals classified as working	59
Percentage working (59/122)	48%
Add 10% for foreman and personal time (optional)	10%
Adjusted field-rating index	58%

Source: Adapted from Oglesby et al. (1989), p. 175.

At the end of the observation period, the percentage of workers who were performing useful activities is calculated using the following formula:

$$\text{percentage of individuals who are working} \\ = \text{number of individuals working/number of individuals observed} \quad (6\text{-}1)$$

The accuracy of the observations increases if more than one individual is recording observations and if there are an adequate number of observations. After the percentage of individuals working has been calculated, an additional 10% is added to the subtotal to account for the foreman and for personal time, such as restroom and coffee breaks. If union workers are working on a project, union agreements stipulate the number of foremen required to supervise the work crews.

Field ratings below 60% indicate an inadequate level of productivity. Productivity rates for specialty crews should be higher than 60%, whereas productivity rates for casual labor or workers cleaning up after other activities might be considerably lower (Oglesby et al. 1989, p. 176). Table 6-3 provides a sample of a completed field-rating report.

Field ratings should be verified by performing repeat studies on different days. A sufficient amount of individual observations should be recorded for the results to be statistically significant. One hundred observations is considered sufficient to identify work-process problems, but 400 observations provide more reliable data (Oglesby et al. 1989).

5-Minute Ratings

The purpose of 5-min ratings is to "1) provide management with information about delays and to indicate their order of magnitude, 2) measure the effectiveness of a crew, and 3) indicate where thorough, detailed observations or planning might result in cost savings" (Oglesby et al. 1989, p. 181). The 5-min-rating technique is not as accurate as the field-rating method, but it is an effective method for making a general work evaluation.

The ratio of not working or a work delay to the total observed amount of time is recorded, and if during the observation time the work delay is more than 50% of the time, then the individual effort for that block of time is classified as a delay. If the observer notes that a delay is less than 50%, then the block of time for that individual is marked as being effective. Observations for small units of time are made to evaluate a crew, and all members are rated as effective or not effective or the activity is rated as being delayed or not delayed. Productivity rates should be reported as percentages to assess the level or pace of activities. A highly efficient worker would receive a rating of 80–90%, and less efficient workers might be rated at 50–60%.

Five-minute ratings are a method for rapidly evaluating the productivity of individual workers, but they only provide information on whether a worker is idle or productive during the period under observation. These ratings are conducted over short periods, usually 5 min, as the name of the sampling method implies. When five or fewer workers are being observed, 5 min is an adequate amount of time for the data-collection sequence, but if there are more than five workers, the total amount of observation time should be equal to or exceed the total number of workers.

Before an observer starts to record observations, he or she should record general information about the job, including the job title, the date, the type of crew being observed, the time interval of the observations (such as 1 min), the type of operation, and weather conditions. A title, such as ironworker, welder, carpenter, or pipe fitter, is listed for each worker, and each worker is assigned a worker number. It is helpful to also note a method for identifying each worker, such as the color of shirts or hard hats or other identifiable features.

Each observation should be recorded by observation number and time of day. At each designated time interval, the observer checks whether each worker is performing **effective work** and notes the duration of any delays. After all of the observations have been recorded, an effectiveness ratio is calculated for each worker. The formula for the effectiveness ratio is as follows:

$$\text{effectiveness ratio} = \text{total number of effective observations}/\text{maximum total number of observation periods} \quad (6\text{-}2)$$

An effectiveness ratio should be calculated for each individual worker and also for all of the workers during the total observation period. Notes should be added that indicate the type of task performed for each observation period. Figure 6-2 shows a blank 5-min-rating form.

Figure 6-3 is an example of a completed 5-min-rating report for the lifting and moving of precast concrete wall panels by a crane and a crew that is positioning the panels and temporarily fastening them. The top of the table in Figure 6-3 indicates the individuals involved in the work activities and their respective trades. In this study, a crew of six was being observed, and the observer divided the time into units of 1-min intervals. This was a 12 min per cycle operation. The effectiveness rating was calculated for each worker and for the total crew.

Work Sampling

Work-sampling forms are used to collect and summarize data on the effectiveness of the work being performed by crew members. These data are then used to calculate **labor utilization factors**. Before data are gathered using work-sampling techniques,

5 Minute Rating

Date:__________
Page: _____ of _____

Analysis Crew: __________
Job: __________
Time Interval: __________
Weather: __________

Observ. #	Time of Observ.	Man 1	Man 2	Man 3	Man 4	Man 5	Man 6	Operations:
1								
2								
3								
4								
5								
6								
7								
8								
9								
10								

	Man 1	Man 2	Man 3	Man 4	Man 5	Man 6	Total
Maximum Total	10	10	10	10	10	10	60
Total Effective	0	0	0	0	0	0	0
Effectiveness Ratio	0	0	0	0	0	0	0

5 Minute Rating

Date:__________
Page: _____ of _____

Analysis Crew: __________
Job: __________
Time Interval: __________
Weather: __________

Observ. #	Time of Observ.	Man 1	Man 2	Man 3	Man 4	Man 5	Man 6	Operations:
1								
2								
3								
4								
5								
6								
7								
8								
9								
10								

	Man 1	Man 2	Man 3	Man 4	Man 5	Man 6	Total
Maximum Total	10	10	10	10	10	10	60
Total Effective	0	0	0	0	0	0	0
Effectiveness Ratio	0	0	0	0	0	0	0

Figure 6-2. Blank 5-Min-Rating Form

Source: Courtesy of Dr. Fred Rahbar.

the observer should record information about the project, including the types of crews being analyzed, the job title, the date, and the weather conditions. The data collected are recorded by observation number and time, usually in 1-min or 5-min intervals.

The observations recorded during each time internal should include

- The number of workers involved in direct work,
- The number of workers performing indirect work,
- The number of workers handling materials,

5 Minute Rating

Analysis Crew:	Precast Panel Erection	Date:	8/1/2007
Job:	University Dormitory Structure	Page:	1 of 1
Time Interval:	1 minute		
Weather:	Sunny	□	=Effectiv Work

Observ. #	Time of Observ.	Man 1 Iron Worker	Man 2 Iron Worker	Man 3 Carpent.	Man 4 Carpent.	Man 5 Carpent.	Man 6 Welder	Operations:
1	10:13	□						Crew waiting for panel hoisting
2	:14	□	□	□	□	□		Landing panel/welder waiting to tack
3	:15	□	□	□	□	□		Landing panel/welder waiting to tack
4	:16		□	□	□	□		Install upper brace bolts
5	:17		□		□	□		Install braces
6	:18			□	□	□		Align panels
7	:19			□	□	□		Align panels
8	:20			□	□	□		Align panels
9	:21	□	□	□				Unhook crane
10	:22	□	□	□				Unhook crane
11	:23						□	Welder tacks rebar, crew waits for
12	:24						□	next panel to be hoisted
13	:25						□	" "
								Total
Maximum Total		13	13	13	13	13	13	78
Total Effective		5	6	8	7	7	3	36
Effectiveness Ratio		38%	46%	62%	54%	54%	23%	46%

Figure 6-3. Completed 5-Min-Rating Form

Source: Ogimachi et al. (n.d.).

- The number of workers who are idle,
- The number of workers involved in miscellaneous ancillary activities, and
- The type of work being performed during each time interval.

At the conclusion of the observation period, each of the previously listed categories is summed up to provide the total number of workers involved in direct work, indirect work, material handling, being idle, or miscellaneous activities. The percentage of the total represented by each activity is calculated using the following formula:

$$\text{percentage of total} = \text{total number of workers in category} / \text{maximum total summation of all categories of workers} \quad (6\text{-}3)$$

Figure 6-4 is a work-sampling form completed for a safety-cable-laying and concrete-decking operation.

Effective Work, Essential Work, and Ineffective Work

Productivity ratings are used to determine whether the work being performed is effective, essential contributory, or ineffective work. These ratings might be affected by site conditions, the type or nature of the work trades involved in activities, a lack of appropriate support, or environmental influences. Therefore, poor performance may not necessarily be caused by any one particular craft or individual.

Effective work includes activities that are directly involved in the work process or effort under evaluation. Movement of materials or labor within the immediate work area and activities essential to achieving the objective of the activity are considered

WORK SAMPLING

Page 1 of 9

Analysis Crew: Decking and Laying Safety Cables

Job:

Date

Weather: Clear 65 degrees F

MS= Miscellaneous Activity
DW= Direct Work
IO= Indirect Work
IA= Ineffective Work
MH = Material Handling

Observation #	Time of Observation	1 DW	2 IO	3 MH	4 IA	5 MS	6	Operations: Type of Work
		Number of people observed in each category designated						
1	8:25			2		2		North Decking, 2-Laying Safety Cables.
2	8:26			2		2		North Decking, 2-Laying Safety Cables.
3	8:27			2		2		North Decking, 2-Laying Safety Cables.
4	8:28			2		2		Lift moving down, 2 - Material Handling
5	8:29			2		2		Lift moving down, 2 - Material Handling
6	8:30				2			Restroom visit, No Deck work yet.
7	8:31				2			Restroom visit, No Deck work yet.
8	8:32	2	2					Decking started, Cable laying resumed
9	8:33	2	2					2-Decking, 2-Laying Safety Cables.
10	8:34	2	2					2-Decking, 2-Laying Safety Cables.
								Total
Maximum Total		6	6	10	4	10	0	36
% of Total		17%	17%	28%	11%	28%	0%	100%
Labor Utilization Factor =			28%					

Figure 6-4. Completed Work-Sampling Form

Source: Ogimachi et al. (n.d.).

effective work. Examples of effective work are as follows (Oglesby et al. 1989, pp. 176–177):

- Painting,
- Laying bricks,
- Nailing wallboard,
- Transporting soil from an excavation,
- Mixing mortar for bricks,
- Welding, threading pipe,
- Cutting boards to be used in the work activity.

Essential contributory work includes activities that do not directly contribute to the main work activity but are essential to finishing a unit of work. Examples of essential contributory work are as follows (Oglesby et al. 1989, p. 177):

- Handling materials, reading plans, and waiting while another member of a crew is doing productive work;
- Movements that are near where an activity is being performed, such as within a radius of 35 ft [10.7 m];
- Building scaffolding to serve as a work platform;
- Measuring a pipe;
- Placing a pipe in a machine preparatory to cutting and threading;
- Spotting an empty truck;
- Moving materials within an area of 10 ft [3.1 m] to 35 ft [10.7 m] from the primary work function.

Ineffective work is work that is not useful; this term typically represents idle workers. Examples of ineffective work are as follows (Oglesby et al. 1989, p. 177):

- Walking empty-handed,
- Carrying anything farther than 35 ft [10.7 m] from the work position,
- Taking coffee breaks,
- Waiting for trucks,
- Riding in trucks,
- Correcting errors,
- Returning to the shop for tools,
- Discussing non-work-related items.

The difference between essential contributory work and ineffective work is hard to distinguish unless adequate information is provided regarding the necessity of each task. To determine whether work is contributory or ineffective, consider whether

- Crew members are part of a work crew and handling the same materials as everyone else,
- A crew is properly sized and balanced for a task, and
- There are alternative activities that a worker could be doing (Oglesby et al. 1989, p. 177).

Workers who are not part of a crew should be classified as ineffective workers.

Ineffective work includes idle workers or workers performing tasks that are not necessary to complete an activity. Work tasks that are performed incorrectly or with the wrong tools are considered ineffective work. Rework performed incorrectly is also ineffective work. Work is effective if it contributes to the completion of an activity (Oglesby et al. 1989).

Figures 6-5 and 6-6 show how graphics such as pie charts or area charts might be used to summarize installation durations for activities. Figure 6-7 is a pie-chart showing the percentage of time expended on effective work, ineffective work, indirect work, and handling materials for a work-sampling activity.

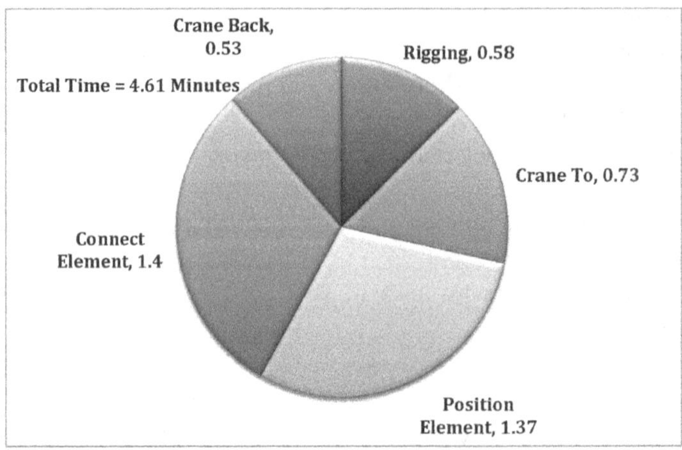

Figure 6-5. Summary of Installation Activity Times

Source: Ogimachi et al. (n.d.).

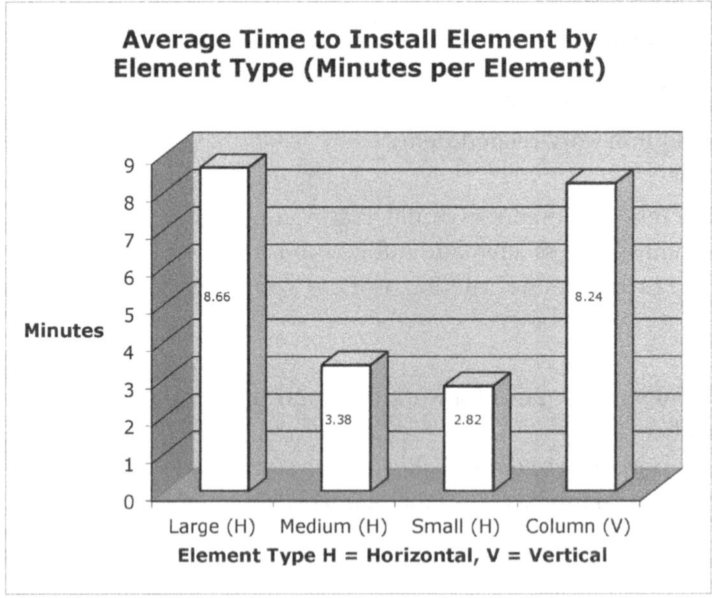

Figure 6-6. Summary of Element Installation Times

Source: Ogimachi et al. (n.d.).

Labor Utilization Factors

One of the objectives of productivity improvement studies is to increase the amount of effective work, but some types of essential contributory work are required to complete activities. Therefore, allowing for essential contributory work in reports of overall performance is a common practice. Labor utilization factors include essential

Concrete Project – Summary of Observations for Decking/Connecting/Laying Safety Cables

Data	Number of Observations					Labor Utilization Factor
	Direct Work	Indirect work	Material Handling	Ineffective Work	Total	
DD/MM/YY	149	18	21	92	280	57%
DD/MM/YY	56	59	0	2	117	60%
Total	205	77	21	94	397	58%
	52%	19%	5%	24%		

k

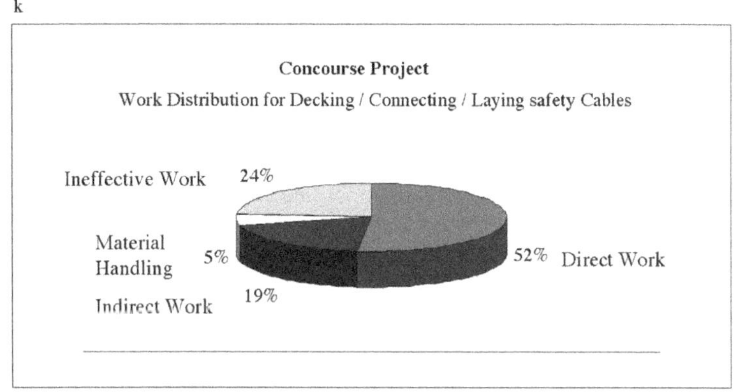

Figure 6-7. Summary of Work-Sampling Observations

Source: Ogimachi et al. (n.d.).

contributory work. They are calculated by adding the number of observations of effective work to one-fourth the number of observations of contributory work and dividing this sum by the total number of observations. The formula for labor utilization factors is as follows (Oglesby et al 1989, p.180):

$$\text{labor utilization factor} = [\text{effective work} + 0.25(\text{essential contributory work})]/\text{number of total observations} \quad (6\text{-}4)$$

where

$$\text{number of total observations} = \text{effective work} + \text{essential contributory work} + \text{ineffective work} \quad (6\text{-}5)$$

Work-Process Flow Diagrams

Work-process flow diagrams are pictorial representations that demonstrate the layout of the jobsite, the location of equipment relative to work-operation areas, transportation sequences, worker locations, existing structures, and other site-specific information. Figure 6-8 illustrates the flow of activities for a crane operation erecting precast concrete panels, and Figure 6-9 shows a work-process flow diagram that indicates the layout of the work; the spacing or relative location of the work in progress, including staging; and the location of the observations.

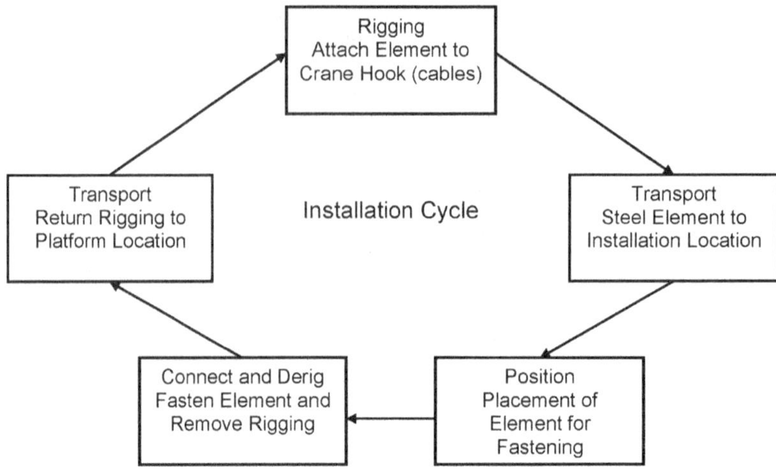

Figure 6-8. Work-Process Flow Diagram for Crane Operation Showing Installation Cycle

Source: Ogimachi et al. (n.d.).

Work-Process Charts

Work-process charts are used to record specific operations, the tasks workers are performing, and the operation of equipment over a specified time interval. In addition, work-process charts are used to record delays that occur during operations and the number of workers involved in each task. Because work-process charts are used to deconstruct operations into specific tasks, they provide productivity improvement team members with an opportunity to propose alternative work methods for performing either one or more tasks or transportation schemes.

Work-process charts use five symbols to represent what is occurring when an observer is recording work activities. Figure 6-10 contains the five symbols used in work-process charts and the corresponding activities that each symbol represents. The symbols used in Figure 6-10 were adopted for use in process-analysis procedures by the **American Society of Mechanical Engineers (ASME)** (2008).

Work-process charts should include the title of the project, the location of the project, the work item under review, the date, the weather conditions, the name of the evaluator(s), and the evaluation number. The number for each element being moved, the locations of each element and the levels in the structure where they are located, and the element type should also be included in work-process charts. A brief description of each task sequence is included next to the appropriate symbol, along with the number of workers and the time of the observation. Notes should be added for each task to help clarify the sequencing of the operations.

Figure 6-11 is a work-process chart for the installation of a concrete platform on a beam frame, and Figure 6-12 is a work-process flow diagram for steel erection at an airport terminal for the installation of a frame concrete platform beam. Work-process charts are used to show both existing and proposed work processes, to compare existing processes to proposed processes, and to demonstrate potential time savings if new alternative processes are implemented for a task.

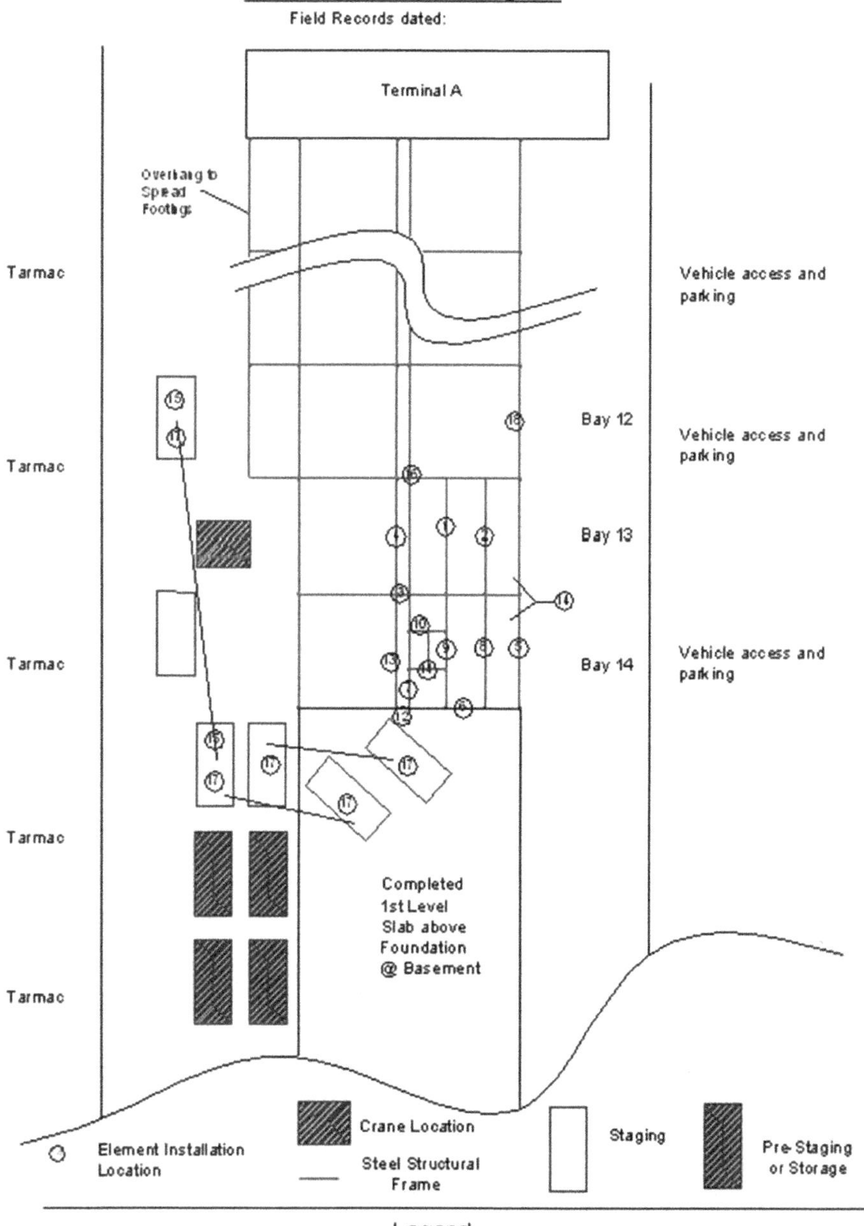

Figure 6-9. Work-Process Flow Diagram for Crane Operation Showing Jobsite Layout

Source: Ogimachi et al. (n.d.).

Symbol	*Operation*
○	Operations
⇨	Transportation
▢	Inspection
D	Delay
▽	Storage

Figure 6-10. Work-Process Chart Symbols

Source: Adapted from ASME (2008).

Crew Balance Charts

Crew balance charts are used to record the activities of each crew member during a designated period. The main purpose of crew balance charts is to provide a record of when workers are working effectively, what task they are performing, and when they are idle. The data in crew balance charts are used by productivity improvement team members to determine whether all of the crew members are necessary and whether using a different combination or a smaller number of crew members would reduce the amount of idle time, thus increasing productivity.

Histograms are used to record the operations performed by each worker. The *y*-axis of the histogram chart lists the time when each observation was recorded on the chart. The *x*-axis lists the types of workers involved in the operation under or above each histogram. If an activity is continuous, it is marked by completely filling in the histogram next to the time of the observation. Idle times are designated on the histogram by blank spaces. If appropriate, a short description of each activity is provided next to the location on the histogram where each activity is recorded using abbreviations.

Crew balance charts are also used to record effective work, contributory work, and ineffective work. These categories are marked on the histograms, each represented by a different color. After all of the work activities have been recorded for the designated period, the total number of minutes for each specific work task is calculated for each worker and for the total work cycle. Crew balance charts do not demonstrate the levels of effectiveness or efficiency for operations, nor do they provide information on whether a worker is working rapidly or using proper work techniques. Examples of some of the types of activities that are recorded in crew balance charts are listed in Table 6-4.

Crew balance chart histograms are also used to determine whether one or more workers might be eliminated from the crew and whether their work assignment might be performed by another crew member. Potential alternative crew configurations should be evaluated by redrawing the crew balance chart with the tasks from the eliminated worker distributed to other crew members. Figure 6-13 contains a crew

Process Chart Worksheet

Project: Sample Project **Study Location**:__________

Work Item: Installation of Frame Concrete Platform on Beam Frame **Prepared By**: Group 2

Chart No.: 1 **Date**: ____________

No.	Flow Sequence	Operation	Transportation	Inspection	Delay	Storage	Time	Notes
1.	Connecting crane cables to frame	●	⇨	□	D	▽		
2.	Lifting for adjustment	○	⇨	□	◗	▽		
3.	Checking level and C.G.	●	⇨	□	D	▽		
4.	Putting the platform back	○	➡	□	D	▽		
5.	Adjusting cables	○	⇨	□	◗	▽		
6.	Lifting again for checking of level	○	➡	□	D	▽		
7.	Taking the platform to the place	○	➡	□	D	▽		
8.	Adjustments for putting in place	○	⇨	□	◗	▽		
9.	Bolting to the beam	●	⇨	□	D	▽		
10.	Welding the sheet metal	●	⇨	□	D	▽		
11.	Pouring concrete in the open parts	●	⇨	□	D	▽		
12.	Vibrating the poured concrete	●	⇨	□	D	▽		
13.	Finishing of concrete	●	⇨	□	D	▽		

Summary	Evaluation		Proposed		Difference	
Functions	**No.**	**Time**	**No.**	**Time**	**No.**	**Time**
○ Operations						
⇨ Transportation						
□ Inspection						
D Delays						
▽ Storage						

Figure 6-11. Work-Process Chart for Installation of Concrete Platform

Source: Ogimachi et al. (n.d.).

balance chart for a concrete placing and finishing operation. The original crew configuration included nine crew members.

If several categories of workers are being observed at the same time, it becomes difficult to record their actions at 1-min intervals. Recording work activities with a digital camera, a digital video camera, or a time-lapse camera provides visual documentation of the work process and a repetitive method for evaluating the work sequences. Figures 6-14 and 6-15 show two alternatives developed to improve on the original crew configuration for placing and finishing concrete in Figure 6-13. In the first alternative, the general laborer was eliminated and his work was consolidated

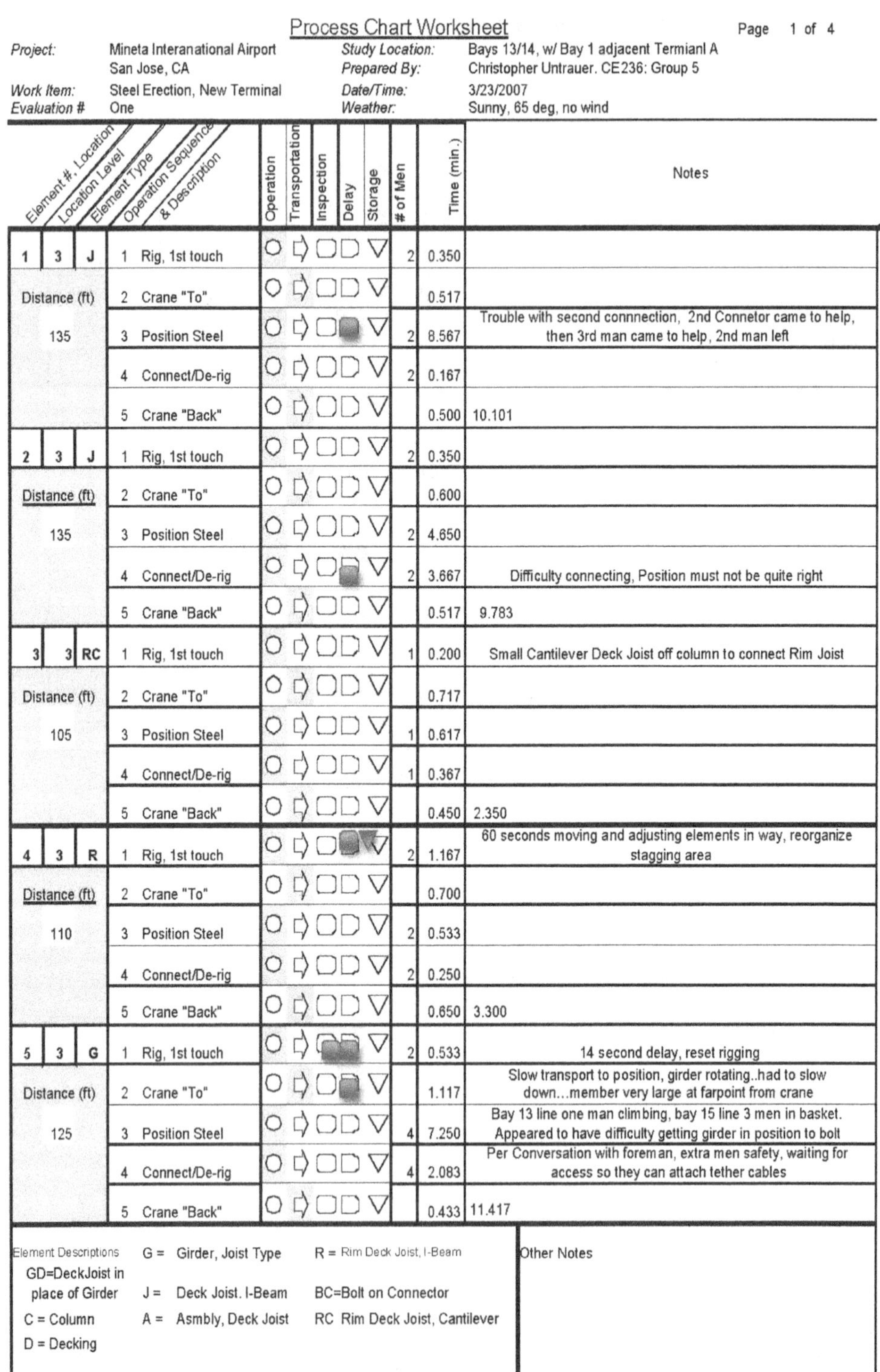

Process Chart Worksheet

Page 1 of 4

Project:	Mineta Interanational Airport San Jose, CA	*Study Location:*	Bays 13/14, w/ Bay 1 adjacent Termianl A
		Prepared By:	Christopher Untrauer. CE236: Group 5
Work Item:	Steel Erection, New Terminal	*Date/Time:*	3/23/2007
Evaluation #	One	*Weather:*	Sunny, 65 deg, no wind

Element #, Location	Location Level	Element Type	Operation Sequence & Description	Operation	Transportation	Inspection	Delay	Storage	# of Men	Time (min.)	Notes
1	3	J	1 Rig, 1st touch						2	0.350	
Distance (ft)			2 Crane "To"							0.517	
135			3 Position Steel				■		2	8.567	Trouble with second connnection, 2nd Connetor came to help, then 3rd man came to help, 2nd man left
			4 Connect/De-rig						2	0.167	
			5 Crane "Back"							0.500	10.101
2	3	J	1 Rig, 1st touch						2	0.350	
Distance (ft)			2 Crane "To"							0.600	
135			3 Position Steel						2	4.650	
			4 Connect/De-rig				■		2	3.667	Difficulty connecting, Position must not be quite right
			5 Crane "Back"							0.517	9.783
3	3	RC	1 Rig, 1st touch						1	0.200	Small Cantilever Deck Joist off column to connect Rim Joist
Distance (ft)			2 Crane "To"							0.717	
105			3 Position Steel						1	0.617	
			4 Connect/De-rig						1	0.367	
			5 Crane "Back"							0.450	2.350
4	3	R	1 Rig, 1st touch				■	■	2	1.167	60 seconds moving and adjusting elements in way, reorganize stagging area
Distance (ft)			2 Crane "To"							0.700	
110			3 Position Steel						2	0.533	
			4 Connect/De-rig						2	0.250	
			5 Crane "Back"							0.650	3.300
5	3	G	1 Rig, 1st touch			■	■		2	0.533	14 second delay, reset rigging
Distance (ft)			2 Crane "To"				■			1.117	Slow transport to position, girder rotating..had to slow down...member very large at farpoint from crane
125			3 Position Steel						4	7.250	Bay 13 line one man climbing, bay 15 line 3 men in basket. Appeared to have difficulty getting girder in position to bolt
			4 Connect/De-rig						4	2.083	Per Conversation with foreman, extra men safety, waiting for access so they can attach tether cables
			5 Crane "Back"							0.433	11.417

Element Descriptions			Other Notes
GD=DeckJoist in place of Girder	G = Girder, Joist Type	R = Rim Deck Joist, I-Beam	
	J = Deck Joist. I-Beam	BC=Bolt on Connector	
C = Column	A = Asmbly, Deck Joist	RC Rim Deck Joist, Cantilever	
D = Decking			

Figure 6-12. Work-Process Flowchart for Airport Steel Erection

Source: Ogimachi et al. (n.d.).

Table 6-4. Examples of Crew Balance Chart Activities

Crew Balance Chart Activities
Doing nothing
Doing rework or repairs
Holding materials while someone else installs the materials
Being idle
Doing inefficient work
Doing other activities of value
Doing productive work
Waiting for another worker to finish his or her work because of space conflicts
Standing by for a purpose but unproductive
Transporting materials
Waiting for instructions
Waiting for material deliveries
Waiting for tools or machines
Doing work paced by machines

Source: Adapted from Oglesby et al. (1989), p. 223.

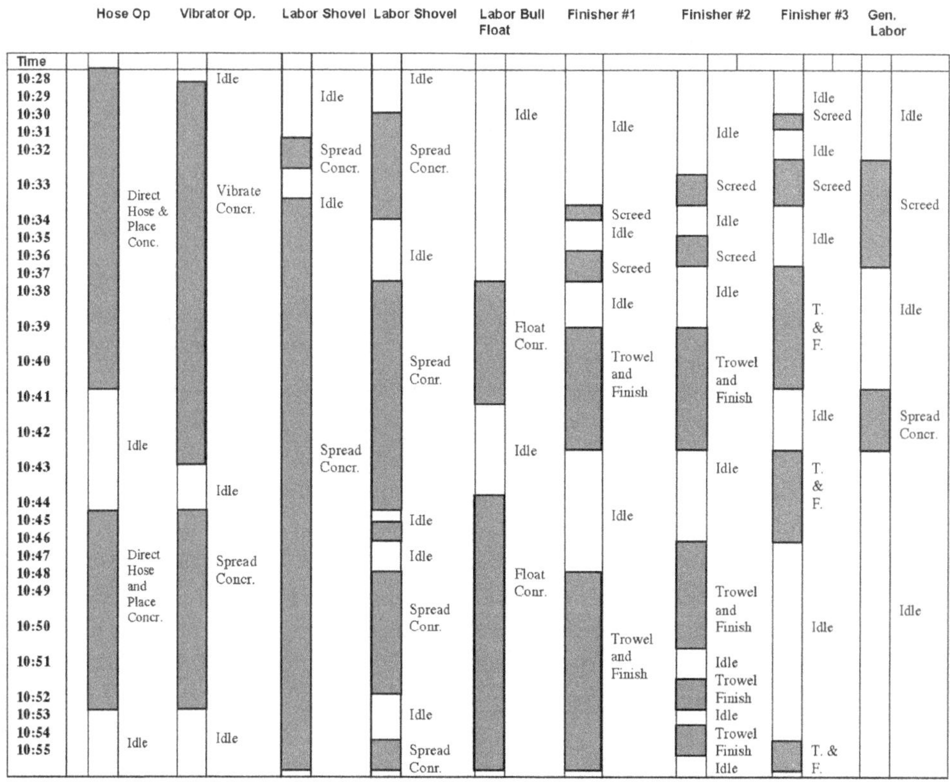

Figure 6-13. Crew Balance Chart: Existing Method for Concrete Placement

Source: Ogimachi et al. (n.d.).

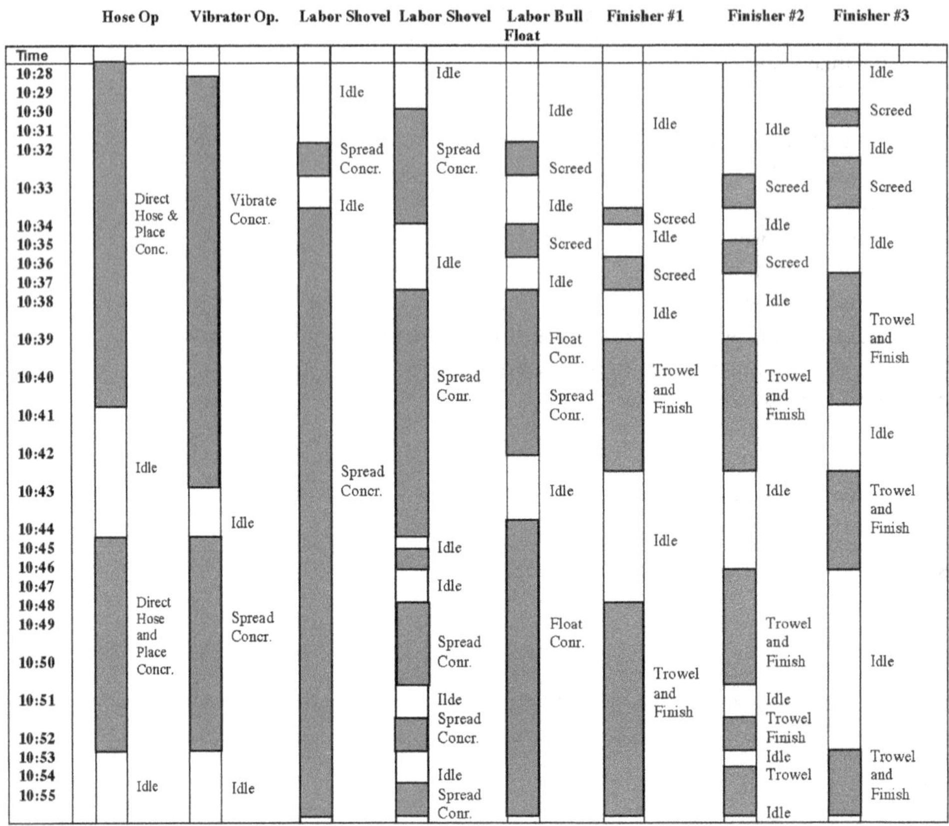

Figure 6-14. Crew Balance Chart: Alternative Method 1 for Concrete Placement

Source: Ogimachi et al. (n.d.).

and distributed to the other three finishers. In alternative two, the crew was reduced to two finishers.

The information provided in crew balance histograms is used to develop alternative crew sizes, to consolidate tasks, to reduce idle time, to clarify information about crew assignments, to address how efficient the workers are at completing tasks, and to determine how activities might be completed more efficiently.

Data collected on work processes during a productivity improvement study should be analyzed to determine whether a task might be completed with fewer workers, whether materials are being moved unnecessarily, whether power is sufficient and equipment is properly sized, whether alternative methods are available for performing tasks, and whether digital video recordings adequately demonstrate how workers are performing the tasks.

Methods Time Measurement

In addition to all of the techniques for recording productivity data mentioned in the previous sections, there are other methods for conducting studies that deconstruct

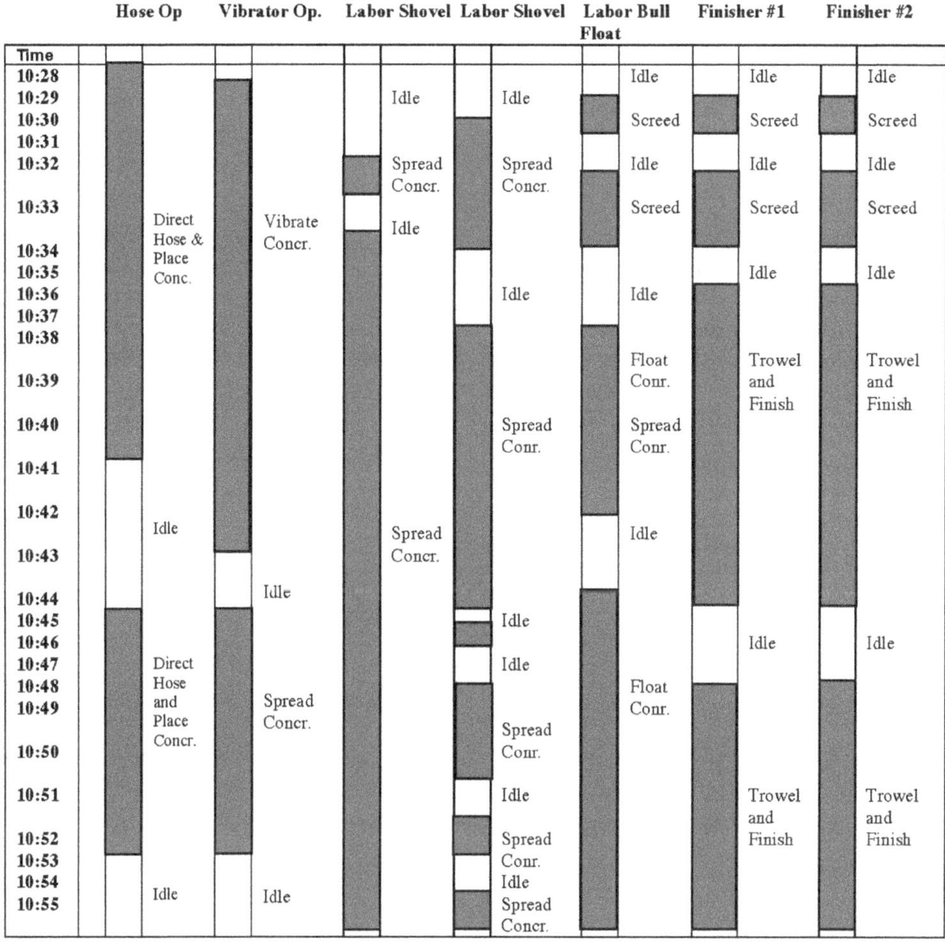

Figure 6-15. Crew Balance Chart: Alternative Method 2 for Concrete Placement

Source: Ogimachi et al. (n.d.).

work activities into even smaller units. These smaller units are referred to as **time measurement units** (TMUs). The system that uses TMUs is called Methods Time Measurement (MTM), and it has been in use in the United States since the 1940s. Time measurement units are recorded using the following units (Adrian 2004, p. 197):

1 TMU = 0.00001 hour = 0.000583 minute = 0.035 second
1 hour = 100.000 TMU
1 minute = 1.6667 TMU
1 second = 27.78 TMU

Examples of TMUs used for particular body motions are listed in Table 6-5.

Methods Time Measurement is mostly used by members of public agencies, by the U.S. Navy, and by industrial companies (Adrian 2004). The TMUs are determined

Table 6-5. Methods Time Measurement Units

MTM Motion	Number of TMUs
Reaching or moving 1 in. (2.54 cm)	2 TMU
Reaching or moving 2 in. (5.08 cm)	4 TMU
Grasping	2 TMU
Regrasping	6 TMU
Simple foot motion	10 TMU
Foot motion with pressure	20 TMU
Leg motion	10 TMU
Kneeling on both knees	80 TMU
Arising from knees	90 TMU
Standing	50 TMU
Turning	6 TMU
Turning and applying pressure	20 TMU
Turning body	20 TMU
Walking per pace	17 TMU

Source: Adapted from Methods Time Measurement Association for Standards and Research (1987), p. 95; Adrian (2004), p. 197.

for each body motion required for a worker to complete a specific task, and then, for each worker, total TMUs are calculated for all of the body motions necessary to complete that task. Alternative body movements are evaluated by computing the total TMUs required, and that value is compared with the original number of TMUs to determine whether the proposed alternative requires more or fewer TMUs than the original operation.

Kaizen

Another method for implementing productivity improvements is ***Kaizen***, which is a Japanese term that means "continuous improvement" in English. The productivity improvement approaches mentioned in previous sections are considered *revolution approaches*, which attempt to implement improvements all at one time. Kaizen productivity improvement methods evaluate work processes in the same manner as other methods, but continuous prioritized or controlled steps are implemented gradually to meet overall goals or objectives for improving processes (Salmon 2007).

Using an incremental approach for introducing work-process improvements is not as disruptive to the day-to-day activities necessary to maintain business as introducing completely new processes. A longer period is required to achieve the desired results, but it is easier to persuade employees and management personnel to implement new work-process techniques when they are introduced slowly. The Japanese management philosophy is based on consensus decision making, which is a process where all of the participants agree on a plan before it is implemented, and the *Kaizen* approach allows time for new work processes to become acceptable before they are fully implemented on projects.

6.5 Productivity Rating and Work-Process Analysis Reports

Productivity rating reports provide management personnel, foremen, and superintendents with information that helps them determine which tasks might be performed more efficiently or at a lower cost. Productivity rating reports highlight ineffective work layouts, inappropriate or inconvenient material placement or handling, and poorly sized or unbalanced crews. Productivity rating reports should also include written details about observations to help clarify the ratings.

Work-process analysis reports summarize the results of work-process studies and include the layout of jobsites, summaries of the data collected during productivity improvement studies, visual representations of work-task sequences, records showing which workers performed each task, crew balance charts, and a listing of the amount of material moved during each task. These reports also contain tables that indicate the amount of time lost during the performance of each activity. All of the work-process analysis methods described in the previous sections are included in work-process analysis reports along with information on

- The procedure used for conducting the study;
- The data collected;
- Summaries of the analysis results, including charts and graphs;
- A written analysis of the results; and
- Recommendations for work-process improvements.

Work-process analysis reports are used to communicate to management personnel the benefits that might be achieved by implementing productivity improvement recommendations, such as increased productivity, time savings, cost savings, and increased profits.

6.6 Work-Distribution Analysis

Work-distribution analysis (WDA) is another technique used to select activities that would benefit from the implementation of improved work processes. Jobs are divided into specific tasks, distributed within an organization, and evaluated to determine the most effective work distribution. Work functions, tasks, personnel, positions, work counts, and work-hour distributions are all factors that should be recorded and analyzed to facilitate the drafting of proposed work-process revisions.

A variety of indices or ratios are used to compare productivity over distinct periods. Samples of these productivity indices and ratios are provided in Table 6-6.

6.7 Recording Work-Process Operations

There are several options for recording work-process activities that help productivity improvement team members illustrate and evaluate productivity rates by providing

Table 6-6. Samples of Productivity Ratios and Measures

Productivity Ratios and Measures
Output/input
Actual production/estimated production
Actual work hours/estimated work hours
Current year sales/previous year sales
Cost, quantity, and number of orders/previous year cost, quantity, and number of orders
Efficiency and response time
Market share
Units produced or units sold
Revenue/payroll costs
Revenue/production costs
Selling price per unit/direct and indirect costs

visual representations of work activities. Work-process activities should be recorded for full days or for a set period. The advantage of recording work-process activities for an entire day is that it provides data on the daily fluctuations of production rates and on which times during the day workers are most productive. It also allows for observations of how each activity interacts with other activities. Daily recordings highlight problems and help facilitate the development of strategies for preventing the recurrence of specific problems. Methods for recording activities, including periodic, moving-average, and cumulative affect measurements, are discussed in the following sections.

Periodic Recordings

Periodic recordings do not provide as much detailed data as **daily recordings** do, but they are easier to analyze because they require fewer calculations. Periodic recordings are not useful for evaluating production-rate fluctuations over long periods because they provide data only for short time intervals, but they are useful for analyzing and determining spot production rates. Periodic recordings are less expensive than daily recordings and provide visual summaries of work-crew configurations, jobsite layouts, material storage, and original and proposed work processes to upper-level management.

Moving-Average Recordings

Moving-average recordings provide daily feedback over a period of several days. These recordings provide data that takes into account the variations in production rates that occur during different days of the week. This approach is used to analyze short-term trends in production rates.

Cumulative Recording

Cumulative recording documents all of the tasks required to complete an activity and determines the percentage of time required for each task. The cumulative-recording

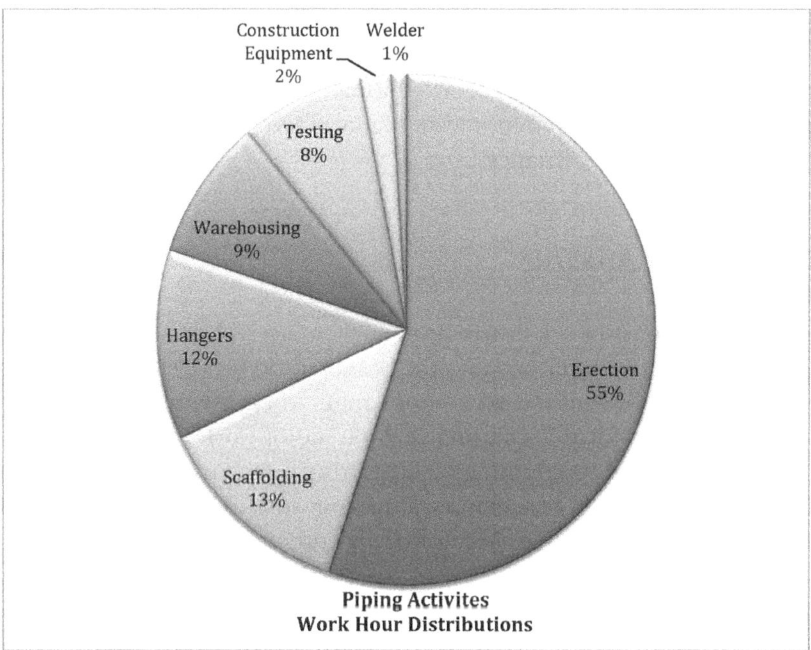

Figure 6-16. Summary of Cumulative-Recording Result for Piping Installation

approach is used for critiquing overall progress because it provides information on the percentage of time required for each task before and after productivity-improvement techniques are implemented on activities. Figure 6-16 provides a summary of cumulative-recording results for a piping-installation operation.

Time-Lapse Photography

Time-lapse photography is used to record work processes over long periods. Special time-lapse photography cameras are required that are used to take photos for a preselected number of frames, rather than as cumulative video recordings. When time-lapse recordings are played, they are played either at normal speeds, which make the actions in the photos seem disjointed, or at slower speeds, which allow observers to analyze specific tasks and work crews.

Time-lapse photography is not used as frequently as other recording methods because of the high cost of time-lapse cameras. If short time intervals are being analyzed, digital cameras are used to mimic time-lapse photography by the user taking photographs at predetermined, equal time intervals and then splicing the photographs together on the computer to create a time-lapse video presentation of the still photographs.

Digital Video Recordings

Digital video recordings are useful for recording all activities over time intervals of 1–2 h, depending on the recording speed and the recording times available on the recording media. Unfortunately, most digital video recorders hold a charge for only a few hours; therefore, an electrical power source or multiple batteries are required if work-process activities are to be recorded for an entire workday. Special equipment or software programs are required to view digital video recordings at different speeds or to print individual video frames.

6.8 Work-Count Analysis

As mentioned previously, work-distribution analysis is a technique used for determining the type of work accomplished and the best method for dividing the work among workers. Work-count analysis is used to supplement work-distribution analysis because it provides techniques for determining the volume of work in terms of measurable units. The data obtained by counting or measuring work provide a basis for developing improved methods and procedures, adjusting work assignments, evaluating performance, eliminating areas that slow down work processes, demonstrating personnel requirements, and stimulating interest in work activities.

To obtain an accurate measure of the volume of work being performed, identical tasks should be counted for the same types of activities. These tasks should be tangible and representative of the work being performed during an activity. Examples of the activities or items that could be included in a work-count analysis for an office are listed in Table 6-7.

Tasks that could be counted at construction jobsites include

- Cubic feet of concrete placed per day,
- Lineal feet of formwork built per day,
- Cubic yards of soil excavated per day,
- Number of bricks installed per day,
- Number of workers employed per day,
- Number of truck deliveries per day, and
- Amount of material distributed per day.

The flow of work at construction jobsites has certain key points that might be used as milestones to measure work progress: for example, when work is first assigned

Table 6-7. Examples of Activities for Work-Count Analysis

Activities	Activities
Number of e-mails sent	Log of telephone calls received
Number of e-mails received	Log of trips made by office personnel
Log of telephone calls initiated	Decision log
Log of visitors to the office	Log of materials ordered
Number of memos sent	Log of materials received
Number of memos received	Log of meetings and meeting attendees

to workers, when a decision is made or an action is taken, when a process is completed, or when materials are stored in a warehouse or a lay-down yard. Work-counting methods help to highlight situations where work progress declines and negatively affects subsequent activities. Data collected through work counting helps to identify procedures that require further study using work-process analysis techniques.

The method used for counting tasks depends on the type of operation being analyzed in a study. For example, the number of personnel working on a specific task is recorded on time cards by activity number. Bar codes and computers are useful tools for tracking work counts. Bar codes are scanned using handheld devices, and the data are instantly recorded and stored in computer files.

Work-count analysis techniques provide data that are used to help schedule work after it has been determined how long it takes to complete each task in relation to other tasks. Work assignments are then adjusted to accommodate actual conditions. Work counting is also used to determine the relationships between tasks if, in some circumstances, several individuals are performing unrelated tasks. Work counting assists in determining whether or not these tasks should be redistributed so that workers are performing related tasks. However, to do this, the volume of work being performed should be large enough to warrant specialization.

A work count should be used to determine methods for distributing work more equitably to workers, which results in a better balance of the workload for the workers. Work counts are used to determine when the work slows down because a delay in the flow of work may indicate a need for readjusting procedures or reassigning personnel. Work counts also help to determine whether there are understaffed or overstaffed organizational units. Most workers are interested in the accomplishment of their work tasks in relation to the accomplishments of other units engaged in similar work. A work count is useful for showing comparative production in order to increase interest in work and to help foster a competitive spirit.

Work-counting techniques help to determine the volume of work being completed by workers. They are used as part of **work-distribution charts**, they provide a formal appraisal of the division of work, and they provide information that could indicate a need for further investigations into how work processes might be revised to increase production rates.

6.9 Work-Measurement Methods

This section provides information on the different types of work-measurement methods used in construction and during productivity improvement studies.

Units Completed

Units completed is a measurement method that is used to determine the total number of units accomplished by workers. It provides detailed information that is easily verified by counting or measuring the number of units again. It is used for measuring items such as masonry, doors, windows, and electrical outlets.

Percent Complete

Percent complete is a measurement method that estimates what percentage of an activity has been completed at a specific point in time. The work completed is measured, and that volume is divided by the total volume to determine the percent complete. Percent complete may be applied to operations such as constructing formwork, excavating and backfilling excavations, and placing concrete.

Level of Effort

Level of effort is used when activities involve overlapping subtasks. The subtasks have to be measurable or their status easily defined by some type of measure. This method is most useful when there are a large number of similar commodities and the work is performed for extended periods. Level of effort provides more detail and objectivity than merely estimating how much work has been completed, and it is less expensive than counting or measuring the number of units completed. Level of effort measurements would be used in situations such as installing flooring, constructing formwork, and excavating soil.

Level of effort relies on predetermined **rules of credit** that establish appropriate credit for partially completed work performed in several stages. This method is often used for bulk commodity items. For example, the stages involved in constructing formwork are fabricating, erecting, aligning, tightening, oiling, stripping, and cleaning. Table 6-8 provides an example of how rules of credit are applied to wall-form construction subtasks.

Incremental Milestones

The **incremental milestone** measurement technique is used when a few items involved are hard to evaluate using any of the other measurement methods or are erected over long periods. Levels of completion are measured each time a predetermined milestone activity is reached during the activity. This method is used to measure work progress in areas such as structural steel erection, equipment installation, and heating, ventilating, and air-conditioning duct installation. Figure 6-17 shows how a percentage of the fee would be paid at each incremental milestone for the installation of heavy equipment.

Table 6-8. Rules of Credit Proportions for Wall-Form Subtasks

Subtask	Unit of Measure	Rules of Credit (percentage)
Erect initial wall form	Square feet (Square meter)	40
Erect second wall form	Square feet (Square meter)	40
Final bracing and plumbing	Square feet (Square meter)	10
Strip and clean	Square feet (Square meter)	10
Total task	Square feet (Sqaure meter)	100

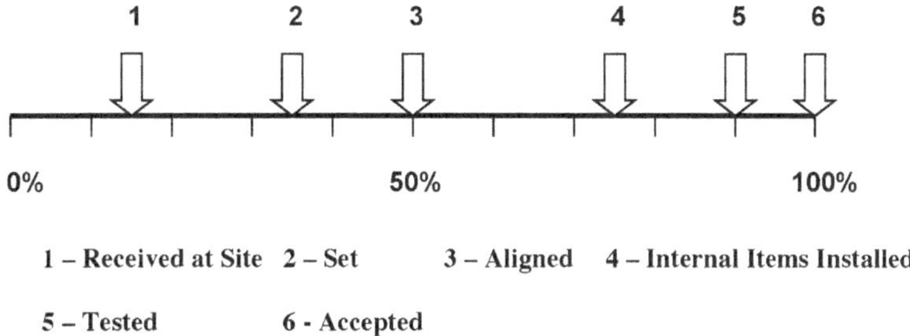

Figure 6-17. Allocation of Payments Using Incremental Milestones

Start-to-Finish Percentage

Start-to-finish percentage is a measurement method whereby an activity is divided into individual tasks that are either not started or not completed. Tasks may not have been started yet or they may be only partially complete and this is recorded for each task when progress is being evaluated during construction. This method is used in areas such as demolition, salvage operations, and erecting cranes.

Construction Specification Institute Codes

In addition to the methods for measuring productivity presented in the previous sections, members of firms also monitor performance by tracking forecasted work hours versus actual work hours in **performance evaluation reports**. Activities are tracked by account number or by **Construction Specification Institute** (CSI) Master Format **codes**, which are part of a system for organizing the information contained in construction contracts. There were originally 16 divisions and as of 2013, there were 49 CSI divisions.

The CSI Master Format is included in the ***CSI Master Format*** (CSI 2012a) and the ***CSI Manual of Practice*** (CSI 2012b). The CSI Master Format is used by over 85% of the large construction firms in the United States. Firms that do not use this format usually have their own unique format, which is followed on all of their projects.

A five-digit CSI code includes a broad scope, a medium scope, and a narrow scope designation for each item of work. The first part of the CSI code designates the CSI division that an item would fall under in a contract or specification, and the medium and narrow scope classifications designate specific work items within the 16 divisions. The original 16 CSI divisions were

Division 1—General Requirements
Division 2—Site Work
Division 3—Concrete
Division 4—Masonry
Division 5—Metals
Division 6—Wood and Plastics

Division 7—Thermal and Moisture Protection
Division 8—Doors and Windows
Division 9—Finishes
Division 10—Specialties
Division 11—Equipment
Division 12—Furnishings
Division 13—Special Construction
Division 14—Conveying Systems
Division 15—Mechanical
Division 16—Electrical (CSI 2012b, p. 3)

The CSI Master Format was expanded in 2004 and 2005 to include 49 divisions (CSI 2005):

General Requirement Subgroup

Division 1—General Requirements
Division 2—Existing Conditions
Division 3—Concrete
Division 4—Masonry
Division 5—Metals
Division 6—Wood, Plastic, and Composites
Division 7—Thermal and Moisture Protection
Division 8—Openings
Division 9—Finishes
Division 10—Specialties
Division 11—Equipment
Division 12—Special Construction
Division 13—Conveying Equipment
Division 14—Reserved for Future Expansion
Division 15—Reserved for Future Expansion
Division 16—Reserved for Future Expansion
Division 17—Reserved for Future Expansion
Division 18—Reserved for Future Expansion
Division 19—Reserved for Future Expansion

Facility Services Subgroup

Division 20—Reserved for Future Expansion
Division 21—Fire Suppression
Division 22—Plumbing
Division 23—Heating, Ventilating, and Air Conditioning
Division 24—Reserved for Future Expansion
Division 25—Integrated Automation
Division 26—Electrical

Division 27—Communications
Division 28—Electronic Safety and Security
Division 29—Reserved for Future Expansion

Site and Infrastructure Subgroup

Division 30—Reserved for Future Expansion
Division 31—Earthwork
Division 32—Exterior Improvements
Division 33—Utilities
Division 34—Transportation
Division 35—Waterway and Marine
Division 36—Reserved for Future Expansion
Division 37—Reserved for Future Expansion
Division 38—Reserved for Future Expansion
Division 39—Reserved for Future Expansion

Process Equipment Subgroup

Division 40—Process Integration
Division 41—Material Processing and Handling Equipment
Division 42—Process Heating, Cooling, and Drying Equipment
Division 43—Process Gas and Liquid Handling, Purification, and Storage Equipment
Division 44—Pollution Control Equipment
Division 45—Industry-Specific Manufacturing Equipment
Division 46—Water and Wastewater Equipment
Division 47—Reserved for Future Expansion
Division 48—Electrical Power Generation
Division 49—Reserved for Future Expansion

The *CSI Manual of Practice* and the *Master Format* list the broad scope headings under each division, and the medium and narrows scope headings are customized for each project. The specifications are available under each of the appropriate specification sections. The CSI Master Format makes it easier to locate information on specific items because they are consistently included in the same specification sections in any contract.

The *Master Format* is a useful reference for writing specifications because it provides a standard format for organizing specification sections. The *CSI Manual of Practice* also includes information on writing contracts and specifications. The last section of the *CSI Manual of Practice* includes an extensive list of key words for every item used in construction projects and the corresponding CSI code for each of the key words. This list is used to locate the CSI code for an item, and the CSI code is then used to locate the division and specification section where the item is described in the specifications.

Performance Evaluation Reports

Performance evaluation reports track activities by CSI code or company account number, and they include the budgeted work hours and the actual work hours for each period and to date, the budgeted quantities and the actual quantities for each period and to date, the budgeted unit rate, and the **performance factor**. Table 6-9 provides a sample performance evaluation report. The formulas used in the productivity evaluation report are in the note at the bottom of Table 6-9.

The performance factors listed in Table 6-9 measure whether an activity is on schedule, behind schedule, or ahead of schedule relative to work hours. They also indicate whether an activity is over budget, under budget, or on budget. A performance factor of over 1.0 indicates that the performance of an activity is above the budgeted performance to date. A performance factor that is below 1.0 indicates that an activity is behind the budgeted performance to date. The formula for performance factors is the following:

$$\text{performance factor for work hours} = \text{actual work hours accomplished to date} / \text{budgeted work hours to date} \tag{6-6}$$

$$\text{performance factor for quantities} = \text{actual quantities to date} / \text{budgeted quantities to date} \tag{6-7}$$

Performance evaluation reports are used to track the performance of individual activities during a designated period, such as each week, and for the period to date. After managers have reviewed these reports, they are able to adjust resources accordingly to balance the performance of activities that are performing at rates better than the budgeted rates and to apply resources to activities that are not performing as well as the budget projections.

6.10 Systems Engineering Concepts

Systems engineering is a process used to analyze a complete system rather than its individual components. The definition of systems engineering is "a group of elements acting as a whole and moving towards achieving some common goal or end" (Martin 1991, p. 2). Systems engineering is used to solve complex problems and requires extensive communication and interaction among the various engineering disciplines. Systems engineering processes should be applied to a component if any of the following conditions exist (Martin 1991, p. 2):

- The component is complex.
- The component is not available off-the-shelf.
- The component requires special materials, services, techniques, or equipment for development, production, deployment, test, training, support, or disposal.
- The component cannot be designed entirely within one engineering discipline.

Table 6-9. Example of Performance Evaluation Report

			Work Hours				Quantities				Budgeted Performance Total			
			Budget	Period	To Date	Percentage	Budget	Period	To Date	Percentage	Unit Rate	Period	To Date	Work Hours
Account	*Activity*	*Unit of Measure*	(1)	(2)	(3)	(4)	(5)	(6)	(7)	(8)	(9)	(10)	(11)	(12)
01735	Piping	Linear ft	13,500	6,850	18,240	135	10,630	1,970	9,170	86	1.27	0.38	0.64	21,440
05121	Str. Steel	Piece	9,630	2,280	7,820	81	6,420	1,580	5,170	81	1.50	1.04	0.99	9,710
16832	Cable	Linear ft	62,280	7,300	10,600	17	692,020	96,120	143,270	21	0.09	1.19	1.22	50,200

Note: Column (4) = [Column (3)/Column (1)] x 100; Column (8) = [Column (7)/Column (5)] x 100; Column (10) = Column (9)/[Column (2)/Column (6)]; Column (11) = Column (9)/[Column (3)/Column (7)]; Column (12) = Column (3) x [Column (5)/Column (7)].

The different types of systems include **closed systems** and **open systems**. In closed systems, a business or firm is completely isolated from the environment, and management has control over all system components. In open systems, businesses interact with the environment. All social systems are categorized as open systems. Open systems have permeable boundaries. Systems are divided into programs, projects, and tasks.

Systems Engineering Processes and Methods

A project manager is expected to set objectives, establish plans, organize resources, staff the organization, implement controls, issue directives, motivate personnel, find innovative solutions and alternatives, and remain flexible (Kerzner 2005). A **systems engineer** is expected to analyze and evaluate entire processes and systems. A systems engineer conducts research and develops and analyzes systems. When systems engineering methods are implemented, the following questions are addressed (Martin 1991, p. 5):

- *Need*: What needs are going to be fulfilled? What is wrong with the current situation? Is the need clearly articulated?
- *Operational Concept*: Who are the intended users? How will the products be used? How is this different from the present situation?
- *Functional Requirements*: What specific service will be provided, and at what level of detail will it be provided? Are element interfaces well defined?
- *System Architecture*: What is the overall plan of attack? What elements are involved in the overall approach? Are these complete, logical, and consistent?
- *Allocated Requirements*: Which elements address which requirements? Is the allocation appropriate? Are there any unnecessary requirements?
- *Detailed Design*: Are the details correct? Do they meet the requirements? Are the interfaces satisfied?
- *Implementation*: Will the solution be satisfactory in terms of cost and schedule? Can existing pieces be revised?
- *Test*: What provides evidence of success? Will the result chosen meet the needs of users?

Systems Engineering Approach to Life-Cycle Phases

According to James Martin (1991), a systems engineering approach might be used during life-cycle phases to analyze alternatives for meeting objectives. The systems approach to life-cycle phases includes the following (Martin 1991, p. 7):

- Planning, data gathering, and procedures;
- Analytical studies and basic engineering;
- Major reviews;
- Detail engineering;
- Construction;
- Testing and implementation.

The systems engineering approach also involves the following (Martin 1991, p. 13):

- Reviewing interrelationships between subsystems,
- Management of the dynamic process,
- Integrating into the total system,
- Systematically assembling into a unified process,
- Seeking optimal solutions or strategies.

Systems engineering concepts are similar to productivity improvement methods. In both techniques, the participants and work activities are analyzed to determine alternative work processes that improve productivity or in some manner benefit the system.

6.11 Summary

This chapter discussed a variety of data-analysis techniques that are useful when quantifying work-process operations at construction jobsites. The techniques presented in this chapter were work sampling, activity sampling, field rating, 5-min rating, work-process charts, and work-process diagrams. In addition, productivity measurement methods were introduced, and an explanation was provided for the types of construction activities where each of the productivity measurement techniques could be used on construction projects. Performance factors and the CSI Master Format were discussed in relation to how to measure whether activities are ahead of or behind their expected duration or performance. Systems engineering concepts were discussed to demonstrate the differences and similarities between systems engineering and productivity improvement methods.

Productivity improvement team members should determine which of the techniques presented in this chapter are applicable to the activities they are analyzing and incorporate the appropriate techniques into their data-analysis process. Chapter 7 provides case studies that demonstrate how the different methods presented in this chapter are used during productivity improvement studies.

6.12 Key Terms

American Society of Mechanical Engineers (ASME)
closed systems
Construction Specification Institute
Construction Specification Institute codes
crew balance charts
CSI Manual of Practice
CSI Master Format
cumulative recording
daily recordings
effective work
essential contributory work
field ratings
5-min ratings
incremental milestone
ineffective work
Kaizen
labor utilization factors
level of effort
Methods Time Measurement

moving-average recordings
open systems
percent complete
performance evaluation reports
performance factor
periodic recordings
productivity rating report
rules of credit
start-to-finish percentage
systems engineer
systems engineering
time-lapse photography
time measurement units
units completed
work-count analysis
work-distribution analysis
work-distribution charts
work-measurement analysis
work-process analysis
work-process charts
work-process flow diagrams
work sampling

6.13 Exercises

6.1 Discuss whether implementing productivity improvement methods without productivity having been directly measured first is a viable technique for increasing productivity?

6.2 What should be accomplished in productivity improvement meetings?

6.3 What should be included in a work-process analysis plan?

6.4 What are three analysis tools that could be used for analyzing work processes?

6.5 What are the six types of forms, charts, and diagrams that should be used to summarize the results of work sampling?

6.6 How does the data-recording process differ using work-sampling forms versus field-rating forms?

6.7 What is the advantage in using 5-min ratings instead of the other data-analysis techniques discussed in this chapter?

6.8 What types of items are shown on work-process flow diagrams?

6.9 What are the five categories monitored on work-process charts for each task?

6.10 What are process charts used for during work sampling?

6.11 Explain what type of information should be shown on crew balance charts?

6.12 Explain the six techniques that should be used for recording work-process data?

6.13 Explain the difference between productivity improvement studies and systems engineering.

6.14 Explain how to calculate a field-rating index.

6.15 Perform a productivity improvement study of a construction operation using each of the data-analysis techniques introduced in this chapter.

References

Adrian, J. (2004). *Construction productivity measurement and improvement*, Stipes, Champaign, IL.

ASME. (2008). *Work process chart symbols,* New York. <https://www.asme.org/search.aspx?searchText=work%20process%20chart%20symbols&#page=1,category=> (March 2014).

Construction Specification Institute (CSI). (2005). *Master Format 2004 edition,* Alexandria, VA.

Construction Specification Institute (CSI). (2012a). *Master Format 2012 edition,* Alexandria, VA.

Construction Specification Institute (CSI). (2012b). *The project resource manual—CSI manual of practice,* Alexandria, VA.

Kerzner, H. (2005). *Project management,* Wiley, New York.

Martin, J. (1991). *Systems engineering guidebook,* Construction Research Council Press, New York.

Methods Time Measurement Association for Standards and Research. (1987). *Methods-time-measurement MTM application data,* Park Ridge, IL.

Ogimachi, N., Spangrud, J., Untrauer, C., Champaneria, V., Bahman, B., and Alturi, D. (n.d.). "Productivity improvement study for structural steel erection on the north concourse project at the Norman Y. Mineta San Jose International Airport." *Course project,* San Jose State University, San Jose, CA.

Oglesby, C., Parker, H., and Howell, G. (1989). *Productivity improvement in construction,* McGraw Hill, New York.

Salmon, J. (2007). "A Kaizen approach to data management, part one." *Professional Surveyor,* October, 48–49.

CHAPTER 7

Evaluating Productivity Improvement Alternatives: Case Studies

The objective of productivity improvement programs is to implement alternative work processes with the expectation that the proposed productivity improvement techniques will reduce costs (inputs), increase profits, or provide a safer working environment. Productivity improvement team members propose alternative work processes, along with justifications for each alternative, and management personnel review the proposed alternatives and select an alternative or a combination of alternatives to be implemented on the project.

This chapter presents information on how to develop and then select appropriate productivity improvement alternatives to be implemented on projects after analyzing current project information. It also provides information on evaluating and prioritizing productivity improvement alternatives and includes four detailed case studies that demonstrate how to collect and analyze productivity improvement data and provide recommendations for improving production rates.

7.1 Evaluating and Prioritizing Productivity Improvement Alternatives

Various methods for comparing productivity improvement alternatives are available, and the selection of the method to be used depends on the variables involved in the work processes and the type of study used to conduct the productivity improvement analysis. Some methods involve cost accounting, balancing work crews, and comparing the potential results of the recommended improvements with the original production rates and costs. Background information on each alternative proposed should be provided, along with details of the recommended alternative work processes, to illustrate how each alternative was analyzed and evaluated by members of the productivity improvement team.

Different analysis techniques are used for comparing the financial viability of productivity improvement schemes. For example, present worth or equivalent uniform annual worth analysis methods may be used to determine the present worth of costs and income. These techniques show which alternative will yield the highest net present worth or equivalent uniform annual worth. For public agencies, minimizing costs to the public is equivalent to maximizing profits for public clients; therefore, net present worth or equivalent uniform annual cost analysis are methods used to determine which alternative will result in the lowest net present cost or equivalent uniform annual cost associated with the alternatives.

As mentioned in Chapter 6, work crews are evaluated and different crew configurations tested to see which crew configuration increases production rates. Work crews are evaluated to determine whether reducing or increasing the size and types of crews enhances workflow processes, maximizes production, or reduces overhead. Problems related to workflow, production, and high overhead may be caused by idle workers, the use of incorrect trades, or inexperienced crews. Alternative work layouts are modeled to determine the most efficient flow of work and availability of materials. It is also important to evaluate whether training programs should be implemented before workers start performing repetitive work tasks and whether they would help motivate workers and improve worker retention.

Existing equipment schedules are evaluated and used to create alternative schedules that reflect the optimal use of equipment and minimize the time equipment is idle. These changes should increase the production of each piece of equipment. To justify the use of expensive tools, there should be a method in place for forecasting increased productivity rates resulting directly from the use of the new tools. To that end, schemes should be developed for sharing expensive or scarce tools and equipment between crews to improve efficiency. Methods for selecting appropriate equipment and the right size equipment for a job should be established to help improve production rates. Before using more expensive, larger capacity equipment an analysis should be performed that demonstrates that the equipment will decrease the time required to complete work tasks by increasing production rates and possibly result in higher profit margins.

When conducting a productivity improvement study, team members should know whether the cost of materials is the most important factor in their selection or whether other considerations would influence selection decisions. For example, paving-operation alternatives should be compared by evaluating their impact on the environment, the quality of the materials, and the timeliness of their delivery. The delivery systems used for placing asphalt and concrete could affect the quality and efficiency of work. Selecting manufacturers, material suppliers, or fabricators solely on the basis of cost, with no consideration for the travel distance required to deliver materials, might not produce the highest quality end product, the most optimal schedule, or the highest possible profit margins. Similarly, using crews with more experience might increase costs but also lead to higher profit margins. Alternative work layouts should be compared to determine which minimize congestion and optimize accessibility. Complying with labor contracts, safety measures, and labor laws might increase overhead costs, but these costs are recoverable from clients. Overhead costs should be evaluated to determine whether investing in additional overhead would create longer term benefits.

To evaluate the previously mentioned items, data need to be collected to support or refute proposed changes in work processes. Chapter 6 introduced techniques for the systematic collection of productivity improvement data, and demonstrated the applicability of these methods. Four case studies are included in the following sections that illustrate how to implement a productivity improvement study. Each of the case studies explains the projects studied and the data-collection procedures used during each of the case studies. The case studies also include the data collected and recommendations for improving productivity resulting from the analysis of that data.

7.2 Case Study 1: Concrete Parking Structures and Police Station

(This case study was conducted by Dr. Basel M. Abulfateh, Assistant Undersecretary for Services for the Royal Court in the Kingdom of Bahrain.)

For this productivity improvement case study, specific work processes were analyzed for the construction of two freestanding parking garage structures and a new police department office building. The smaller of the two parking structures is located on approximately 1.2 acres (0.486 hectares) that was previously a parking lot. It has a capacity of 422 cars and has two subsurface levels and one surface level with a total area of 136,000 ft^2 (12,634.8 m^2). The larger of the two parking structures is located on a site of approximately 5 acres (2.02 hectares) that was also previously used as a parking lot. The new one-story parking structure has a capacity of 930 cars and a total area of 273,000 ft^2 (25,362.5 m^2). The parking structures were constructed using **cast-in-place concrete** for two reasons: (1) the two subsurface levels required the construction of retaining and foundation walls and (2) the amount of land designated for the project limited the type of equipment that could be used for this structure. It would have been difficult to use **precast concrete** members because they are hard for a crane to lift and maneuver in constricted locations (Abulfateh n.d.).

A three-story police department building is connected to the larger parking structure by a pedestrian bridge. This building includes offices, dispatch rooms, computer rooms, shops, locker rooms, a laboratory, and a darkroom for a total of 20,350 ft^2 (1,890.6 m^2). The police structure was constructed using precast concrete members.

The project consisted of the following major elements of work:

- *Utilities*: Extending all of the utilities—including the sanitary and storm sewers; the gas, power, water, and steam lines; and telephone lines—from the street or right-of-way to the structure.
- *Earthwork*: Earthmoving, excavating, grading, compacting, excavating for drainage, importing soil, and fine-grading.
- *Pavement*: Placing concrete and asphalt, paving, and painting parking space lines.
- *Landscaping*: Installing trees, shrubs, annuals, perennials, ground cover, turf grass, and an irrigation system.
- *Parking structures*: Installing caissons, grade beams, foundation beams, columns, spandrels, double-tee floor slabs, poured concrete decks, post-tensioned concrete slabs, steel framing, bar joists with poured concrete on metal decking, and steel roof members.

- *Police station office building*: Installing caissons, grade beams, foundation walls, retaining walls, precast-prestressed concrete beams, columns, spandrels, double-tee floor slabs, poured concrete decks, and steel decks.
- *Parking structure exteriors*: Installing cast-in-place concrete, precast concrete panels, native sandstone, structural steel, metal doors and frames, built-up bituminous roofing, and vitreous clay tile roofing.
- *Parking structure interiors*: Building an exposed structural system.
- *Police station office building interior*: Installing gypsum wallboard on metal studs, suspended acoustical tile in a metal-tee grid, ceramic tile, carpet, and raised computer floors.
- *Conveying systems*: Installing hydraulic elevators.
- *Mechanical system*: Installing a ventilation system, a carbon-monoxide monitoring system, a fire-protection sprinkler system, parking-control equipment, and vertical circulation by the hydraulic elevator.
- *Electrical system*: Installing power and lighting, emergency power, site lighting, telephone and data conduits, a video system, and fire alarm and fire detection systems.

The projects were scheduled to be completed within nine months. All three of the structures were constructed by one construction firm. Several different superintendents were in charge of the construction sites at different times. The parking structures had two superintendents, and the police building had three superintendents. The turnover in the supervisory personnel was caused by vacations, firing, and relocation to other projects.

After the site was excavated, the next stage of construction was the erection of the topping slabs and then backfilling and site utilities. At the second site, major activities involved construction of the ground walls, decks, the tower, and the installation of utilities.

Data-Collection Techniques

Similar activities that occurred at both sites were studied to compare the productivity rates between the two sites. In particular, two tasks were analyzed: (1) the on-site concrete-casting process for the foundations of the two parking structures and (2) the process of erecting precast concrete members for the police station office building. Field ratings were used at Site 1 to study the process of erecting the precast members for the police office building. Data were collected during several visits over a four-month period. The specific productivity improvement techniques used for this study were interviews, productivity ratings, 5-min ratings, process charts, and process flow diagrams.

Construction Tasks

The process of casting the concrete for the parking structure foundations required concrete trucks and a space where the trucks could **queue** (wait) for the trucks in front of them to empty their loads. Concrete pump trucks were used for the casting of the concrete. Work crews were responsible for placing, vibrating, leveling, and applying the curing requirements for the concrete. The foreman was responsible for testing the concrete using slump tests and for filling cylinders and cubes for the stress tests conducted before the concrete was placed in the foundations. When one crew member failed to complete his or her assigned task on time, the entire process was delayed because every step of the work process depended on the completion of the preceding activities.

The second process analyzed during the case study was the erection of the precast concrete members for the police station office building. This process included (1) transporting the precast concrete members on trucks, (2) unloading the precast concrete members from the trucks and transporting them to their position in the structure using one or more cranes, and (3) fixing the precast concrete members. During the study, the weather was dry, and the temperature varied between 50°F (10°C) and 75°F (23.9°C), within the acceptable temperature range for placing concrete. According to company policy, the workers were not allowed to work more than 45 h per week; therefore, fatigue was not a major factor affecting labor productivity rates at these sites. Figures 7-1 and 7-2 summarize the results of the productivity study for Sites 1 and 2. The data were collected under similar conditions, and then averages were calculated for each site. The averages are shown in Figures 7-1 and 7-2.

Table 7-1 provides the results obtained from the data-collection process on effective work, supportive work, and ineffective work for Sites 1 and 2.

Figures 7-3 and 7-4 contain a summary of the results of the 5-min ratings conducted on five randomly selected workers who were part of the construction teams at Sites 1 and 2. Table 7-2 summarizes the results of the 5-min ratings for the concrete casting of the foundations at Sites 1 and 2.

Figures 7-5 and 7-6 represent the relationship between the sample number and the average percentage of productivity for the field ratings for Sites 1 and 2, and

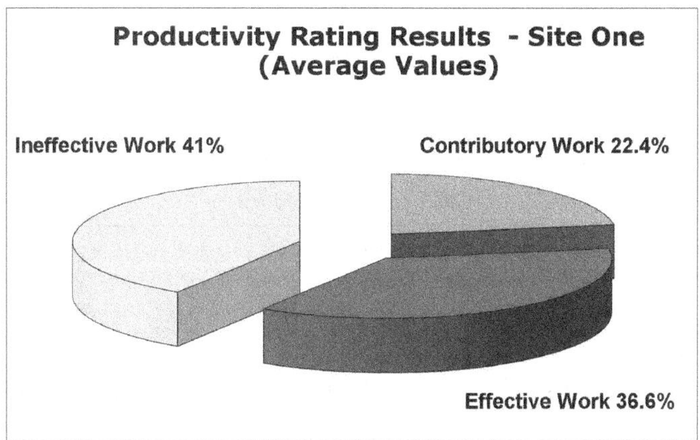

Figure 7-1. Productivity Rating for Site 1

Source: Abulfateh (n.d.), Figure 2; reproduced with permission.

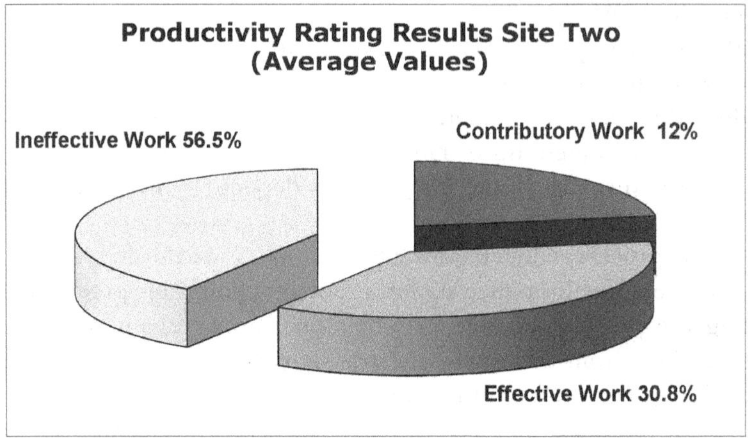

Figure 7-2. Productivity Rating for Site 2

Source: Abulfateh (n.d.), Figure 3; reproduced with permission.

Table 7-1. Summary of Effective, Supportive, and Ineffective Work for Sites 1 and 2 (Case Study 1)

Site 1 Productivity Study Results			Site 2 Productivity Study Results		
Effective Work	Supportive Work	Ineffective Work	Effective Work	Supportive Work	Ineffective Work
7	8	14	16	5	24
7	6	17	15	3	23
9	6	13	9	3	30
10	7	11	9	4	29
11	6	14	9	9	26
16	8	6	19	8	16
18	6	6	19	8	16
7	5	14	9	3	29
Total = 85/232 = 36.6%[a]	Total = 52/232 = 22.4%[b]	Total = 95/232 = 41%[c]	Total = 105/341 = 30.8%[a]	Total = 43/341 = 12.6%[b]	Total = 193/341 = 56.6%[c]

Source: Abulfateh (n.d.), Table 1; reproduced with permission.

[a] Percentage of effective work = effective work total observations/(effective work + supportive work + ineffective work) [Eq. (6-1)].

[b] Percentage of supportive work = supportive work total observations/(effective work + supportive work + ineffective work) [Eq. (6-2)].

[c] Percentage of ineffective work = ineffective work total observations/(effective work + supportive work + ineffective work) [Eq. (6-3)].

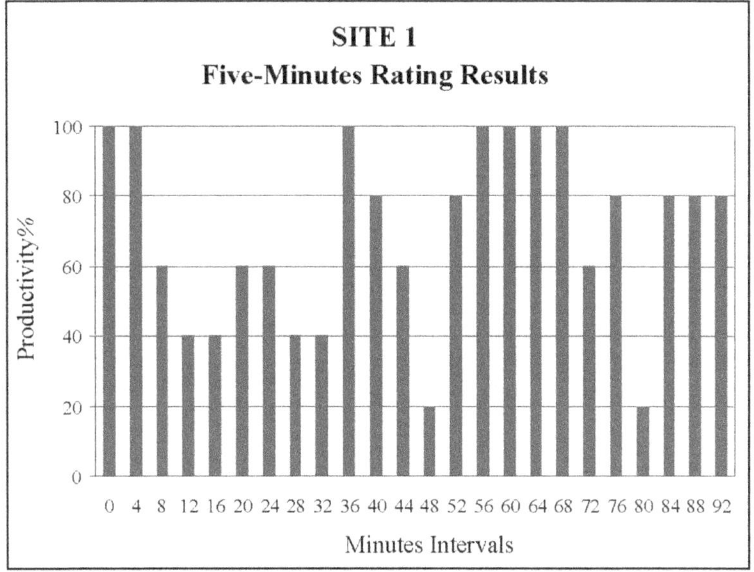

Figure 7-3. 5-Min Rating for Site 1

Source: Abulfateh (n.d.), Figure 4; reproduced with permission.

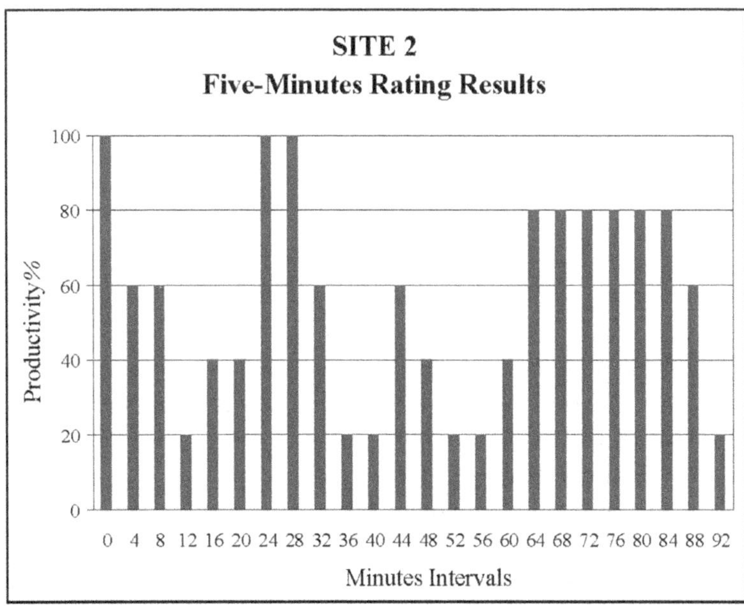

Figure 7-4. 5-Min Rating for Site 2

Source: Abulfateh (n.d.), Figure 5; reproduced with permission.

Table 7-2. Summary of Productivity Ratings at Different Time Intervals for Sites 1 and 2 (Case Study 1)

Time Intervals	Productivity Percentage (Site 1)	Productivity Percentage (Site 2)
0	100	100
4	100	60
8	60	60
12	40	20
16	40	40
20	60	40
24	60	100
28	40	100
32	40	60
36	100	20
40	80	20
44	60	60
48	20	40
52	80	20
56	100	20
60	100	40
64	100	80
68	100	80
72	60	80
76	80	80
80	20	80
84	80	80
88	80	60
92	80	20

Source: Abulfateh (n.d.), Table 2; reproduced with permission.

Figure 7-7 shows the sample size and average percentage of productivity for the erection of the precast concrete members for the police station office building.

Figure 7-8 shows the flow process for the concrete-casting sequence for Site 1, and Figure 7-9 shows the process chart worksheet for this operation. Figure 7-10 shows the flow process for the concrete-casting sequence for Site 2, and Figure 7-11 is the process chart worksheet for this operation at Site 2.

Figure 7-12 is the flowchart for the erection of the precast concrete members for the police station office building, and Figure 7-13 is the flowchart worksheet for this operation.

Figures 7-14 to 7-18 provide proposed recommended revisions to the work-process flow for Sites 1 and 2.

Figures 7-19 to 7-29 contain photos of the work areas that were studied for this case study.

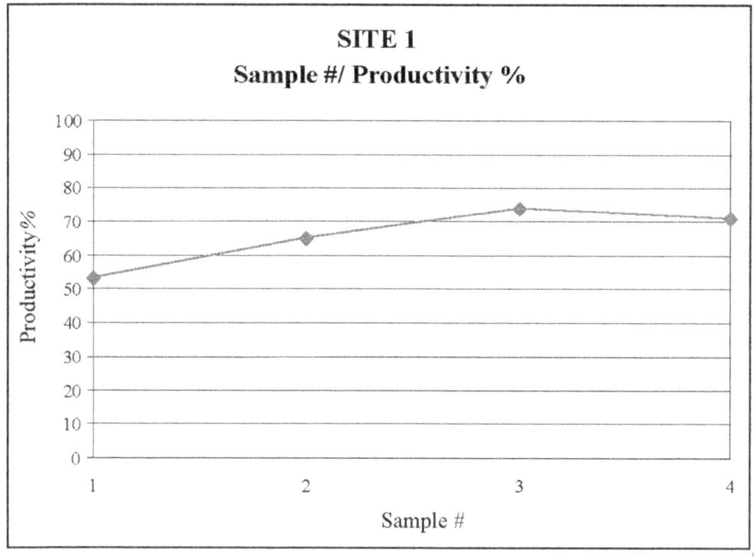

Figure 7-5. Sample Numbers versus Productivity Percentages for Site 1

Source: Abulfateh (n.d.), Figure 6; reproduced with permission.

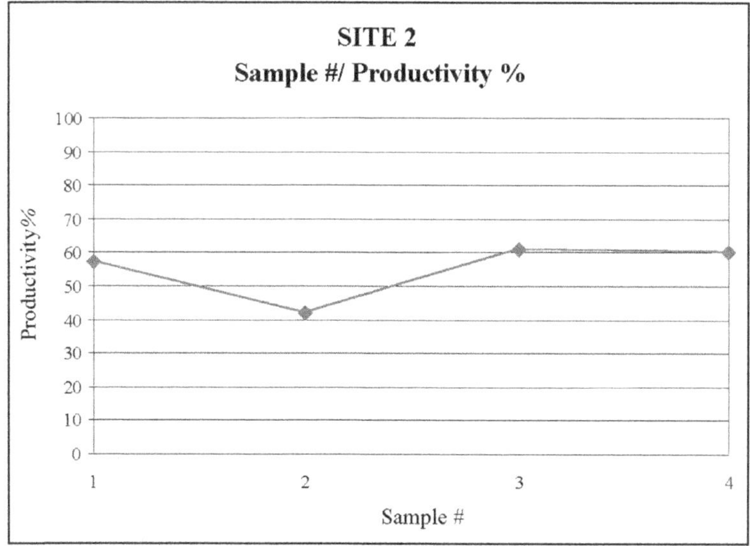

Figure 7-6. Sample Numbers versus Productivity Percentages for Site 2

Source: Abulfateh (n.d.), Figure 7; reproduced with permission.

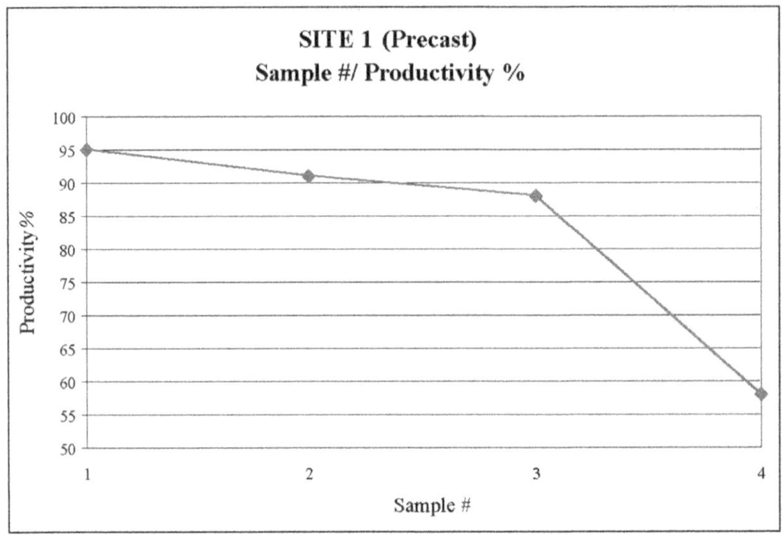

Figure 7-7. Sample Numbers versus Productivity Percentages for Site 1 Precast

Source: Abulfateh (n.d.), Figure 8; reproduced with permission.

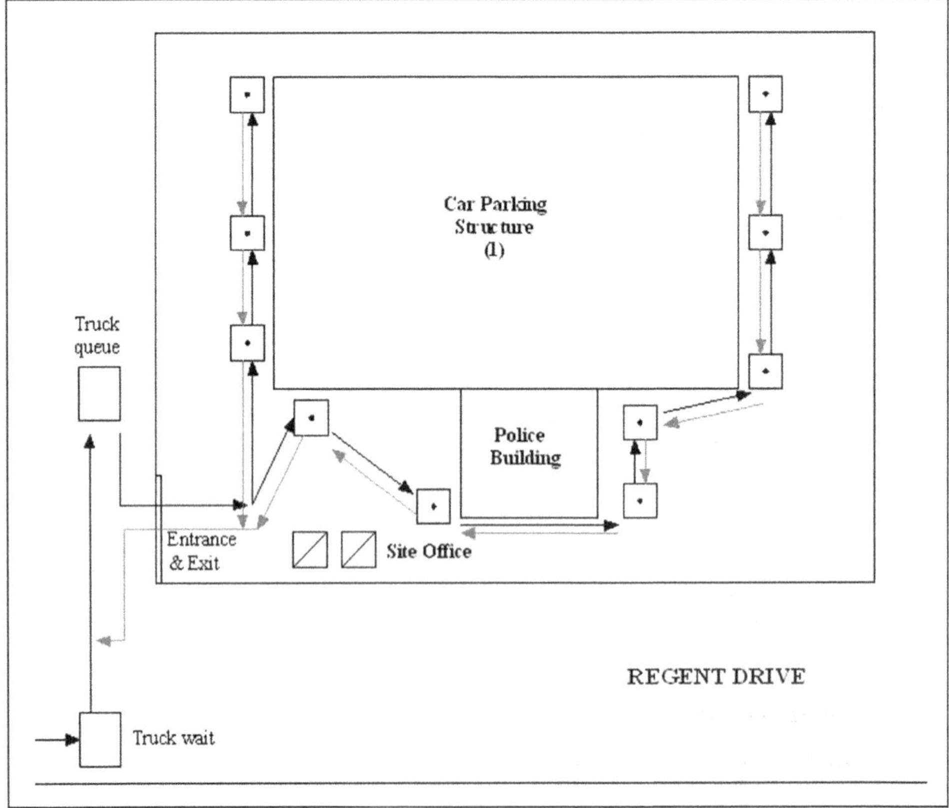

Figure 7-8. Flowchart of Casting Concrete for Site 1

Source: Abulfateh (n.d.), Figure 9A; reproduced with permission.

Process Chart Worksheet

Project:	Car parking	Study Location:	Site 1
Work Item:	Concrete (foundation)	Prepared By: ______	Date: ______
Evaluation No.	1		

No.	Flow Sequence	Operation / Transportation / Inspection / Delay / Storage	Quantity	Distance in Feet	Time	Notes
1.	Concrete truck trans.	○ ➡ □ D ▽		C	C	From supplier to site
2.	Wait at street	○ ⇨ □ ◗ ▽			10	
3.	Travel to unloading location	○ ➡ □ D ▽		300	1	
4.	Wait in queue	○ ⇨ □ ◗ ▽			45	Average time
5.	Truck enters site	○ ➡ □ D ▽		700	5	Where concrete has to be placed
6.	Slump and stress tests	○ ⇨ ■ D ▽			13	
7	Casting Concrete	● ⇨ □ D ▽	4		30	
8.	Delivery approval	○ ⇨ ■ D ▽			2	
9.	Truck leaves the site	○ ➡ □ D ▽		700	6	
		○ ⇨ □ D ▽				
		○ ⇨ □ D ▽				*C = Constant Value
		○ ⇨ □ D ▽				
		○ ⇨ □ D ▽	4	1700	112	

Summary		Evaluation		Proposed		Difference	
Functions		No.	Time	No.	Time	No.	Time
○	Operations						
⇨	Transportation						
□	Inspection						
D	Delays						
▽	Storage						
Distance Traveled (in feet)							

Figure 7-9. Process Chart Worksheet for Site 1 Parking Garage Concrete Foundation

Source: Abulfateh (n.d.), Figure 9B; reproduced with permission.

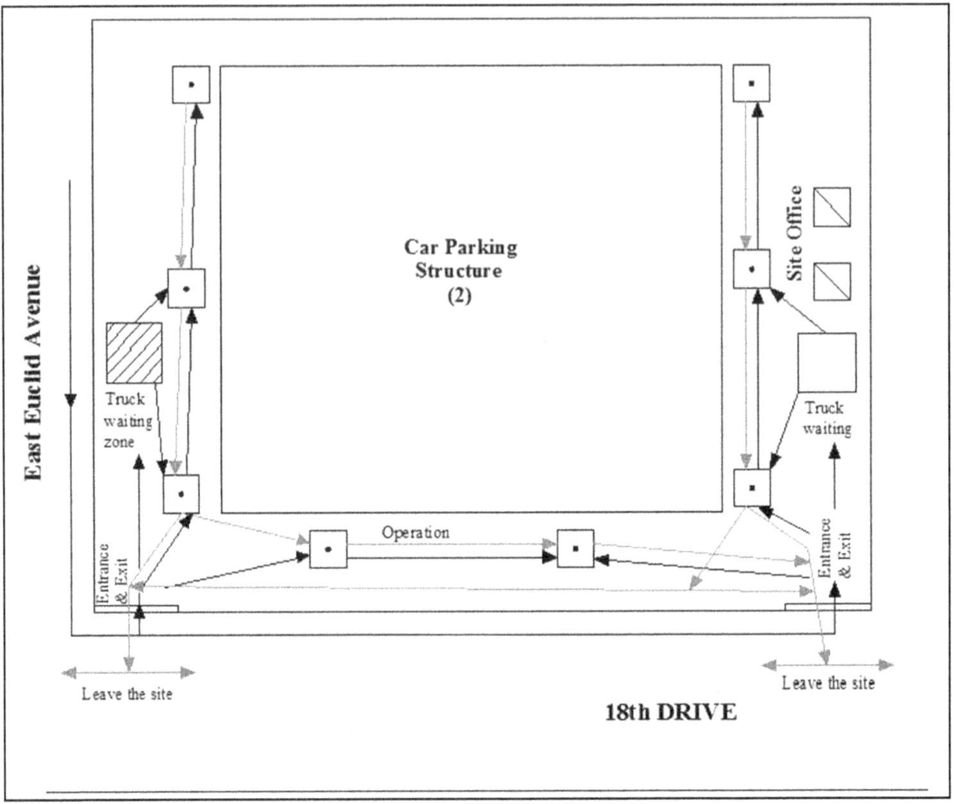

Figure 7-10. Flowchart for Casting Concrete for Site 2

Source: Abulfateh (n.d.), Figure 10A; reproduced with permission.

Analysis and Results of Interviews with Jobsite Personnel

After the data were collected at the jobsites, it was analyzed, and the results were examined to identify the factors that either positively or negatively affected construction productivity. The following sections summarize the results of the analysis.

One of the major problems observed by all of the project participants, including the project manager, was that the project schedule was too short. According to the schedule, all three of the structures had to be constructed in nine months. There were **liquidated damages** of $25,000 for each structure not finished by the end of the ninth month. In addition, there was a penalty of $2,500 per delay day per structure. The contractor did not provide enough workers or equipment to complete the job on time. Management personnel thought that the delays occurred because of the poor productivity of the workers, and as a result, one of the superintendents was fired from the job. Because the second superintendent joined the project after it had started, it was difficult for him to quickly learn what had occurred on the project before he took over.

Process Chart Worksheet

Project:	Car parking	Study Location:	Site 2		
Work Item:	Concrete (foundation)	Prepared By:		Date:	
Evaluation No.	2				

No.	Flow Sequence	Operation / Transportation / Inspection / Delay / Storage	Quantity	Distance in Feet	Time	Notes
1.	Concrete truck trans.	○ ➡ □ D ▽ (Transportation)		500	5	From one gate to waiting area
2.	Wait in queue	○ ⇨ □ ◗ ▽ (Delay)			45	
3.	Travel to unloading location	○ ➡ □ D ▽ (Transportation)		1300	15	Travel to casting place anywhere in the site
4.	Slump and stress tests	○ ⇨ ■ D ▽ (Inspection)			13	
5.	Casting Concrete	● ⇨ □ D ▽ (Operation)	4		30	
6.	Delivery approval	○ ⇨ ■ D ▽ (Inspection)			2	
7.	Truck leaves the site	○ ➡ □ D ▽ (Transportation)		1200	15	
		○ ⇨ □ D ▽				
		○ ⇨ □ D ▽				
		○ ⇨ □ D ▽				
		○ ⇨ □ D ▽				
		○ ⇨ □ D ▽				
		○ ⇨ □ D ▽	4	3000	125	

Summary	Evaluation		Proposed		Difference	
Functions	No.	Time	No.	Time	No.	Time
○ Operations						
⇨ Transportation						
□ Inspection						
D Delays						
▽ Storage						
Distance Traveled (in feet)						

Figure 7-11. Process Chart Worksheet for Site 2 Parking Garage Concrete Foundation

Source: Abulfateh (n.d.), Figure 10B; reproduced with permission.

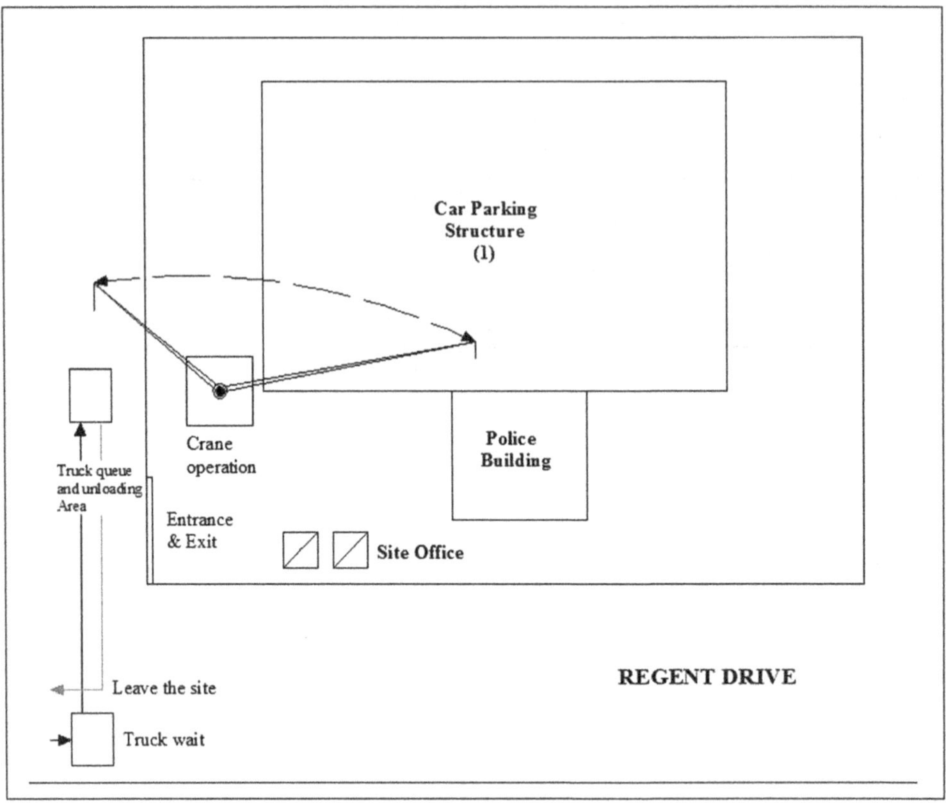

Figure 7-12. Flowchart of Erecting Precast Members for Police Building

Source: Abulfateh (n.d.), Figure 11A; reproduced with permission.

The project manager indicated that he knew how to alleviate some of the problems causing delays, but he would not suggest solutions to the contractor because if his suggestions failed he thought he might be liable. The contractor claimed that the project manager misled him into thinking that he could request additional time and money. This indicated that the communication between the project manager and the contractor needed to be improved so that the project manager could effectively direct and assist the contractor.

Several superintendents indicated that a major problem at both of the construction sites was vandalism and the theft of tools. During the project, several individuals were charged with theft and vandalism while they were under the influence of alcohol. As a result, the contractor increased security at an additional cost to the company. The superintendents all agreed that the police failed to provide sufficient security for the two jobsites. One superintendent stated, "It is a shame that the police cannot protect their own headquarters."

The superintendents further mentioned that a more thorough plan for the project should have been provided by the engineer. For example, at Site 2 there

Process Chart Worksheet

Project:	Police Building	Study Location:	Site 1
Work Item:	Erecting Precast	Prepared By:	______ Date: ______
Evaluation No.	3		

No.	Flow Sequence	Operation / Transportation / Inspection / Delay / Storage	Quantity	Distance in Feet	Time	Notes
1.	Precast truck Travel	○ ➡ □ D ▽ (Transportation)		*C	C	From supplier to waiting street
2.	Wait in street	○ ⇨ □ D ▽ (Delay)			7	
3.	Travel to queue zone	○ ➡ □ D ▽ (Transportation)		300	1	
4.	Wait in queue zone	○ ⇨ □ D ▽ (Delay)			120	
5.	Check and approve delivery	○ ⇨ ■ D ▽ (Inspection)			10	
6.	Unload the precast member	● ⇨ □ D ▽ (Operation)			120	Truck stays in place until the precast members are unloaded
7.	Truck leaves the site	○ ➡ □ D ▽ (Transportation)		300	2	
		○ ⇨ □ D ▽				
		○ ⇨ □ D ▽				
		○ ⇨ □ D ▽				
		○ ⇨ □ D ▽				*C = Constant Value
		○ ⇨ □ D ▽				
		○ ⇨ □ D ▽		600	260	

Summary	Evaluation		Proposed		Difference	
Functions	No.	Time	No.	Time	No.	Time
○ Operations						
⇨ Transportation						
□ Inspection						
D Delays						
▽ Storage						
Distance Traveled (in feet)						

Figure 7-13. Process Chart Worksheet for Erecting Precast Concrete for Site 1

Source: Abulfateh (n.d.), Figure 11B; reproduced with permission.

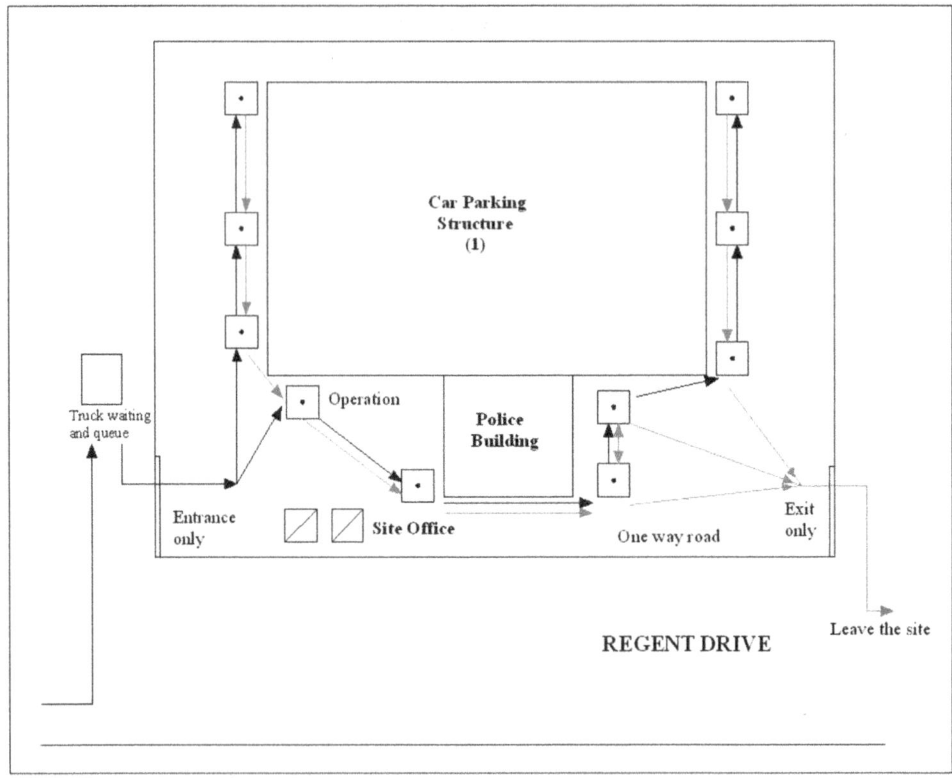

Figure 7-14. Proposed Revised Flow Diagram for Site 1

Source: Abulfateh (n.d.), Figure 22A; reproduced with permission.

was a high groundwater table, and water had to be pumped out of the excavation area for the foundation ditches. The superintendent had to determine where to drain the water. He did not anticipate that the managers working for the construction company would not plan for this type of situation. The contractor did not have permission from the city to dump the pumped water into the local drainage system, and as a result, the work was delayed until permission was issued by the municipality.

The lack of coordination between the project schedule and the fire department authorities also caused a problem. At Site 1, the work was delayed because the construction crews could not continue with their work until sprinkler systems and other requirements were inspected and approved by the fire department. Management personnel working for the contractor failed to arrange for an inspection appointment compatible with the project schedule. The superintendent of Site 1 stated, "I cannot imagine that management arranged a meeting with those people (fire department personnel) ten days after we actually needed them."

Preliminary interviews were conducted with jobsite personnel, and Table 7-3 provides the interview guide questions used for this project. One of the results of the

Table 7-3. Interview Guide Questions for Case Study 1

1. *Tools*
   a. What tools were used the most often?
   b. Where were the tools located when not being used during the project?
   c. If two people needed the same tool, who decided who would use it first?
   d. Where were the tools stored at the jobsite?
   e. Who repaired the tools when they broke during the project?
   f. Where and how did workers acquire new or more tools?
   g. Who ordered the tools for the workers?
   h. What types of tools could be ordered?
   i. What types of orders were you able to approve?
   j. How long did it take for the tools that you ordered to be delivered to the jobsite?
2. *Management Practices*
   a. Planning practices
      (1) How did you plan the work being conducted for the project?
      (2) Have you been trained to plan work?
      (3) How much were you allowed to change a method already being used?
      (4) How much latitude did you have in varying the methods used on the project?
   b. Milestones and priorities
      (1) Did you have milestone dates for the completion of activities?
      (2) Who provided the milestone dates to you?
      (3) Did you have a list of priorities?
      (4) Who provided the list of priorities?
3. *Information Requirements*
   a. Information for planning
      (1) How did you acquire the information used in the plan?
      (2) Was there a better way to acquire the information than the method used?
      (3) How did you know which activity to work on next?
      (4) How did you know what resources you were able to use?
      (5) How did you determine what resources were available at the jobsite for use?
      (6) How did you locate resources at the jobsite?
   b. Feedback systems
      (1) What was the unit cost of the operation?
      (2) Was the cost above or below the budgeted cost?
      (3) How did the performance achieved compare to the scheduled performance?
      (4) How much rework was required?
      (5) Was the quality of the work adequate?
      (6) Did you ever have access to the control reports during construction?
      (7) Did you understand the control system being used on this project?
      (8) Have you ever been trained to use the control system?
   c. Problem solving and decision making
      (1) How were problems recognized on the project?
      (2) How were solutions to the problems developed?
      (3) Were you ever involved in recognizing problems and determining solutions for them?
      (4) How were the solutions to the problems evaluated?
      (5) Who approved the use of the solutions to the problems?
      (6) Were all of the solutions developed implemented?
      (7) Who checked to verify that the solutions were successful?

Source: Adapted from Sanvido and Paulson (1991).

Process Chart Worksheet

Project:	Car parking	Study Location:	Site 1
Work Item:	Concrete (foundation)	Prepared By:	________ Date: ________
Evaluation No.	1 (proposed)		

No.	Flow Sequence	Operation / Transportation / Inspection / Delay / Storage	Quantity	Distance in Feet	Time	Notes
1.	Concrete truck transportation	○ ➡ □ D ▽		C	C*	From supplier to site
2.	Wait in queue	○ ⇨ □ ◗ ▽			45	
3.	Concrete truck enters site	○ ➡ □ D ▽		700	3	
4.	Slump, stress tests, and delivery approval	○ ⇨ ■ D ▽			13	
5.	Casting concrete	● ⇨ □ D ▽	4		30	
6.	Truck leaves the site	○ ➡ □ D ▽		1100	3	
		○ ⇨ □ D ▽				
		○ ⇨ □ D ▽				
		○ ⇨ □ D ▽				
		○ ⇨ □ D ▽				
		○ ⇨ □ D ▽				
		○ ⇨ □ D ▽				*C = Constant Value
		○ ⇨ □ D ▽	4	1800	94	

Summary		Evaluation		Proposed		Difference	
Functions		No.	Time	No.	Time	No.	Time
○	Operations	1	30	1	30	-	-
⇨	Transportation	4	12	3	6	-	6
□	Inspection	2	15	1	13	1	2
D	Delays	2	55	1	45	1	10
▽	Storage						
Distance Traveled (in feet)		1700		1800		-100	

Figure 7-15. Proposed Revised Process Chart Worksheet for Site 1

Source: Abulfateh (n.d.), Figure 22B; reproduced with permission.

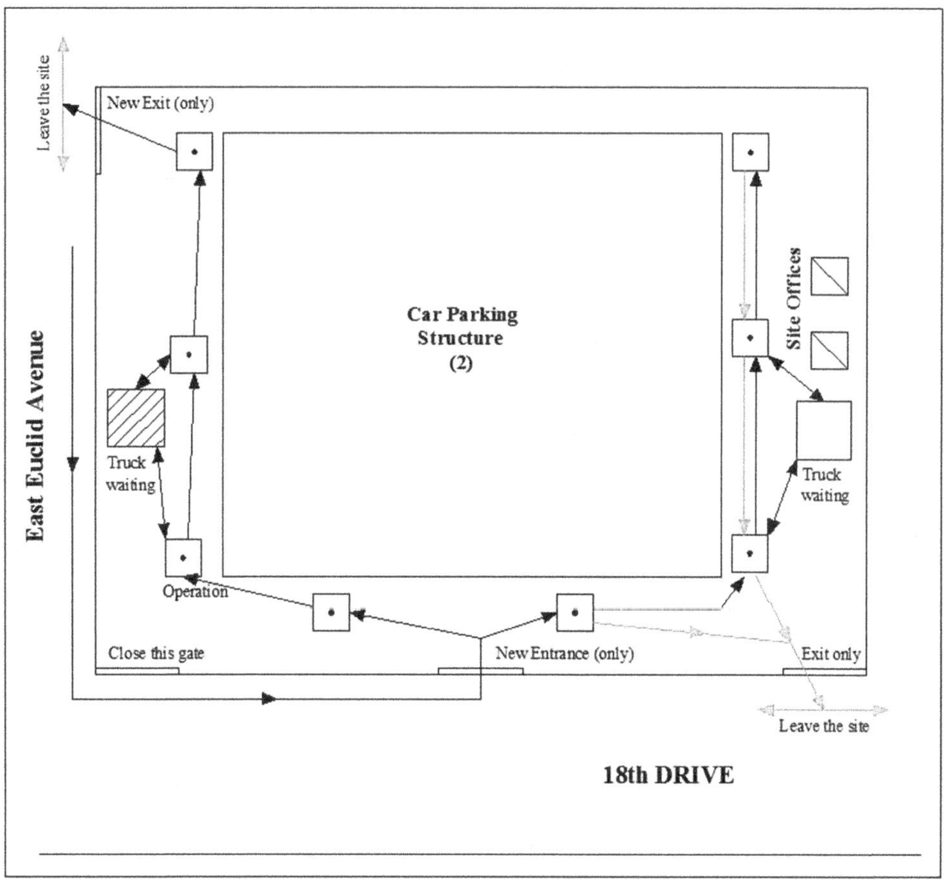

Figure 7-16. Proposed Revised Flow Diagram for Site 2

Source: Abulfateh (n.d.), Figure 23A; reproduced with permission.

preliminary interviews was the development of a series of questions that helped to highlight the responsibilities of the different parties involved with the project. The project manager, the superintendent, the foremen, and selected crew members were interviewed for this case study.

Analysis of Field Ratings

The field productivity percentages of the samples recorded are shown in Figures 7-5, 7-6, and 7-7. For the concrete-casting work processes at Sites 1 and 2, productivity rates decreased between 10:30 a.m. and 12:00 p.m. The workers had a break from 10:00 to 10:15 a.m. and from 12:00 to 12:30 p.m. As the work continued for the same phase of an activity, the productivity increased, especially at Site 1, where it increased by 20%.

Process Chart Worksheet

Project:	Car parking	Study Location:	Site 2
Work Item:	Concrete (foundation)	Prepared By:	________ Date: ________
Evaluation No.	2 (proposed)		

No.	Flow Sequence	Operation / Transportation / Inspection / Delay / Storage	Quantity	Distance in Feet	Time	Notes
1.	Concrete truck transportatioin	○ ➡ ☐ D ▽ (Transportation)		700	3	From gate to waiting area
2.	Wait in queue	○ ⇨ ☐ D ▽ (Delay)			45	
3.	Travel to cast	○ ➡ ☐ D ▽ (Transportation)		1300	5	Travel to where concrete has to be placed
4.	Slump, stress tests, and delivery approval	○ ⇨ ☐ D ▽ (Inspection)			13	
5.	Casting concrete	● ⇨ ☐ D ▽ (Operation)	4		30	
6.	Truck leaves the site	○ ➡ ☐ D ▽ (Transportation)		1200	6	
		○ ⇨ ☐ D ▽				
		○ ⇨ ☐ D ▽				
		○ ⇨ ☐ D ▽				
		○ ⇨ ☐ D ▽				
		○ ⇨ ☐ D ▽				
		○ ⇨ ☐ D ▽				
		○ ⇨ ☐ D ▽	4	3200	102	

Summary		Evaluation		Proposed		Difference	
Functions		No.	Time	No.	Time	No.	Time
○	Operations	1	30	1	30	-	-
⇨	Transportation	3	35	3	14	-	21
☐	Inspection	2	15	1	13	1	2
D	Delays	1	45	1	45	-	-
▽	Storage						
Distance Traveled (in feet)		3000		3200		-200	

Figure 7-17. Proposed Revised Process Flowchart Worksheet for Site 2

Source: Abulfateh (n.d.), Figure 23B; reproduced with permission.

Process Chart Worksheet

Study

Project:	Police Building	Location:	Site 1
Work Item:	Erecting Precast	Prepared By:	______ Date: ______
Evaluation No.	3 (proposed)		

No.	Flow Sequence	Operation / Transportation / Inspection / Delay / Storage	Quantity	Distance in Feet	Time	Notes
1.	Precast truck travel	○ ➡ □ D ▽ (Transportation)		300	1	Directly to waiting zone
2.	Wait in queue	○ ⇨ □ ◗ ▽ (Delay)			120	
3.	Check and approve for delivery	○ ⇨ ■ D ▽ (Inspection)			0	Because it could be done while the truck is waiting
4.	Unload the precast members	● ⇨ □ D ▽ (Operation)				
5.	Truck leaves the site	○ ➡ □ D ▽ (Transportation)		300	2	
		○ ⇨ □ D ▽				
		○ ⇨ □ D ▽				
		○ ⇨ □ D ▽				
		○ ⇨ □ D ▽				
		○ ⇨ □ D ▽				
		○ ⇨ □ D ▽				*C = Constant Value
		○ ⇨ □ D ▽				
		○ ⇨ □ D ▽		600	123	

Summary		Evaluation		Proposed		Difference	
Functions		No.	Time	No.	Time	No.	Time
○	Operations	1	120	1	120	-	-
⇨	Transportation	3	3	2	3	1	-
□	Inspection	1	10	1	0	-	10
D	Delays	2	127	1	120	-	7
▽	Storage						
Distance Traveled (in feet)		600		600		-	

Figure 7-18. Proposed Second Revised Process Flowchart Worksheet for Site 2

Source: Abulfateh (n.d.), Figure 23C; reproduced with permission.

Figure 7-19. Site 1 Car Parking Structure

Source: Abulfateh (n.d.), Figure 12; reproduced with permission.

Figure 7-20. Site 1 Police Station Office Building

Source: Abulfateh (n.d.), Figure 13; reproduced with permission.

The labor productivity rates were negatively affected by breaks, especially when the laborers had not finished a certain part of their work tasks before they took a break. For example, they would stop working on tasks that could have been completed in only 5 min in order to take a break. Productivity decreased more than 30% during the last half hour of the workday, between 4:30 and 5:00 p.m. The reason for

Figure 7-21. Precast Members Transported by Trucks

Source: Abulfateh (n.d.), Figure 14; reproduced with permission.

Figure 7-22. Limited Parking Area for Trucks Unloading Precast Members

Source: Abulfateh (n.d.), Figure 15; reproduced with permission.

this reduction was the nature of the work being performed during this time interval. It took approximately a half hour to completely erect each precast concrete member: therefore, if the workers thought that they could not fix a certain member within the remaining half hour of work, they simply did not start the erection sequence. The contractor was not paying for overtime work shifts, and as a consequence, none of the workers would work one minute after 5:00 p.m.

Figure 7-23. Lifting Precast Concrete Member by Crane

Source: Abulfateh (n.d.), Figure 16; reproduced with permission.

Figure 7-24. Fixing Precast Concrete Member

Source: Abulfateh (n.d.), Figure 17; reproduced with permission.

Productivity Ratings and 5-Min Ratings

Figures 7-1 and 7-2 include the average values of the 5-min ratings. The productivity rates at Site 1 were higher than the productivity rates at Site 2 (59% and 43%). For Site 2, ineffective work averaged 56.5% of the total work. If the average number of laborers at Site 2 was 45, then 25 of them were not working at any one time. Although the samples collected at Site 1 indicated an average total for ineffective work of 41%, the productivity rate for Site 1 was higher than this value (59%). There were signifi-

Figure 7-25. Site 2 Car Parking Structure

Source: Abulfateh (n.d.), Figure 18; reproduced with permission.

Figure 7-26. Limited Work Area around Site 2

Source: Abulfateh (n.d.), Figure 19; reproduced with permission.

cant differences in the production rates at Sites 1 and 2. The production rate at Site 2 was low, the site was poorly organized, and its layout was inefficient.

Process Charts and Flow Diagrams

After reviewing the process charts and flow diagrams, the following recommendations were developed:

Figure 7-27. Foundation Work at Site 2

Source: Abulfateh (n.d.), Figure 20; reproduced with permission.

Figure 7-28. Groundwater Being Drained by Hydraulic Pumps

Source: Abulfateh (n.d.), Figure 21; reproduced with permission.

1. For the foundation work at Site 1, a new flow diagram and process chart were developed to help improve productivity rates, and they are shown in Figures 7-14 and 7-15. The modifications introduced were the following:
   - By making different arrangements with the concrete supplier, the waiting period for the concrete trucks outside the jobsite could be reduced or eliminated.

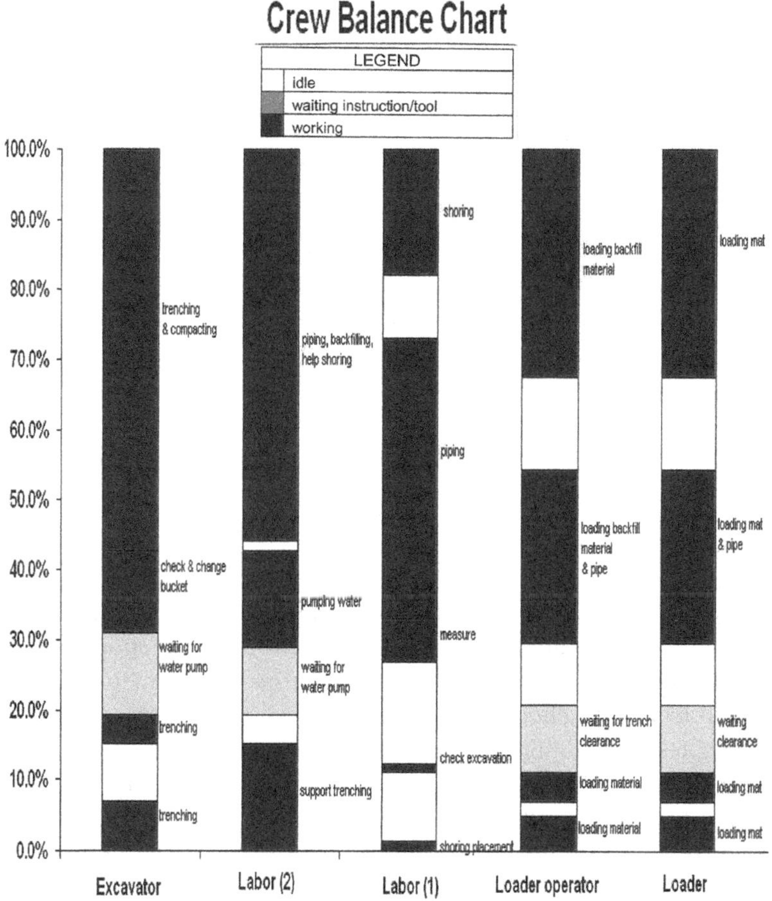

Figure 7-29. Overall View of Parking Garage and Police Headquarters Building

Source: Abulfateh (n.d.), Figure 22; reproduced with permission.

- The process of conducting the concrete and slump tests and filling the cubes and cylinders with concrete for the stress tests could be combined with the process of concrete delivery approval.
- The delivery path could be changed by opening a new gate on the other side of the jobsite. This recommendation was adopted by the superintendent after the productivity improvement team members suggested it to him. The street parallel to the jobsite should have been a one-way street during construction, as this would have significantly improved the movement of the trucks and reduced the time required for the trucks to travel into and out of the site.

2. For the foundation work at Site 2, the recommendations were summarized as follows:
   - Combining the process of conducting the concrete tests and slump tests and filling the cubes and cylinders with concrete for stress tests with the process of concrete delivery approval.

- Changing the direction of the phases within the site was accomplished by opening a new gate at the southwest corner of the site. Having traffic move in the direction shown in the revised flowchart would increase the travel distance; however, the travel time would be decreased because the trucks would not have to wait for other trucks to leave the site or move to another location to allow other trucks access.

3. For the precast concrete work at Site 1, the only change to the flow diagram and the process chart was to eliminate the truck waiting period by arranging with the supplier the intervals at which the trucks would be released to drive to the jobsite. The limited area provided within and around the jobsite had a significant negative effect on productivity rates. It was not only a problem for the movement of trucks, but it also prevented the contractor from having on-site storage for materials. All of the materials had to be delivered when they were needed, which meant that if one truck was late, the work stopped until the materials on the truck arrived at their proper location.

Subcontractor Performance

The poor performance of the subcontractors reduced productivity rates at both sites. The superintendents indicated that the subcontractors did not cooperate with them. When construction was behind schedule at Site 2, the subcontractors had to change their schedule to fit the schedule of the general contractor. However, some of the subcontractors had work obligations at other projects; therefore, it was difficult for them to provide the required amount of resources, labor, and equipment, and productivity rates declined rapidly. The subcontractors argued that they had provided the required resources on time, but because the project was behind schedule, they had moved their resources to other sites where they were also working on projects.

Conflicts between the subcontractors, especially at Site 2, also resulted in delays. Due to poor management, some subcontractors did not or would not change their plans and schedule. This caused some activities to conflict with other activities performed by different subcontractors. For example, at Site 2, a conflict between two subcontractors occurred because one subcontractor wanted to have his concrete trucks cast a section of the retaining walls; however, a second subcontractor had excavated the entrance to erect drainage pipes. The engineer working for the general contractor was consulted, and he solved the problem by bringing in a concrete-pumping truck at the expense of the general contractor to allow the first subcontractor to cast the concrete without entering the site.

Technical Difficulties

Another problem that reduced productivity rates, especially at Site 1, was technical difficulties, and two situations occurred at Site 1. First, a precast girder did not fit in the space available for it because one of the two retaining walls designed to support the girder had inclined slightly after it had been backfilled. The entire operation was halted to remove the backfill and the misaligned retaining wall, and then the wall was erected again and backfilled with soil. It was discovered that the surveyor checked the alignment of the retaining wall only when it was erected. He failed to check the alignment after the backfill was put in place. This might have been avoided if a checklist had been used to verify the technical aspects of the job and if a training program had been implemented for the technicians.

The second situation that arose was attributable to incorrect signs and directions written on a precast concrete member. After the workers had erected the concrete member, they were not able to connect the joints. The superintendent asked that the member be removed and rotated 180°, which was the right position for it. This situation suggests that before precast members were unloaded from the trucks, the superintendent should have verified that the directions on them were correct to ensure their accurate placement.

Conclusions

Studying two similar structures allowed the productivity rates of different crews to be compared during construction. Having one project constructed at two sites by one contractor allowed for a comparison of productivity rates at the two sites. Factors that caused differences in the productivity rates were the type of concrete used (precast or cast in place), the area available around the jobsites, and the experience of the superintendents.

7.3 Case Study 2: Steel Erection

[This case study was conducted by Dr. Fred Rahbar, Project Consultant to Aramco Overseas Company (AOC).]

The second case study evaluated the steel erection processes used on a two-story steel-framed building designed to allow for upward expansion in the future. The cladding is clay brick with an extensive glass storefront on the south wall. The boilers and refrigeration compressors are housed in a heavily reinforced penthouse on the south portion of the roof, and the air handlers are mounted on the roof. The roof is lightweight concrete built up with a bituminous ply system. The second floor has an extensive computer room with a raised floor for cooling. There is an elevator near the south center portion of the building. The transformer bank pad is mounted near the northeast corner of the building. The water and sewer lines were tied into the municipal system networks (Rahbar n.d).

All of the work was performed by subcontractors and supervised by a construction manager. The job superintendent had 30 years of building construction experience, and his deputy had five years of work experience. Quality control was primarily the responsibility of the construction manager.

At the time of the case study, the number of subcontractors on the site varied between six and seven. Major operations were the following:

- Concrete work on the roof slab and the penthouse slab,
- Metal stud framing of the exterior walls,
- Steel framing of the storefront wall,

- Fireproofing of the structural steel on the second floor, and
- Masonry work on the north exterior wall.

The productivity analysis methods used during this case study included

- Field ratings,
- Productivity analysis,
- Crew balance,
- Video analysis, and
- Process flow analysis.

Site observations of the steel-erection processes were recorded and are shown in (1) the work-sampling chart in Figure 7-30, (2) the 5-min crew rating summarized in Figure 7-31; (3) the crew balance chart for erecting a beam in Figure 7-32; (4) the crew balance chart for erecting a column in Figure 7-33; and 5) the process chart for erecting a column, a girder, and a beam in Figure 7-34.

Work Sampling and 5-Minute Crew Rating

Twenty observations at 5-min intervals were recorded for 420 workers. Only 180, or 43% of the workers, were actively engaged in a work activity, and only 41% of the steel-erection crew was working, as shown in Figure 7-30.

An overall index of less than 60% indicates that the productivity of a job activity is unsatisfactory. A field rating of this size is merely an indication of probable conditions, and additional ratings are necessary. Also, even though the work sample showed that people were working, they were not necessarily doing productive work. The data from this rating are shown in Figure 7-31. Ineffective work included (1) locating the proper beams, (2) reversing beam number 1, (3) reading the plans, and (4) cutting beams.

Crew Balance Charts

After an examination of the items highlighted in the 5-min ratings, alternative methods were developed that could result in savings in both time and cost. The proposed alternatives were (1) conduct field measurements before installation; (2) rearrange columns and beams in the storage area so that they are lifted according to the planned sequence, thus avoiding delays in checking the plans; (3) relocate bolts closer to the work-assembling area; (4) not tightening the column bolts permanently before the beams and girders are set; and (5) marking the proper location of the girders and beams to indicate how they will be assembled, thus avoiding having to reverse operations.

The crew balance chart shown in Figure 7-32 for erecting a beam indicates that several of the crew members were idle or waiting during the operation. A revision of the chart eliminated two crew members without any significant effect on the other crew members. A second revision indicated that the time required to perform the operation could be reduced by 39%. The crew balance chart for the column erection

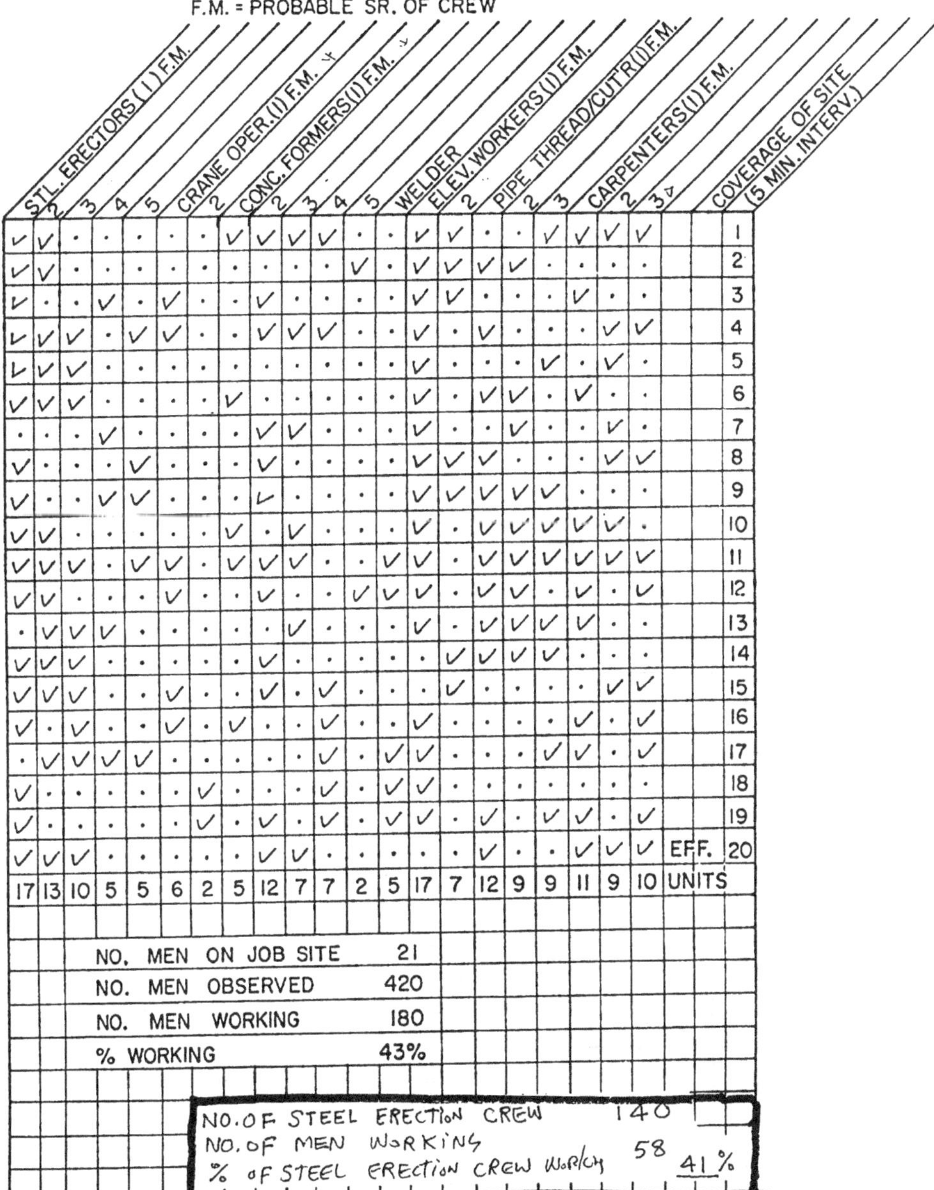

Figure 7-30. Work-Sampling Chart for Steel-Erection Crew

Source: Rahbar (n.d.), Figure 1; reproduced with permission.

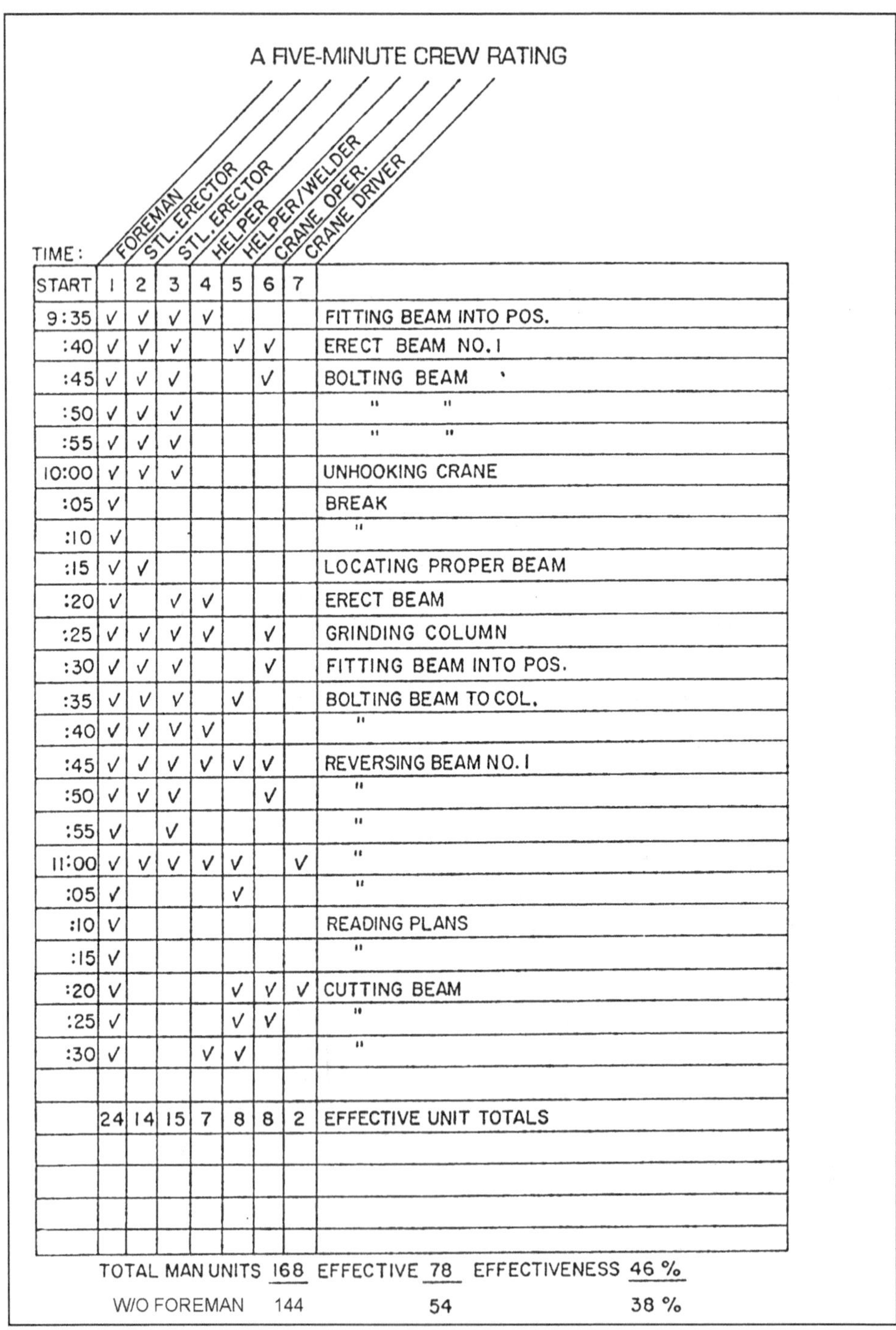

A FIVE-MINUTE CREW RATING

TIME: START	1 FOREMAN	2 STL. ERECTOR	3 STL. ERECTOR	4 HELPER	5 HELPER/WELDER	6 CRANE OPER.	7 CRANE DRIVER	
9:35	√	√	√	√				FITTING BEAM INTO POS.
:40	√	√	√		√	√		ERECT BEAM NO. I
:45	√	√	√			√		BOLTING BEAM
:50	√	√	√					" "
:55	√	√	√					" "
10:00	√	√	√					UNHOOKING CRANE
:05	√							BREAK
:10	√							"
:15	√	√						LOCATING PROPER BEAM
:20	√		√	√				ERECT BEAM
:25	√	√	√	√		√		GRINDING COLUMN
:30	√	√	√			√		FITTING BEAM INTO POS.
:35	√	√	√		√			BOLTING BEAM TO COL.
:40	√	√	√	√				"
:45	√	√	√	√	√	√		REVERSING BEAM NO. I
:50	√	√	√			√		"
:55	√		√					"
11:00	√	√	√	√	√		√	"
:05	√				√			"
:10	√							READING PLANS
:15	√							"
:20	√				√	√	√	CUTTING BEAM
:25	√				√	√		"
:30	√			√	√			"
	24	14	15	7	8	8	2	EFFECTIVE UNIT TOTALS

TOTAL MAN UNITS 168 EFFECTIVE 78 EFFECTIVENESS 46 %

W/O FOREMAN 144 54 38 %

Figure 7-31. 5-Minute Rating for Steel-Erection Crew

Source: Rahbar (n.d.), Figure 2; reproduced with permission.

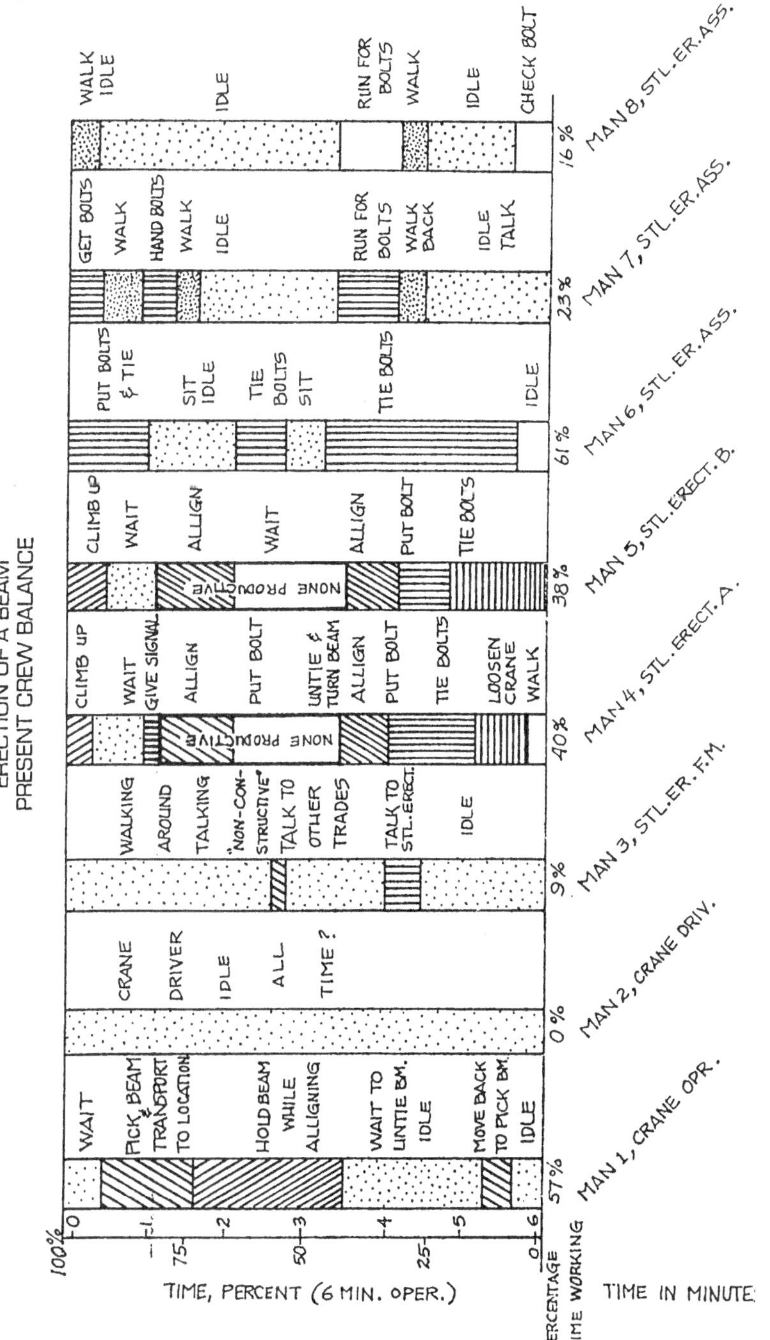

Figure 7-32. Crew Balance Chart for Erecting Beam

Source: Rahbar (n.d.), Figure 3; reproduced with permission.

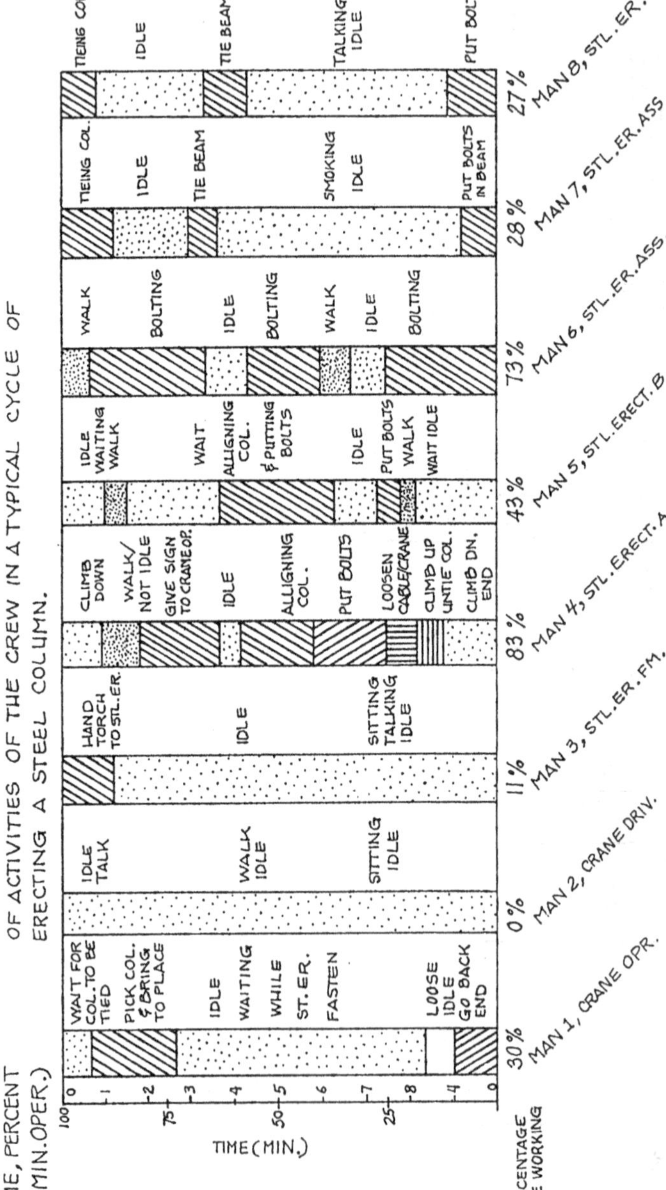

Figure 7-33. Crew Balance Chart for Erecting Column

Source: Rahbar (n.d.), Figure 4; reproduced with permission.

PROCESS CHART

Erection of 1 Column, 1 Girder, and 1 Beam. —

	Symbol	Activity	Transport. Dist.
1	●	Tie & Hook Column to lift, 1 man	
2	➡	Lift column to be placed	60 ft.
3	■	Align & Level Column, 2 men	
4	●	Bolt column, 2 men	
5	●	Unhook Column from crane, 1 man	
6	D	See plans for correct girder, 2 men	
7	●	Tie & hook girder to lift, 1 man	
8	➡	Look for correct bolts, 3 men	120 ft.
9	➡	Lift girder into place	60 ft.
10	■	Align and place girder, 2 men	
11	●	Bolt girder, 2 men	
12	●	Unhook girder from crane, 1 man	
13	D	See plan for correct beam, 2 men	
14	●	Tie and hook beam to lift, 1 man	
15	➡	Lift beam into place	60 ft.
16	■	Align & place beam, 2 men	
17	●	Bolt beam, 2 men	
18	●	Unhook beam from crane, 1 man	

Figure 7-34. Process Chart for Erecting Column, Girder, and Beam

Source: Rahbar (n.d.), Figure 5; reproduced with permission.

in Figure 7-33 was also revised to eliminate two workers. Postponing the permanent tightening of columns reduced the overall time required for this work process by 10%.

Process Charts

The process chart in Figure 7-34 shows nine operations, four movements of materials, and three inspections. A revised process chart indicated that the travel distance could be reduced by 120 ft (35.6 m) along with eliminating two major delays.

Other work-process alternatives that could be investigated include (1) using a second, lighter crane to lift the beams and girders while the main crane is dedicated to column erection; (2) assembling columns and girders on the ground and lifting the frame into place; and (3) using larger wrenches to bolt the columns and beams.

Methods Time Measurement for the Welding Operation

For this part of the project the total time for changing a welding rod on a 400-amp welding rod holder was 207.70 time measurement units (TMU), or 7.48 s. For activities 6 and 8 in this work process, the trigger could be redesigned so that workers do not have to hold the welding holder in their right hand in order to reach the trigger. This would reduce the time by 11.2 TMU. In activity 1, if the welding shield was positioned in a fixed location, the time for this activity would be 13.1 TMU, not 18.6 TMU. The rod holder could be redesigned so that the next rod is automatically fed. Also, activities 28 and 29 could be performed simultaneously. These changes would result in a reduction of at least 27.30 TMU, or a 13% savings in time.

Field Ratings

Twenty field-rating observations of four separate operations were made, and the field-rating index varied from a low of 67 to a high of 100, with the average being 88.40. The field ratings were performed on a sunny day with a temperature of 80°F (26.7°C). A factor of 10% was added to the percent working subtotal to account for the foreman and personal time. The crew working on the floor plans for the first-floor slabs and beams was the most productive, with a field-rating index of over 90, whereas the indices for the other operations, such as digging footings and setting up formwork, were 76 to 89. Although the results are not statistically significant for any one crew, the total number of observations embodied in the analysis for all the crews was 20 (219 man observations), which provides a useful indication of the aggregate level of productivity on site.

Concrete crew. The concrete crew had eight workers. When the concrete crew was being observed, it was finishing a slab pour on the roof and tying rebar for a curb adjacent to the penthouse. The bulk of the work appeared to be performed by three concrete finishers, who were kneeling on boards on the set concrete and hand-troweling the finish. The remainder of the crew was engaged in cleaning up and arranging the rebar.

Steel crew 1: This steel crew had seven workers, and they were engaged in welding metal studs on the exterior walls. The men were working fairly rapidly; however, three welders who were on scissor lifts on the outside of the structure were frequently unable to move their lifts due to extremely muddy conditions. This resulted in a delay, but the crew still managed to obtain the highest field rating.

Steel crew 2: This steel crew had eight workers, and they were engaged in welding the metal frames for the glazing on the south wall. There appeared to be some confusion surrounding this operation, which resulted in plan reading and material movement. In addition, some members of the crew were frequently away from the work area for undetermined reasons and were not observable. The relatively low field rating (60%) could be partially explained by the fact that they were just starting and experiencing a learning curve for this custom operation.

Fireproofing crew: The fireproofing crew had seven workers, three of whom were on scissor lifts spraying the structural steel with fireproofing material. In addition, there were three laborers and one foreman. It appeared that the three laborers were attending to the pumps for each of the sprayers; however, it also appeared that the pumps operated automatically. As a result, the three laborers and the foreman were observed to be idle a majority of the time.

Masonry crew: The masonry crew had 11 workers, and they were engaged in both **concrete masonry unit** (CMU) wall construction and clay-brick installation on the northeast portion of the building. This crew had one of the higher field ratings, 80%. Most of the support workers were involved in transporting blocks, bricks, and mortar.

5-Minute Ratings

A 5-min rating was performed on two days at approximately the same time but with different crews and different temperatures. The effectiveness factor ranged from 44% to 58%. The lowest rate of 44% occurred during the unloading of reinforcing steel from a truck to a lay-down area. This figure included the truck driver, who was not doing any physical work. However, another worker also did not do any physical work other than talking to the truck driver during the 5-min rating. The 5-min-rating technique indicated that a more effective plan for the location of the trucks and for setting up wall forms could result in savings because both of these operations had a rating of less than 50%.

Work Sampling

The results obtained from work sampling were converted to labor utilization factors that ranged from as low as 34% to 52%. The lowest figure of 34% was for the forming operation. The figure for the crew unloading reinforcing steel was 41%, and for the ironworker crew working on wall rebar it was 50%. During the study, the operations that involved ironworkers had a higher rating than the work performed by the other trades. Also, formwork operations were ranked low by all three of the analysis methods.

Productivity Analysis

The work was divided into direct work, support work, and delays (or ineffective work). Most of the direct work, accounting for 32% of the effort, was performed by the concrete finishers. An additional 32% of the crew effort was directed toward transporting material, which for the most part was **polyethylene** (PE) sheeting for curing the slab. Over a third of the time, the work being performed was ineffective or delayed (36%). This was because the slab pour had isolated the stairway, which meant that to access the lower floors, the workers had to use a ladder. The cleanup work was slowed down by laborers hauling equipment down the ladder. Also, workers tying rebar were hindered because they were surrounded by fresh concrete with only the ladder to move tools and materials. If there had been a crane available, materials could have been moved more rapidly between the storage yard and the roof.

Concrete crew: The productivity of the concrete crew is shown in Figure 7-35. The concrete crew was conducting effective work less than 40% of the time. The main reason is that they had to wait for the concrete trucks to get into position to place the concrete.

Steel crew 1: The steel crew was primarily welding metal studs to the exterior of the north and west walls. Three workers on scissor lifts worked on the exterior, and the remainder of the crew worked on the second-floor interior. In addition, they had a supply of metal studs and a radial arm saw available on the second floor. Although the crew was observed to be working most of the time, as is shown in Figure 7-36, the work might have progressed faster if a few workers had been measuring and marking studs and one worker had been cutting the materials. Instead, the welders measured, walked to a stud pile, marked the materials, sawed, returned to the wall, and proceeded to weld the piece in place. This contributed to increased travel time, congestion, and some delay. In addition, the area was muddy, which was a detriment to both production and safety because the lifts were fully extended and the workers were

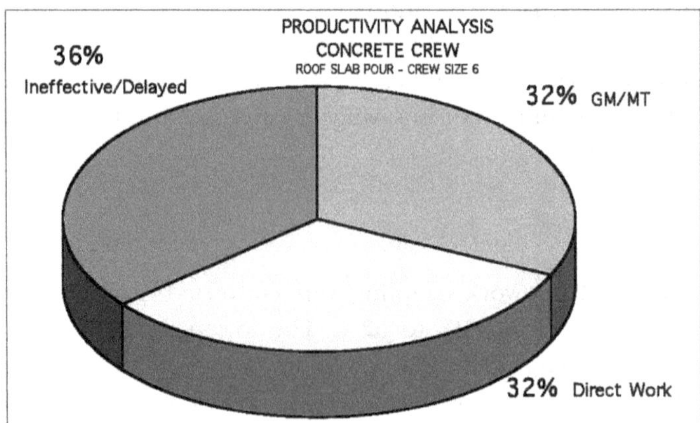

Figure 7-35. Productivity Analysis for Concrete Crew Roof-Slab Pour

Source: Rahbar (n.d.), Figure 6; reproduced with permission.

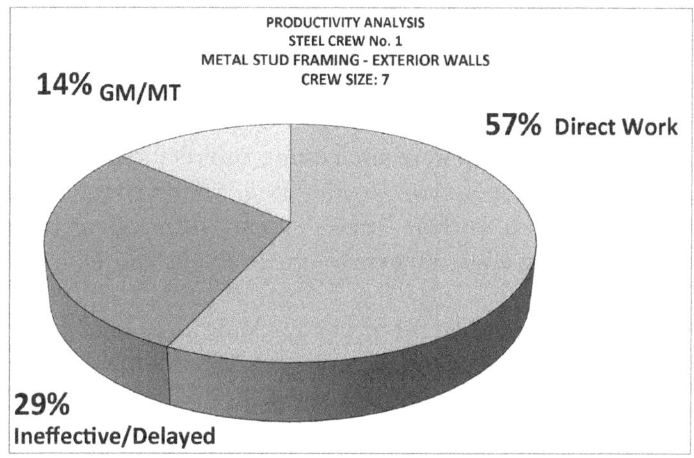

Figure 7-36. Productivity Analysis for Steel Crew 1

Source: Rahbar (n.d.), Figure 7; reproduced with permission.

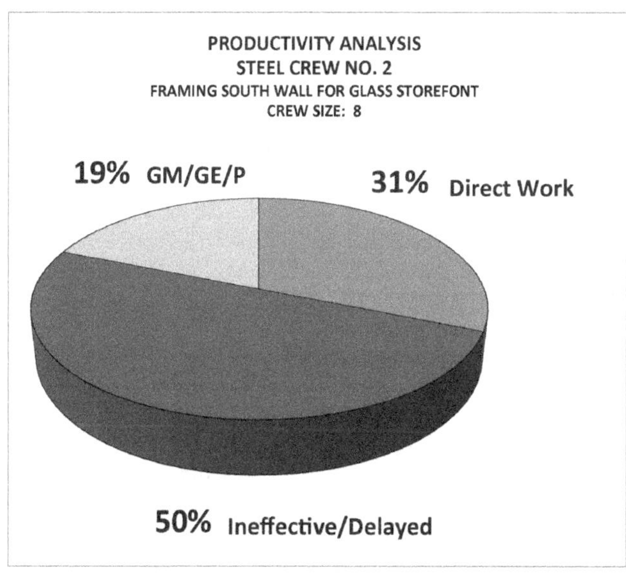

Figure 7-37. Productivity Analysis for Steel Crew 2

Source: Rahbar (n.d.), Figure 8; reproduced with permission.

trying to maneuver them (unsuccessfully) in this configuration. Although a fireproofing crew was working nearby, no interference between the crews was observed; however, the potential certainly existed due to competition for limited electrical power and overspray from the fireproofing.

Steel crew 2: Only 31% of the members of steel crew 2 were working, as is shown in Figure 7-37. The workers were unsure of some aspects of the work, and there were problems with the materials. The prefabricated storefront frames were being dismantled and welded again before being bolted into place, which indicated a flaw in fabrication. In addition, for a crew of eight workers, only a fraction of the workers were visible at any given time, which indicated a need for closer supervision.

Working conditions were congested for this operation. Metal frames placed on sawhorses consumed almost the entire elevated front walkway. As with the first steel crew, workers doing the erection were also cutting and refabricating the frames. One method for improving the operation would be to have three or four workers refabricate the frames while two workers installed them. Although it is likely that the workers were experiencing a learning curve, the problem was the improper fabrication of the materials.

Fireproofing crew: This crew had seven workers, only three of whom were engaged in direct work at any given time, as shown in Figure 7-38. Although the three laborers were available for work, there was no work for them to perform; therefore, this was classified as a supervision delay. The layout of the operation worked well, with the three workers positioned to perform an equal amount of the work with a minimum amount of travel. In addition, bags of the fireproofing material were readily available at the center of the operation. Thus, although the crew was unbalanced, because 50% of the effort was ineffective, the process flow was effective and required no real adjustment.

Masonry crew: When the masonry crew was first observed, it was working on the first course of brick on the north wall of the building in an effort to sheath in the north wall before the cold weather and winter winds. However, as a result of the welding problems experienced by steel crew 1, the masons had to move to the east wall in order for the steel crew to perform some rework. As a result, when the masons

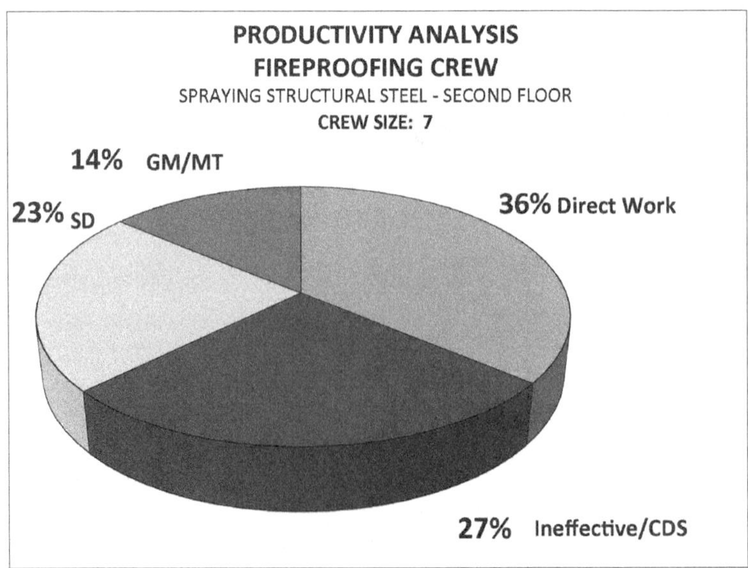

Figure 7-38. Productivity Analysis for Fireproofing Crew

Source: Rahbar (n.d.), Figure 9; reproduced with permission.

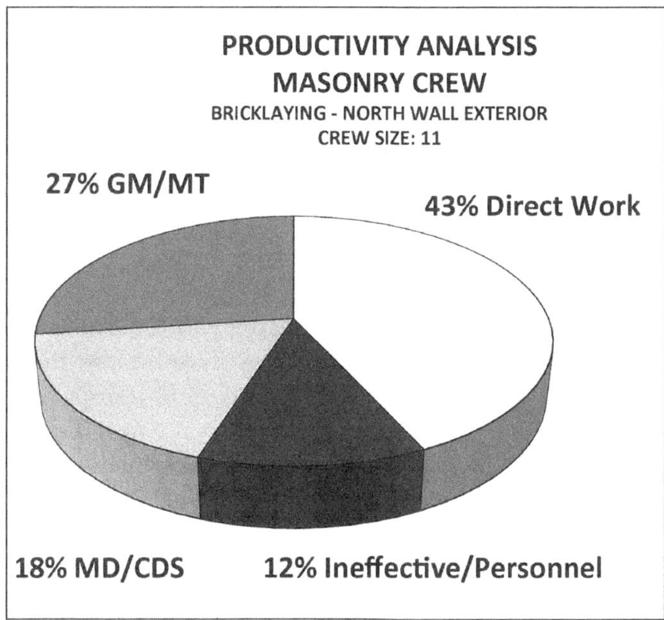

Figure 7-39. Productivity Analysis for Masonry Crew

Source: Rahbar (n.d.), Figure 10; reproduced with permission.

were observed, they were engaged in constructing the protruding northeast section of the building, which was CMU–brick veneer construction. As with the prior observations, the crew appeared to be working diligently, with 43% of the effort directly affecting progress, as is shown in Figure 7-39. However, each mason was responsible for one geographical section of the work, and this resulted in minor delays. As shown in Figure 7-39, material delays were significant, accounting for 27% of the effort lost. This was primarily attributable to the late arrival of mortar and having to wait for blocks to be cut. As a result, the laborers, observed in the original assessment to be idle, became busy.

Analysis of the Digital Video Recordings

Digital video recordings of the masonry crew were conducted on two days, and on both days, the crew was working on the northeast section of the building. During the first phase, the masonry crew had 11 workers. During filming, Workers 9, 10, and 11 were not observed because they were obstructed from view by the completed wall. Workers 9 and 10 were laborers who mixed the mortar and delivered material to the workers staging the work on the north side of the structure. Worker 11 was finishing the brick veneer on this wall, as was Worker 7. Both Workers 7 and 11 were at a higher level than the main group of workers.

The main group of workers consisted of eight workers. Workers 1, 2, 4, and 5 were masons constructing a CMU wall with running bond construction. They were assisted by Worker 3. In addition, a forklift operator, who was Worker 6, kept this crew supplied with mortar, block, and filling for the vertically reinforced cells. Also, the foreman, Worker 8, was present most of the time, as was Worker 7, the brick mason. It appeared that Worker 3 and the foreman were idle a majority of the time. A few minor delays in the work of the masons occurred, and these were usually caused by the late arrival of the concrete for the cells.

To increase production, the mortar mixer should have been moved closer to the work task. Having the mixer placed just below the staging area would have reduced the travel time of the forklift and the mixed mortar. The only additional movement required would be lifting the trough to the staging area once it had been filled with mortar. Travel to the storage yard would still be necessary because the block had been unloaded there. This process could have been sped up by transporting the blocks to the work area at the end of each day in preparation for work the next day. At the end of the day, bricks, blocks, and bags of cement could have been prepositioned; gas tanks for the forklifts and mixer could have been filled; and the mixer could have been charged with material.

During the second phase, two masons were observed filling the top course bond beam for a completed CMU wall, and a forklift operator was delivering the concrete to the masons and assisting them in filling the beam. This operation was inefficient because the two masons had to wait 7 min for the forklift operator to deliver the concrete, and during this time, the masons were idle.

Crew Balance Analysis

During construction of the CMU wall, the masons were working and well supplied with materials, but Worker 3 was not always working. Even when Worker 3 was working to unload the forklift, the operator and usually one of the masons loaded the forklift. The responsibilities of Worker 3 could have been performed by the forklift operator, except it would require that the forklift operator leave the forklift and climb up and down the ladder. A more suitable solution would have been to have more mortar troughs and detachable pallets for the block. Then, the operator could unload, pick up any empty pallets, and immediately return either to the mixer or to the storage yard. Although there was not much room for rearranging or eliminating personnel, the multiple trough and pallet concept could alleviate the problem experienced by masons waiting for the concrete to be delivered to their work area.

Process Flow Analysis

The process chart in Figure 7-34 shows the overall flow of work during construction of the CMU wall. The materials were stockpiled on the scaffolding adjacent to the work area where the masons prepared to lay a course of block by placing horizontal wire reinforcement. Following this operation, each mason laid down a bed of mortar over the previous course with a trowel, placed the block, and finished the joints. Meanwhile, the forklift operator delivered the blocks and mortar, and a laborer unloaded them. After three courses had been placed, the block cores scheduled to receive vertical rebar were filled with mortar, and the rebar was inserted into the center of each block.

Production could have been improved if two or three courses had been worked on simultaneously from one end to the other, in lieu of the broken pattern of all masons working on one section of a course, ensuring that the slower workers controlled the pace. Based on the estimate of 1.12 man min/ft course (3.67 man min/m course), production could have been increased by at least 30% over the present process.

Conclusions

Although the steel-erection crew was working with a complex design and dimensional errors, potential improvements could have been implemented to increase productivity rates. By implementing the recommendations from this case study, productivity rates could have been increased by 50%, with a similar percentage of cost savings. Other alternatives that could have been examined include

- Using a second, lighter crane to lift the beams and girders while the main crane was performing the column erections,
- Assembling columns and girders on the ground and lifting the frame into place, and
- Using larger wrenches to bolt the columns and girders into place.

7.4 Case Study 3: Highway Interchange Construction

(This case study was conducted by Alan Page, Hasan Bas, Matt Frechette, Meenakshi Grandhi, Anwar Beg Mirza, and Nanik Yusuf.)

This section provides information obtained during a productivity improvement case study conducted on a highway interchange construction project. The highway interchange project was designed to ease congestion and provide access to and from a large metropolitan airport. The project included a tunnel, a new bridge over the main highway, and on- and off-ramps that were widened and reconfigured to help ease traffic congestion. The main avenue was widened to six lanes between the airport and the next major intersection, a connector ramp was added from the airport to the highway, and traffic signal improvements were made to the off-ramps and intersections along the main avenue. Underground, there was extensive utility relocation. Over 18,000 ft (5,486.4 m) of new drainage pipe and over 3,000 ft (914.4 m) of new sewer pipe were installed along the project corridor (Page et al. n.d).

The local transportation authority was the administrator for the project and oversaw the **storm-water pollution plan** (SWPP). The state Department of Transportation operates and maintains the highway interchange and the on- and off-ramps, and the city operates and maintains the local streets surrounding the interchange. The approximate cost of the project was $81 million, and the project duration was one and a half years.

Figure 7-40. Map of Case Study Interchange

Source: Page et al. (n.d.), Figure 1; reproduced with permission.

The construction was divided into eight scheduling stages that occurred in four different locations depending on the ongoing traffic flow. The project was divided into these eight stages to keep traffic moving during construction and to complement the schedule for the expansion of the airport. Dividing the project into eight stages created a more complicated schedule. During construction of the drainage pipe system, runs were started, capped, and backfilled while other necessary construction was completed. The runs were then excavated again; therefore, each run could be started and capped multiple times. Figure 7-40 is a map of the construction zone.

Drainage Pipe Installation

The area inside the on-ramp to the highway from the main avenue, north of the geographical intersection of the main avenue and the highway and directly adjacent to the airport, was the location of the construction operations that were observed and used for this productivity improvement study. A subcontractor installed the drainage pipe. The close proximity of the project to the airport and the highway required dust-control measures because the flight path was right over the construction site. The location of the project contributed to communication problems and increased stress on the workers.

Pipe Configurations

There were six separate runs of drainage pipe, and 12-in. (30.48-cm) diameter green polyvinyl chloride (PVC) plastic pipe was used for this project. For the curved pipe sections, 18-in. (45.72-cm) diameter corrugated polyethylene (PE) plastic pipes were used because of their flexibility. The corrugated pipe comes in 20-ft (6.1-m) sections and is then cut to size, as required for each installation.

Identification of the Construction Workers

The crew for this project consisted of a supervisor, a foreman, an excavator operator, a loader operator, and two or three laborers. Each worker wore different colored hard hats, so they will be referred to as red hat, white hat, and yellow hat. The crew had worked together for many years, ranging from 20 years for the excavator operator (red hat) and one year for the foreman (white hat), who usually assisted the loader operator and helped to install pipe in the trenches. Due to the high levels of noise caused by airplanes taking off and landing, the crew members used hand signals instead of verbal communication. The installation rates for the drainage pipe were influenced by the productivity of the heavy construction equipment.

Heavy Construction Equipment

The project used steel shoring, excavators, a loader, a water truck, a groundwater pump, a generator for the pump, and an electronic depth sensor. Two different excavators were used: (1) a Caterpillar C 235B and (2) a Caterpillar C 225. The loader was a John Deere 644G, and the supervisor or the foreman drove the second excavator.

Pipe-Installation Cycle

The contractor installed the drainage pipe using a cycle that included excavation, placing gravel in the trench, spreading the gravel evenly in the trench, installing shoring if the trench is over 5 ft deep, laying pipe, joining the pipes together, backfilling the trench, and compacting the backfill. The contractor excavated the soil using commercial hydraulic excavators. The depth of the trenches for this project ranged from 3 ft (0.91 m) to over 10 ft (3.05 m). The time to excavate depended on the depth of the excavations. For an 8-ft (2.43-m) depth, it took approximately 10 min to excavate for the total length of one pipe. If groundwater was encountered, the representative of the owner was consulted on how to deal with the groundwater. When there was groundwater, a pump was lowered into the trench inside a grooved

Figure 7-41. Excavating Trench for Pipe Installation

Source: Page et al. (n.d.), Figure 2; reproduced with permission.

section of pipe for protection. The pump was then operated during the next phase of the cycle.

When the depth of a trench was over 5 ft (1.52 m), the pipe-laying phase required shoring. The excavator lifted the steel shoring using steel chains attached to the shoring, and then laborers helped to position the shoring in the trench. After the shoring had been placed, the loader obtained a load of gravel and placed it in the trench for the foundation for the pipe. The gravel was then spread evenly in the trench. Next, the loader brought the pipe from the storage pile and lowered it into the trench, where the laborers moved the pipe to its proper location and glued it to the adjacent section of pipe. For the backfill operation, the loader obtained and poured a layer of gravel over the pipe.

If the soil removed from the trench was too moist, dry soil was brought from a stockpile by the loader and placed into the trench. When enough soil was placed, the excavator, fitted with a sheepsfoot roller, compacted the soil. For greater depths, the soil was compacted in levels. All of these operations constituted one cycle. Figures 7-41 to 7-46 show the pipe-installation operations.

Data Collection

Data was collected in three ways: (1) interviews with personnel, (2) on-site observations, and (3) research.

Each of the productivity improvement team members was assigned to observe one, two, or three crew members or a piece of the heavy equipment, either the excavator or the loader. For each minute of observation, the observers recorded the movements and work processes of their assigned crew member or machine. Figure 7-47 provides a list of the symbols used for the data-collection process. So that spread-

Figure 7-42. Installing Pipe in Trench

Source: Page et al. (n.d.), Figure 3; reproduced with permission.

Figure 7-43. Pipe Installed in Trench

Source: Page et al. (n.d.), Figure 4; reproduced with permission.

sheets could be used to summarize the data, these symbols were converted to numbers shown in Table 7-4.

On each form, the time was listed in the left column, and then a pair of columns to the right was used for recording observations of one crew member. In the left column of the pair, the observer entered the symbol that described the work of the crew member. In the right column, the observer recorded comments describing what the crew member was doing at that time. Comments were also used to help

Figure 7-44. Worker Connecting Pipe while in Trench

Source: Page et al. (n.d.), Figure 5; reproduced with permission.

Figure 7-45. Backfilling around Pipe in Trench

Source: Page et al. (n.d.), Figure 6; reproduced with permission.

Table 7-4. Revised Format of Symbols Used for Data Collection in Case Study 3

Symbol Number	Symbol Meaning
1	Waiting for instructions
2	Waiting for materials/tools
3	Other
4	Working productively
5	Working but at less than normal capacity
6	Idle

Source: Page et al. (n.d.), Table 2; reproduced with permission.

Figure 7-46. Backfilling over Pipe in Trench

Source: Page et al. (n.d.), Figure 7; reproduced with permission.

Symbol	*Symbol Meaning*
▲	Waiting for instructions
✞	Waiting for materials/tools
□	Other
✓	Working productively
X	Working but less than normal capacity
O	Idle

Figure 7-47. Symbols Used for Data Collection in Case Study 3

Source: Page et al. (n.d.), Table 1; reproduced with permission.

clarify some of the processes. On-site observations were performed on four different dates.

Field Ratings

Mechanical counters were used for the counts. On most days, the work started at 7:00 a.m. and ended at 3:30 p.m. During the observations, the weather was sunny, and

Table 7-5. Field-Rating Results during Day 1 of Observations in Case Study 3

Observation Number	1	2	3	4	5	6	7	Average
Time of day	12:51	12:52	12:53	12:54	12:55	12:56	12:57	
Number of individuals on site	8	8	8	8	8	8	8	8
Number of individuals observed	6	6	6	6	6	6	6	6
Number of individuals working	2	3	4	4	4	5	5	3.857
Percent working	33%	50%	67%	67%	83%	83%	83%	64%
Field-rating index	33.33	50	66.67	66.67	83.33	83.33	83.33	64.29

Source: Page et al. (n.d.), Table 3; reproduced with permission.

Table 7-6. Productivity-Rating Results for Case Study 3

Worker or Machine	Effective (percentage)	Contributory (percentage)	Idle (percentage)
Yellow hat	71	23	6
White hat	63	27	10
Excavator operator	61	22	17
Red hat	70	24	6
Supervisor	64	13	23
Loader operator	79	6	15
Loader-driver	81	13	6
Average productivity	70	18	12

Source: Page et al. (n.d.), Table 4; reproduced with permission.

four crew members worked on the drainage pipe installation. Table 7-5 provides a sample of a field rating recorded on the first day of the observations. Eq. (7-1) was used to calculate the percent working:

$$\text{percent working} = \text{number of individuals working}/\text{number of individuals observed} \tag{7-1}$$

The results shown in Table 7-5 were calculated for an 8-min time interval. Field ratings were performed on all of the observations for four days. The results obtained were 73.5%, 75.9%, 69.5%, and 64.6%. The average field rating was 70.65% for the crew, which is higher than the normal productivity rate for construction (60%). The adjusted field-rating index was 80.65%.

Productivity Ratings

Productivity ratings are similar to field ratings, except productivity ratings use three categories of work: (1) effective work, (2) essential contributory work, and (3) ineffective (idle) work. The productivity ratings for the pipe-laying operation are shown in Table 7-6.

5-Minute Ratings

The 5-min ratings used the same observation techniques as the field productivity ratings, but the observations were for shorter periods. There were a total of 12 5-min ratings. The crew members were observed for 12 min for each rating. Figure 7-48 contains the results from the 5-min-rating analysis.

5 Minute Rating

Analysis Crew: Group 4
Job: Laying drainage pipe
Time Interval: 11:19am -11:30am
Weather: sunny

Date: 10/13/05
Page: 1 of 6

Observer #	Time of Observer.	Man 1 Bulldozer operator	Man 2 Excavator operator	Man 3 Yellow hat	Man 4 White hat	Man 5 Red hat	Man 6 Supervisor	Operations
1	11.19	x	o	o	o	o	x	delayed due to groundwater
2	11.20	x	o	o	o	o	x	
3	11.21	x	o	o	o	o	x	
4	11.22	x	o	o	o	o	x	
5	11.23	x	o	o	o	o	x	
6	11.24	x	o	o	o	o	x	
7	11.25	x	o	o	o	o	x	
8	11.26	x	o	o	o	o	x	
9	11.27	x	o	o	o	o	x	
10	11.28	x	o	o	o	o	x	
11	11.29	x	o	o	o	o	x	
12	11.30	x	o	o	o	o	x	
								Total
Maximum Total		12	12	12	12	12	12	72
Total Effective		12	0	0	0	0	12	24
Effectiveness Percentage		100.0%	0.0%	0.0%	0.0%	0.0%	100.0%	33.33%

Table 6 **Five Minute rating results for 10/13/05**

Date of observation	Observation number				Daily average
	1	2	3	4	
10/13/2005	33.33%	77.78%	79.17%	90.28%	70.14%
10/14/2005	43.75%	77.08%	81.25%	85.42%	71.81%
10/17/2005	0%	40%	78.33%	76.67%	48.75%

Figure 7-48. 5-Minute Rating for Laying of Drainage Pipe

Source: Page et al. (n.d.), Figure 8; reproduced with permission.

Crew Balance Charts

The crew balance charts are shown in Figures 7-49 to 7-51. The dark lines on the crew balance charts are working times, the shaded areas are waiting for instruction times, and the white areas are idle times. Table 7-7 shows the crew member and equipment idle times extracted from the crew balance charts.

Flow Diagrams and Process Charts

A flow diagram for the work processes used on this project is shown in Figure 7-52. Figure 7-53 contains the process chart symbols used during this case study.

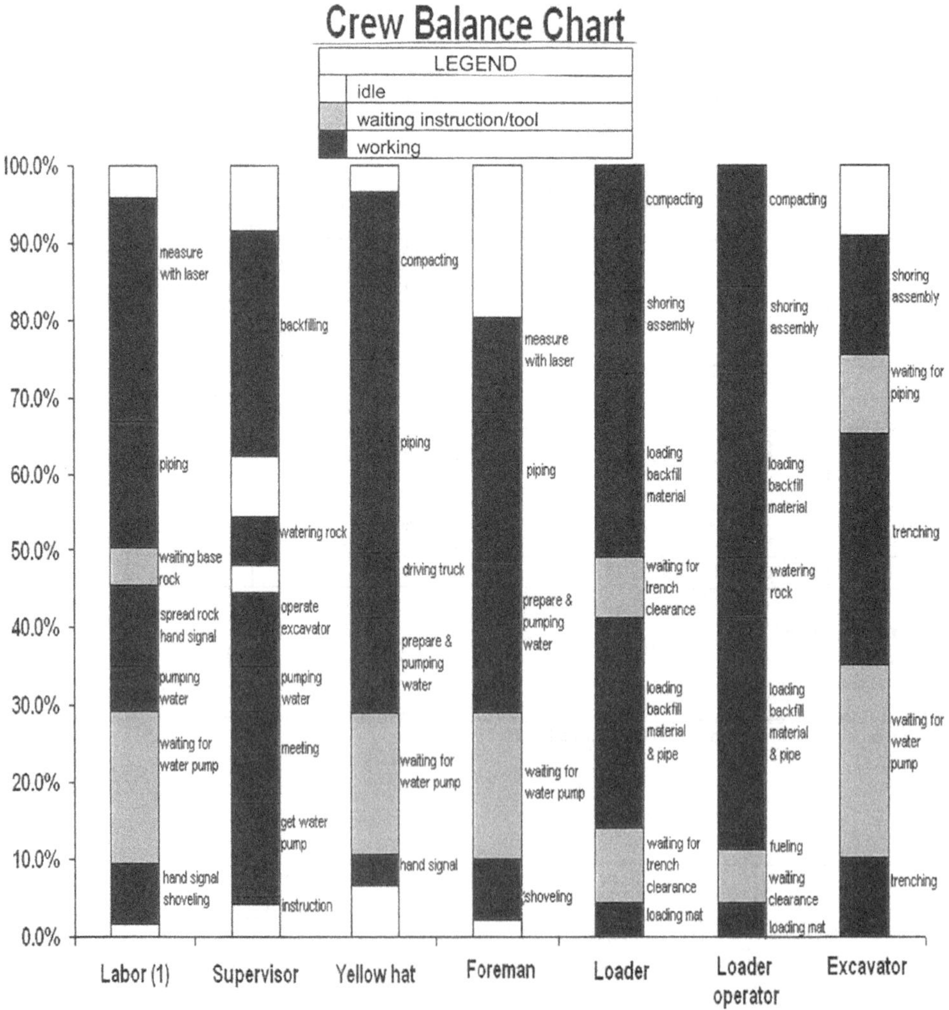

Figure 7-49. Crew Balance Chart for First Observation of Pipe-Installation Operation

Source: Page et al. (n.d.), Figure 9; reproduced with permission.

Project Data

Table 7-8 contains a list of the labor rates used to estimate the cost of the workers on the case study project.

The following is additional information about the case study:

- Project phase duration: 9 days
- Workday: 8 h

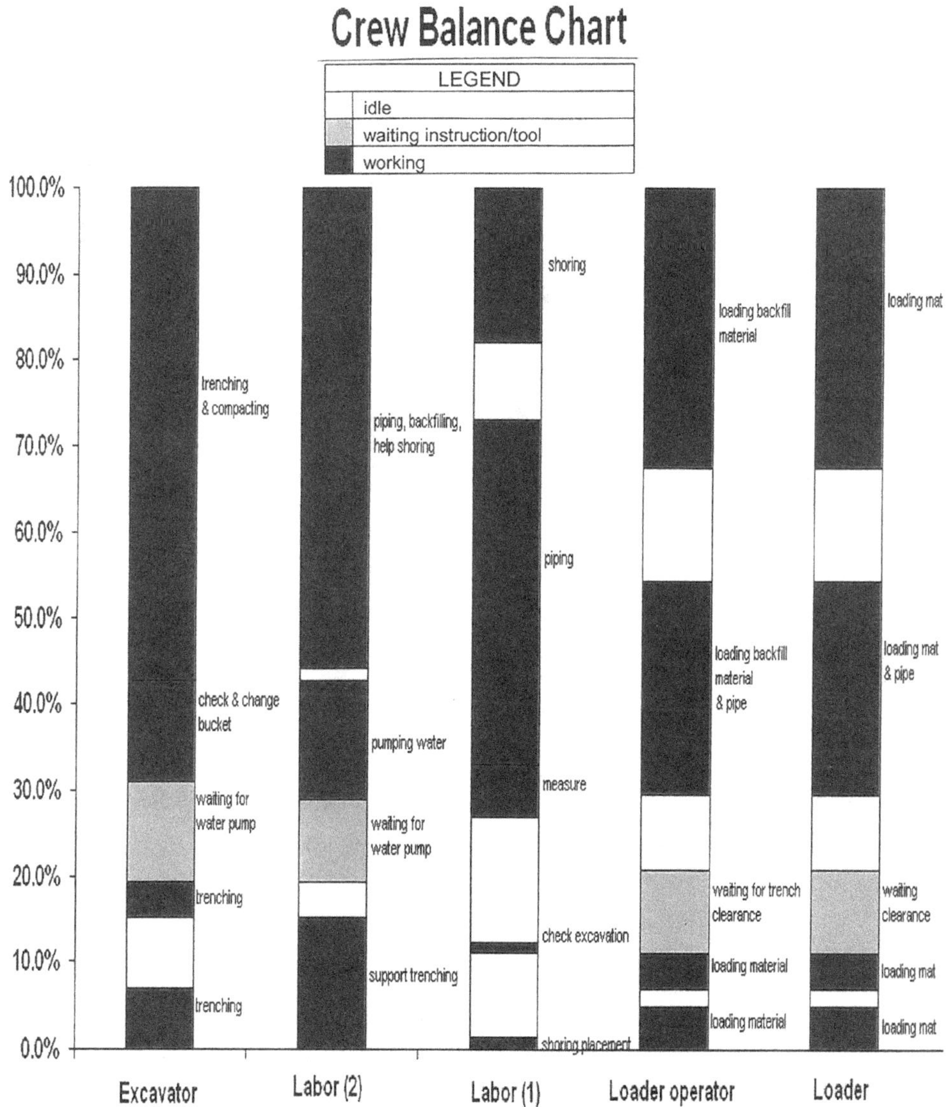

Figure 7-50. Crew Balance Chart for Second Observation of Pipe-Installation Operation

Source: Page et al. (n.d.), Figure 10; reproduced with permission.

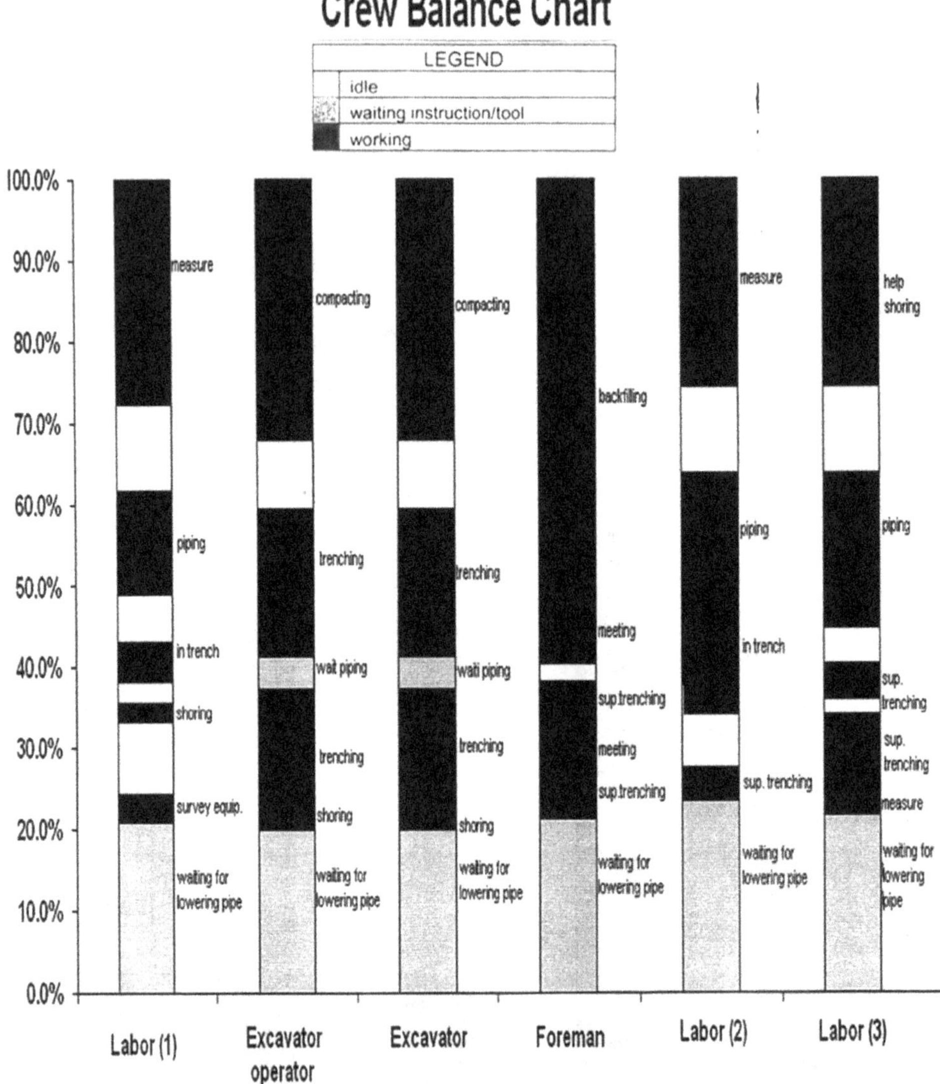

Figure 7-51. Crew Balance Chart for Third Observation of Pipe-Installation Operation

Source: Page et al. (n.d.), Figure 11; reproduced with permission.

- Time wasted by groundwater problems: 40 min per day
- Volume of trenching per cycle: length × width × depth = 37 ft × 4 ft × 9 ft = 1,332 ft^3 [(11.278 m × 1.219 m × 2.743 m) = 37.72 m^3] = 1,322 ft^3/(27 ft^3/yd^3) = 48.96 yd^3
- Number of cycles per day = 3

Table 7-7. Crew Member and Equipment Idle Time Percentages for Case Study 3

Crew Members and Equipment	Idle Time Percentage
Supervisor	23.0
Foreman	5.5
Red hat	20.0
White hat	14.0
Yellow hat	15.0
Loader operator	10.0
Loader	22.0
Excavator operator	7.0
Excavator	8.0

Source: Page et al. (n.d.), Table 5; reproduced with permission.

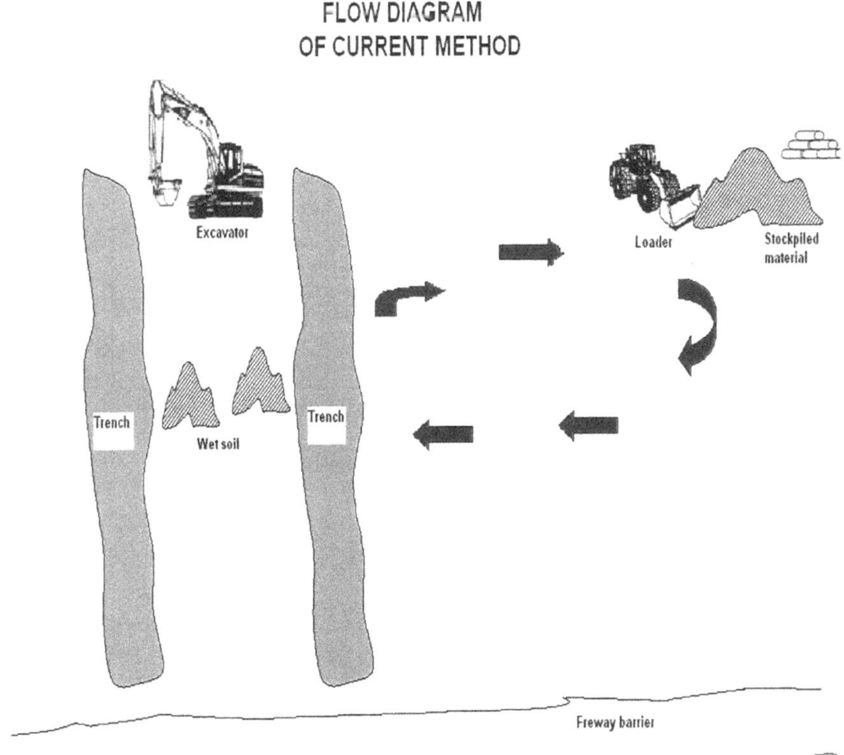

Figure 7-52. Process Flow Diagram for Pipe Installation

Source: Page et al. (n.d.), Figure 12; reproduced with permission.

Symbol	*Symbol Meaning*
●	Foreman signals excavator to start trenching
●	Excavator trenches
■	Valley Transit Authority inspects level of groundwater
◗	Crew is delayed for groundwater inspection
➡	Loader obtains gravel from distant piles
➡	Loader obtains gravel and backfill material from distant piles
●	Groundwater is pumped out of trench
●	Loader dumps bed of gravel into trench
➡	Loader obtains pipe from storage pile
●	Laborers install pipe in trench
●	Loader pours gravel over pipe
●	Loader backfills with soil
●	Excavator compacts with sheepsfoot roller
◗	Sheepsfoot roller is cleaned to enable further compacting, causing delay

Figure 7-53. Process Chart Symbols for Case Study 3

Source: Page et al. (n.d.), Table 6; reproduced with permission.

Table 7-8. Cost of Employees and Equipment for Case Study 3

Employee	Billing Cost/Day (dollars)
Supervisor	432
Foreman	312
Two laborers	592
Two equipment operators	784
Two excavators	1,520
Loader	600
Total	\$4,240/day = \$530/h \$530/h × 72 h = \$38,160/project

Source: Page et al. (n.d.), Table 7; reproduced with permission.

Backfill composition

- Gravel: 10% = 5 yd^3 (3.823 m^3)
- Dry soil from stockpiles: 55% = 27 yd^3 (20.643 m^3)
- Soil from excavation: 35% = 17 yd^3 (13 m^3)

Equipment capacities

- Capacity of John Deere 644G loader bucket: 3.5 yd^3 (2.676 m^3)
- Capacity of Caterpillar C 235B bucket: 2.5 yd^3 (1.911 m^3)
- Capacity of Caterpillar C 225 bucket: 1.5 yd^3 (1.147 m^3)

Loader travel time per cycle (from the video recorded sequences)

- Loading: 22 s
- Travel time (loaded): 35 s
- Unload time: 12 s
- Travel time (empty): 42 s
- Total time: 22 s + 35 s + 12 s + 42 s = 111 s/(60 s/min) = 1 min 51.6 s per cycle

Estimate of proposed dump-truck travel time per cycle

- Loading: 110 s
- Travel time: 22 s
- Dump: 40 s
- Travel time: 18 s
- Total time: 110 s + 22 s + 40 s + 18 s = 190 s/(60 s/min) = 3 min 10 s per cycle
- Dump-truck capacity: 10 yd^3 (7.645 m^3)
- Cost of dump-truck rental: $150 per day

Equipment efficiencies

- Sheepsfoot roller efficiency for the Caterpillar C 225: The sheepsfoot roller head became clogged, and every two passes, the roller had to be shaken to release the wet soil; therefore, the roller was only productive one-third of the time.
- The total compacting time was 39 min per cycle.

Productivity Rating Results

The results of the productivity rating study indicate that the laborers working for the contractor were effective 68% of the time and the contributory effective time was 25%. The laborers were effective 13% more than average productivity ratings. The effective time plus contributory categories was 82%, compared with a 70% average.

The equipment operators were effective 66% of the time and contributory effective 19% of the time. The equipment operators were 28% more effective than average. The field rating was 70.65%. The adjusted field rating was 80.65%, which is well above the acceptable level for construction of 60%.

The 5-min rating for the crew was either high or low depending on whether the crew was waiting for the contractor to approve lowering the pipe run 6/10 of a foot (1.52 cm). The low rating of 40% was at the beginning of the workday. The 44% and the 33% ratings were caused by groundwater delays and waiting for the owner to make a decision about the groundwater.

The high ratings were 78%, 79%, 90%, 77%, 81%, 85%, 78%, and 77%. Of the 12 5-min ratings, eight of them, or two-thirds of the total, were high. The average of the high ratings was 81%, with a variance of 4%.

Even though the production rates on the case study project were high, the work was delayed, and the project exceeded the budgeted cost. The flow of the observed activities was dependent on groundwater and the location of the pipe. The owner provided a 7 1/2% increase in the total payment for drainage pipe installation to compensate for unforeseen site conditions caused by a higher groundwater level than anticipated by the engineer.

Recommendations

This section provides recommendations for improving productivity rates during the installation of the drainage pipes. The areas discussed are safety concerns, groundwater issues, site layout improvements, resource optimization, and reductions in idle time.

The safety issues that affected this particular project were noise and dust. Because airplanes were taking off and landing next to the jobsite, the crew members should have been using ear protection. Jet noise is at a level of 150 dB. According to the **American Speech Language Hearing Association**, "Noise can reduce efficiency in performing daily tasks by reducing attention to tasks. This is a concern of employers when it comes to assuring workers' safety" (Occupational Safety and Health Administration 2006, p. 1). The contractor provided ear protection, but the crew members did not use the ear protection provided for them. The crew members were communicating using hand signals; therefore, the earplugs would not have hindered communication.

The crew members did not wear their dust masks during any of the construction operations. The jobsite had exposed topsoil, and a layer of straw was used to keep the soil from entering the air when it was windy. When the workers climbed into the trenches, additional soil was disturbed and created more dust. The workers were coughing and sometimes covering their faces with their shirts.

Each of the organizations involved with this project assumed different groundwater levels. The owner indicated that the groundwater was higher than the actual level, and the contractor indicated that it was below the actual level. Activities were delayed while the owner decided how to mitigate the differing water levels, and then the contractor filed a claim for the delay. The amount of time wasted and the cost associated with the delay attributable to the groundwater issue were as follows:

$$
\begin{aligned}
\text{time wasted} &= (\text{minutes wasted per day} \times \text{number of days})/60 \text{ minutes per hour} \\
&= (40 \times 9)/60 \\
&= 6\text{ h} \\
&= 6\text{ h}/(8\text{ h/day}) \times 100 = 9x;\ x = 8.33\% \text{ of pipe installation time}
\end{aligned} \tag{7-2}
$$

$$
\begin{aligned}
\text{cost of time wasted} &= \text{time wasted in hours} \times (\text{cost/hour}) \\
&= 6 \times 530 \\
&= \$3{,}180 \\
&= 3{,}180 \times 100 = 38{,}153\,x;\ x = 8.33\% \text{ of pipe installation cost}
\end{aligned} \tag{7-3}
$$

The issue at the jobsite that affected productivity rates the most was the distance between the soil pile and the soil storage area for the trenches. Dry soil was used because the excavated soil was too moist to achieve a 90% compaction level. The time required for the loader to pick up soil and return to the trench was 1 min 51 s. This affected the productivity rate of the crew because during this time the crew members were not performing any tasks. If the soil stockpiles had been closer to the trench, the efficiency of the loader would have improved along with the flow of the work. Figure 7-54 shows a diagram of a proposed alternative that uses a dump truck to move the soil pile closer to the trenches.

The proposed revised work process is as follows:

1. The foreman signals the excavator operator to start trenching.
2. The excavator trenches.
3. The dump truck obtains enough gravel and backfill material for 1 day from the soil pile and moves it closer to the trenches.
4. The groundwater is pumped out of the trench.
5. The loader dumps a layer of gravel into the trench.
6. The loader obtains pipe from the storage pile.
7. The workers install the pipe into the trench.
8. The loader places gravel over the pipe.
9. The loader backfills the trench with soil.
10. The soil is compacted with a sheepsfoots roller.

In the revised process flow diagram, a 10-yd^3 (7.645-m^3) dump truck would be used to move the dry soil to a location closer to the trenching activities. In the present method, two loader round trips are required to deliver the gravel, and eight trips are required to deliver the soil for a total of 10 trips. The total time for 10 loader cycles is 18.5 min. The proposed dump truck would require only one trip to deliver the gravel and three trips to deliver the soil, for a total of three and a half trips for each cycle. The total time for the dump-truck cycle is 11.1 min.

$$
\begin{aligned}
\text{time savings} &= (\text{total time for loader in min} - \text{total time for dump truck in min})/ \\
&\quad 60 \text{ min per hour} \times \text{number of cycles to deliver soil} \times \text{project duration} \\
&= (18.5 - 11.1)/60 \times 3 \times 9 \\
&= 3.33\text{ h} \\
&= 3.33\text{ h}/(8\text{ h/day}) \times 100 = 9x;\ x = 4.625\% \text{ of total pipe installation time}
\end{aligned} \tag{7-4}
$$

Flow Diagram of Proposed Recommendation

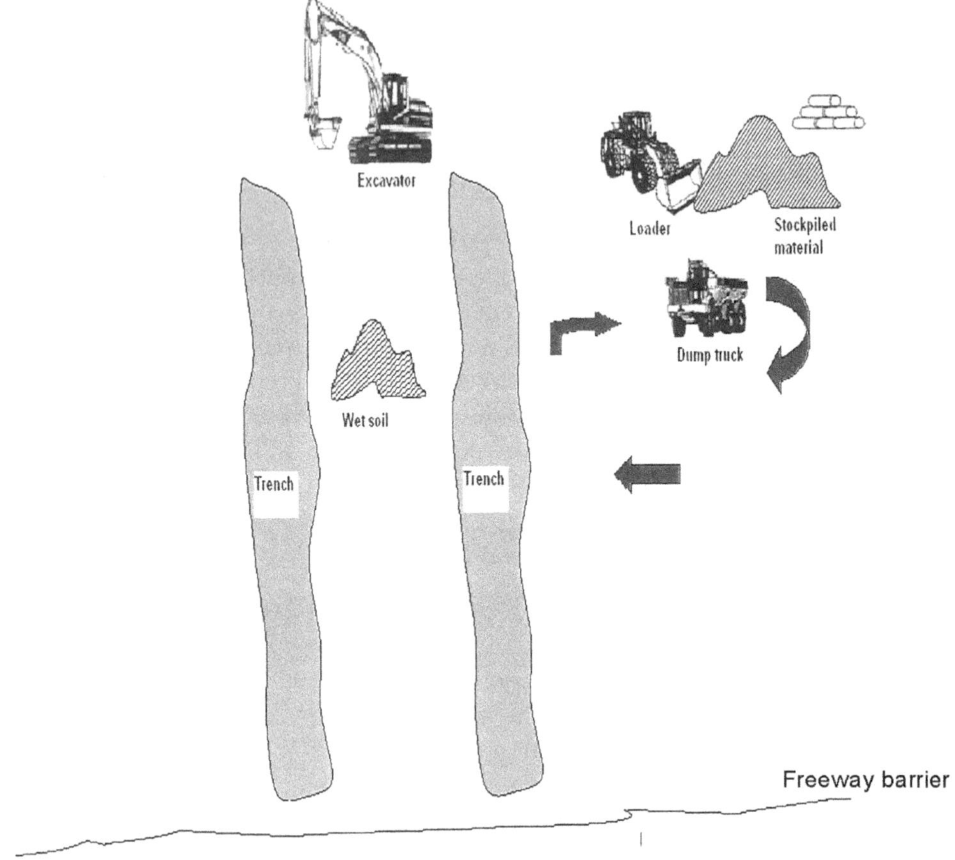

Figure 7-54. Proposed Revised Process Flow Diagram for Pipe Installation

Source: Page et al. (n.d.), Figure 13; reproduced with permission.

$$
\begin{aligned}
\text{cost savings} &= \text{time saved in hours} \times (\text{total cost/hour}) \\
&= 3.33 \times 530 \\
&= \$1{,}764.90 \\
&= 1{,}764.90 \times 100 = 38{,}153x;\ x = 4.625\% \text{ of total pipe installation cost}
\end{aligned}
\tag{7-5}
$$

$$
\begin{aligned}
\text{cost of dump-truck rental} &= \text{cost of dump-truck rental per day} \\
&\quad \times \text{project duration in days} \\
&= 150 \times 9 \\
&= \$1{,}350.00
\end{aligned}
\tag{7-6}
$$

$$
\begin{aligned}
\text{total cost savings} &= \text{cost savings} - \text{cost of dump-truck rental} \\
&= 1{,}765.00 - 1{,}350.00 \\
&= \$415.00 \\
&= 415 \times 100 = 38{,}153x;\ x = 1.088\% \text{ of total pipe installation cost}
\end{aligned}
\tag{7-7}
$$

The loader would also be required to load the dump truck, but for the remainder of the time, its efficiency would be improved, which would in turn improve the efficiency of the crew and equipment. The dump truck could be loaded at the start of the day or at the end of the day.

If the contractor had used two of the newer Caterpillar C 235B excavators instead of one Caterpillar C 225 and one Caterpillar C 235B, time would not have been wasted changing the bucket. Changing the bucket on the older Caterpillar C 225 required 10–12 min, compared with 1 min for the newer Caterpillar C 235B.

The excavator operator had to unclog the sheepsfoot roller by rolling it in dry soil and shaking it, and sometimes a worker would use a rake to remove soil. The total time required for compacting soil was 39 min per cycle, and the noneffective time was 13 min.

$$
\begin{aligned}
\text{time savings} &= (\text{noneffective time in minutes}/60 \text{ minutes per hour}) \times \text{cycles per day} \\
&\quad \times \text{project duration in days} \\
&= (13/60) \times 3 \times 9 \\
&= 5.85 \text{ h} \\
&= (5.85 \text{ h}/8 \text{ h/day}) \times 100 = 9x;\ x = 8.125\% \text{ of total pipe installation time}
\end{aligned}
\tag{7-8}
$$

The proposed process chart eliminates the delay caused by having to clean the sheepsfoot roller. The proposed crew balance chart is shown in Figure 7-55, and for the proposed crew configuration, the white hat worker could be eliminated, resulting in the following cost savings:

$$
\begin{aligned}
\text{cost savings} &= \text{salary per day} \times \text{project duration in days} \\
&= 296.00 \times 9 \\
&= \$2{,}664.00 \\
&= 2{,}662 \times 100 = 38{,}153x;\ x = 6.87\%
\end{aligned}
\tag{7-9}
$$

This worker represents 6.87% of the total crew member and equipment expenses for pipe installation.

Table 7-9 provides a summary of the time and cost savings that could be achieved by implementing the proposed productivity improvement techniques.

7.5 Case Study 4: High-Rise Building Stone-Panel Curtain-Wall Installation

(This case study was provided by Guang Jiang, Amrita Mukherjee, Chakrit Raksamata, Iqbal Jeet Singh, Sorawut Srisakorn, and Milkias Lemma Tefera.)

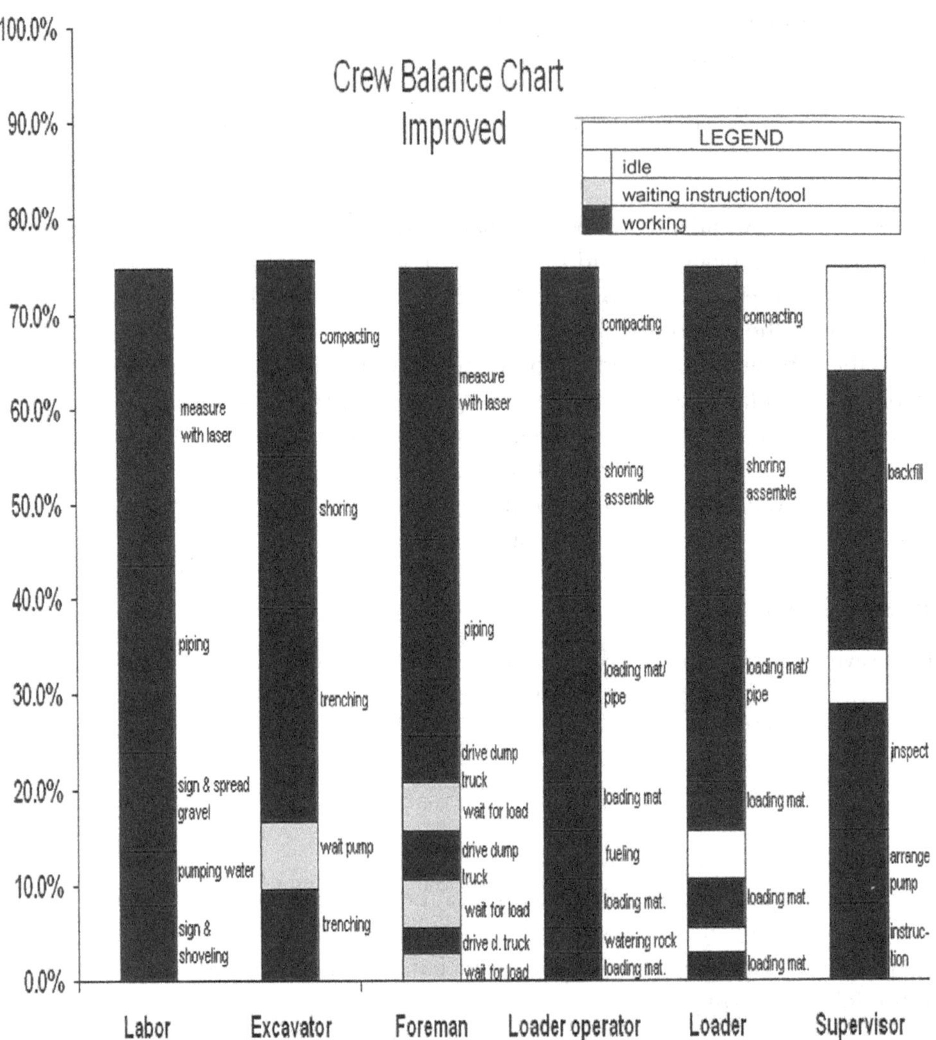

Figure 7-55. Proposed Revised Crew Balance Chart for Pipe Installation

Source: Page et al. (n.d.), Figure 14; reproduced with permission.

Table 7-9. Time and Cost Savings for Case Study 3

Recommendation	Time (percentage)	Cost (percentage)
More effective communication	8.3	8.3
Eliminating one worker	—	7.0
Using a dump truck	4.6	1.0
Changing compactor type	8.1	8.1
Total	21.0	24.4

Source: Page et al. (n.d.), Table 8; reproduced with permission.

Appendix A contains a detailed case study (Case Study 4) of the installation of stone-panel curtain walls on a high-rise building. This case study helps demonstrate how to prepare and conduct a productivity improvement study and includes numerous photos, diagrams, and charts that demonstrate the analysis process and proposed alternative work processes. Case Study 4 evaluated existing work processes, conducted various productivity improvement studies, and developed innovative new work processes that if implemented would have resulted in increased productivity of at least 10% (Jiang el al. n.d).

The case study productivity improvement project was one component of a high-rise building project that included metal, stone-panel curtain walls, windowpanes, and weatherproof sealing. The case study focused on the installation of the stone-panel curtain walls.

Appendix A provides background information on curtain-wall systems. In addition, it includes the productivity improvement project schedule, the methodology used for the case study, and the components of the curtain-wall installation. The stone-panel installation process is described in Appendix A, and information is provided on the floor plan and bay types. A summary of the materials, labor, and equipment used on the project is included, with a listing of the number of stone panels installed per floor.

The case study provides detailed information on the types of platforms used during the project because the platforms were an integral part of the installation process. The data-recording techniques for the installation process are also explained in this case study. In addition, information is provided on the stone-panel curtain-wall installation sequence and cycle. The data-analysis process and the formulas used for the data-analysis phase of the case study are outlined in Appendix A, along with a discussion of the accuracy of the data.

The case study also provides information on the different types of stone-panel curtain walls installed on the project and examines the causes of time differences for the installation of different types of stone-panel curtain walls. An analysis of the sequence of the stone-panel curtain-wall installation is provided in Appendix A, along with productivity index calculations.

At the end of the case study, stone-panel curtain-wall installation work improvement calculations and recommendations for improving the productivity of the installation process are provided. The case study in Appendix A demonstrates how the productivity improvement data collection techniques described in previous chapters could be implemented on construction projects and how to determine the percentage of productivity improvement that might be obtained by altering work processes according to productivity improvement studies.

7.6 Summary

This chapter provided information on how to evaluate and prioritize productivity improvement alternatives. It presented three case studies: one that involved parking structures and a police station, one that analyzed steel-erection processes, and one that reviewed the work processes used for pipe installation at a highway interchange. These three case studies demonstrated how to collect and analyze productivity improvement data and provide recommendations for improving the case study production rates. A fourth productivity improvement case study is provided in Appendix A.

7.7 Key Terms

American Speech Language Hearing Association
cast-in-place concrete
liquidated damages
polyethylene
precast concrete
queue
storm-water pollution plan

7.8 Exercises

7.1 This exercise uses the proposed term project provided at the end of Chapter 1 in Table 1-2, Optional Productivity Improvement Project. Create groups of four or five people. The members of each group will select a construction project and analyze a specific operation according to the project guidelines provided in Table 1-2. At the conclusion of the projects, each group will present its project as though they were presenting to a client.

7.2 In Case Study 1, Site 2, what items either observed or mentioned by the supervisor contributed to low productivity?

7.3 In Case Study 1, Sites 1 and 2, during the field-rating study, what were the two activities that contributed to lower field ratings?

7.4 In Case Study 1, for Site 2 what were the two items that contributed to lower productivity ratings?

7.5 In Case Study 1, for Site 1 what were the recommendations derived from the process chart analysis for improving productivity?

7.6 In Case Study 1, for Site 2 what were the recommendations derived from the process chart analysis for improving productivity?

7.7 In Case Study 2, what were the recommendations for improving productivity that resulted from the crew balance chart analysis?

7.8 In Case Study 2, what were the recommendations for improving productivity that resulted from the process chart analysis?

7.9 In Case Study 2, what were the recommendations that resulted from the 5-min-rating analysis?

7.10 In Case Study 2, what caused the productivity rates during the productivity analysis to be 32%?

7.11 In Case Study 2, what caused the masonry crew to have a productivity rate during the productivity analysis of only 43%?

7.12 In Case Study 2, what were the recommendations for the masonry activity that resulted from the process flow analysis?

7.13 In Case Study 3, why were the productivity ratings higher than they were in Case Studies 1 and 2?

7.14 In Case Study 3, what caused the operation to have a productivity rating of only 40% during the 5-min-rating analysis?

7.15 What could improve the productivity ratings for Case Study 3?

7.16 For the case study in Appendix A, why was it so difficult to install the stone-panel curtain wall?

7.17 For the case study in Appendix A, three different trade unions were involved in the installation of the curtain walls. Would having three unions involved create problems in implementing productivity improvement techniques on this project?

7.18 For the case study in Appendix A, what types of productivity problems were created because the stone panels had to be stored on the roof of an adjacent parking garage? Where else could the stone panels have been stored, and how would this proposed location help to improve productivity?

7.19 Did the size of the stone panels that were being installed in the case study in Appendix A have any effect on installation times, and if so, how?

7.20 How did having digital video recordings of the installation process in Appendix A assist in the evaluation of the work processes and the development of productivity improvement techniques?

7.21 Why were crew balance charts instrumental in analyzing the productivity of the stone-panel curtain-wall installation process in the case study in Appendix A?

7.22 Explain how the number of bays installed before the horizontal platform had to be moved affected productivity in the case study in Appendix A.

7.23 According to the conclusions presented at the end of the case study in Appendix A, could any methods be implemented to help improve the productivity of the stone-panel curtain-wall installation portion of the project, and if so, what were the methods?

References

Abulfateh, B. M. (n.d.). "A study in productivity improvement program on three structures." *Master's Report*, University of Colorado, Boulder, CO.

Jiang, G., Mukherjee, A., Raksamata, C., Singh, I., Srisakorn, S., and Tefera, M. (n.d.). "Stone-panel curtain-wall installation: River Park Tower II project." *Course Project*, San Jose State University, San Jose, CA.

Occupational Safety and Health Administration (OSHA). (2006). *American Speech Language Hearing Association Report*, Washington, DC.

Page, A., Frechette, M., Grandhi, M., Yusuf, N., Mirza, A., and Bas, H. (n.d.). "Coleman Avenue/880 Interchange project." *Course project*, San Jose State University, San Jose, CA.

Rahbar, F. (n.d.). "Steel erection project." *Course project*, Iowa State University, Ames IA.

Sanvido, V., and Paulson, B. (1991). "Site analysis using controller-function charts." *J. Constr. Eng. Manage.*, 117(2), 226–241.

CHAPTER 8

Engineering and Project and Construction Management Productivity Improvement

If engineers and project and construction management personnel are unable to effectively manage their own time, they may find it increasingly difficult to manage their subordinates' time or engineering and construction projects and increase production rates. Unfortunately, it is difficult to measure the productivity of engineers and project and construction management personnel because their work requires analytical reasoning and other types of cerebral activities, before they perform calculations and any type of analysis on paper or on computers. When trying to determine the production rates of engineers and construction managers, one cannot merely measure output, as is done to determine production rates for construction jobsite workers. Rather, a different technique should be used for measuring whether engineers and construction managers are managing their time effectively and working at their theoretical maximum achievable capacity.

One method that helps determine the effectiveness of how engineers and project and construction managers are using their time is a **time-management study** (TMS). Time-management studies are similar to productivity improvement studies in that they measure the amount of time a worker is productive, but they also provide insight into how a worker is using his or her time during a work period. When analyzing the productivity of engineers and other construction team members, TMS provides information on the exact activities performed during specific work periods. Since TMSs are an important aspect of productivity improvement, they are discussed in detail in Section 8.12.

This chapter includes information about engineering and project and construction management productivity variables, distractions, roadblocks, and issues. It also includes a discussion about conducting meetings and resolving conflicts because these activities consume a great deal of the time of engineering and project and construction management professionals. This chapter also covers methods for improving engineering and project and construction management productivity and conducting a TMS.

8.1 Productivity Variables

The productivity rates of engineers and project and construction management personnel are affected by some of the same variables that affect the productivity of construction workers. Among these variables are varying energy levels caused by

- Fatigue,
- Levels of concentration,
- The total amount of work to be performed,
- Motivation,
- Alertness, and
- Interest in a project.

Due to fluctuating blood sugar levels after meals, the peak work hours for most workers are between 9:00 and 11:00 a.m. and between 1:00 and 4:00 p.m. In the construction industry, project and construction managers accomplish more during afternoon peak work hours than they do during morning peak work hours, but this could be due to the work performed in the mornings requiring more immediate problem solving, which detracts from production-related work. Project management personnel are also more productive during Wednesdays and Thursdays, which reflects the planning and organizing activities that occur earlier in the week at construction jobsites.

Engineering productivity is affected by the location of the work being performed because a quieter environment is more conducive to analytical work. Overtime is prevalent among engineers who work in the engineering and construction (E&C) industry, but the overtime is used mainly as a means of performing work in an environment with fewer interruptions and a quiet atmosphere. Engineers work overtime either in the early morning hours or after other employees have left the office at the end of the day. In addition, engineers and construction management personnel often work on Saturdays or Sundays in order to prepare weekly plans in an atmosphere that allows them to concentrate on the task at hand, and not on multiple tasks, as they are required to do during typical workdays.

8.2 Productivity Distractions

In their typical work environment, engineers and project and construction managers encounter a variety of distractions that prevent them from performing their assigned work tasks. In addition to performing their assigned work, engineers and project and construction managers are required to manage the work of others and to be part of a team. The **interdependency** of project teams—that is, the dependence of team members on the work of other team members—makes it important for engineers and project and construction managers to be able to switch between work tasks while they are waiting for others to provide them with information.

The items that most frequently keep engineering and project and construction management team members from completing their assigned tasks on time are (Kerzner 2005, p. 317):

- Incomplete work,
- Rework,
- Delayed decisions,

- Poor communication,
- Telephone interruptions,
- Casual visitors,
- Waiting (for people, decisions, data, memos, information, and so forth),
- Failure to delegate, and
- A poor information retrieval system.

Of the items in this list, the two that cause the most work delays are delayed decisions and poor communication. More information about workplace distractions is available in Roper and Juneju (2009).

It is impossible for employees to be productive 100% of the time, but the productivity of engineering and project management personnel should not fall below 6 h in an 8-h workday (75% efficiency). The profits of firms are based on maintaining high productivity levels, and assigning a monetary value to time helps to illustrate the importance of maintaining engineering productivity rates above 75%. For engineers earning $50,000 a year, with benefits equivalent to 40% of their weekly salary ($20,000), the value of their time is $0.56 per minute {[($70,000/52 weeks/year)/40 hours/week]/60 min/hour}, or $33.65 per hour. At a $100,000 salary (benefits $40,000), the value of their time is $1.12 per minute {[($140,000/52 weeks/year)/40 hours/week]/60 min/hour}, or $67.31 per hour. At a salary of $100,000/for every 2 h of lost productivity per day, the cost is $134.62, or a total of $673.10 per week and $2,692.40 per month. Totaling the cost of the lost productivity for each project team member illustrates that inefficient work hours substantially reduce profits. Since it is difficult to measure the productivity of engineers and project and construction managers, firms might not be able to determine the actual losses caused by the decreased productivity of their personnel.

To reduce the time lost to **productivity distractions**, engineers and project management professionals should investigate the adoption of strategies such as the ones listed in Table 8-1. Additional strategies for producing higher levels of productivity are listed in Table 8-2.

8.3 Productivity Roadblocks

The activities that keep engineers and project and construction management team members from focusing their time on their assigned work tasks are called **productivity roadblocks**. Several examples of productivity roadblocks are listed in Table 8-3.

8.4 Productivity Issues

Engineers and project and construction managers have to deal with unique productivity issues because they are responsible for managing the work processes that occur on E&C projects. In addition to their daily work assignments, they are responsible for making sure that projects are managed as efficiently as possible while not exceeding the budget or missing scheduled completion dates. Project managers spend the majority of their time on four activities (Kerzner 2005, p. 334):

Table 8-1. Strategies for Eliminating Productivity Distractions

Strategies
Ask whether each trip is necessary
Avoid writing or reading useless memos
Control telephone time
Do not procrastinate
Do the hard part first
Follow schedules
Have only relevant people attend meetings
Know your energy cycle
Learn to delegate
Learn to say no
Look ahead
Make decisions quickly
Manage by exception
Refuse to do the unimportant
Send out meeting agendas
Shut off in-house visits for set periods
Travel light
Work at travel stops
Read e-mails only twice a day
Return phone calls only twice a day

Source: Adapted from Mackenzie (1972).

Table 8-2. Strategies for Increasing Productivity

Strategies
Classify activities
Conduct a time analysis
Establish opportunity cost activities
Establish priorities
Focus on opportunities, not problems
Plan solid blocks of time for important matters
Practice calculated neglect
Practice delegation
Train the system (boss, subordinates, and peers)

Source: Adapted from Mackenzie (1972).

1. Meetings (8 hours/week),
2. Productivity Distractions (10 hours/week),
3. Conflicts (12 hours/week),
4. Planning and Replanning (10 hours/week).

Because the time spent on these four activities totals 40 h per week for typical project managers, they likely would not be able to perform all of their assigned work activities without working more than 40 h per week. One way for project managers

Table 8-3. Activities That Reduce Productivity

Activities
Day-to-day administration
Having to explain "thinking" to superiors
Impromptu tasks
Lack of information in a ready format
Lack of sufficient clerical support
Late appointments
Too many levels of review
Too many people working in a small area
Union grievances
Unscheduled appointments or drop-ins

Source: Adapted from Kerzner (2005), p. 33.

to prevent having to work overtime is to reduce the amount of time they spend on these four activities. Other activities that lead to the inefficient use of time at the project management level are the following (Mackenzie 1972, p. 6):

- Lack of planning,
- Lack of self-control,
- Activity traps,
- Managing versus doing,
- People versus task skills,
- Ineffective communication,
- Organizational bottlenecks.

8.5 Conducting Meetings

Engineers and project and construction managers might be able to increase their productivity during the week by conducting more effective meetings and reducing the time required for meetings. Meetings provide an opportunity for people to share information and clarify their performance. They enable participants to provide input into decision-making processes and provide managers with an opportunity to identify members who have the ability to perform tasks or make decisions. They increase the probability that decisions will be implemented and provide a venue for reinforcing organizational values (Kieffer 1988). Not all employees agree that meetings expedite the decision-making process. The most common complaints about meetings are listed in Table 8-4.

Meeting Agendas

When preparing to conduct a meeting, the meeting chair should define the purpose of the meeting and plan what will be accomplished during the meeting. If the chair of the meeting is unclear as to the purpose of the meeting, then he or she should solicit input from potential meeting attendees. Once the purpose of a meeting has

Table 8-4. Common Complaints about Meetings

Complaints
Boring
Too frequent or not frequent enough
Diverted by members with a hidden agenda
Dominated by formal leaders or by a few influential or verbal people
Not focused on important issues
Poorly organized or poorly led
Subverted by members whose behaviors are destructive
Too long

Source: Adapted from Kieffer (1988).

been defined, then an **agenda** should be developed and distributed to everyone who will be attending the meeting. The agenda should include all of the topics that will be discussed at the meeting, with the most urgent topics listed first to ensure that they will be covered during the meeting. Meeting agendas should be distributed to attendees at least one week before the meeting date, unless a meeting is scheduled on short notice.

To ensure that a meeting does not exceed the time allotted for it, the meeting chair should start the meeting on time. Having to wait for late arrivals to a meeting is disrespectful to the people who have arrived on time. Starting on time also sets a precedent that demonstrates that all meeting attendees should be present at the beginning of the meeting. The first activity in a meeting should be a review of the agenda to verify its accuracy and to add items that have surfaced since the agenda was developed and distributed to meeting members.

While a meeting is being conducted, the chair is responsible for ensuring that meeting attendees follow the agenda. If the meeting attendees would like to discuss topics that are not on the agenda, the new items should be discussed as new business toward the end of the meeting. As a meeting progresses, the chair of the meeting should assign responsibility for following up on **action items** to the appropriate person and indicate dates for the completion of each action item. Toward the end of the meeting, the chair should summarize any agreements reached during the meeting and remind attendees of their responsibilities for completing action items. Meetings should be concluded at the scheduled time, as this demonstrates concern for the next commitment of the people attending the meeting. If some agenda items are not addressed by the end of the meeting, they should be included in the agenda for the next meeting, at the beginning of the agenda. Meeting minutes should be distributed in a timely manner to remind meeting attendees of the tasks they were assigned to perform during meetings.

Disruptive Meeting Attendees

One of the most difficult aspects of conducting meetings is dealing with difficult attendees. Meetings might be disrupted if any of the attendees exhibit one or more

Table 8-5. Behaviors That Are Disruptive in Meetings

Behaviors
Challenging attempts to move the group toward decisions
Conducting side conversations
Insisting on a precise, clear definition of each idea to the point that the group does not accomplish anything
Interpreting criticism of ideas as a personal attack
Joking about everything that happens
Talking for the sake of being heard
Urging the group to take action before a problem is clearly identified
Waving off or negating all suggestions or new ideas from others
Talking for long periods

Source: Adapted from Kieffer (1988).

of the behaviors listed in Table 8-5. Some of the disruptive behaviors that are mentioned in Table 8-5 might occur when a meeting is conducted without adequate preparation or when there is no meeting agenda to keep attendees focused on the meeting tasks.

To minimize the time required for meetings, the meeting chair should limit the amount of time commandeered by meeting attendees. When trying to steer meetings back to the original agenda, the chair should adopt strategies that help to neutralize meeting attendees who are being disruptive. The following paragraphs provide suggestions on how to neutralize disruptive behaviors by meeting attendees.

The chair should listen to meeting members but not become involved in debates with them. It is usually not effective to ignore disruptive meeting attendees, as they may merely continue their behavior. Trying to persuade them to join the main discussion sometimes helps to make them feel that the other meeting attendees respect their ideas and contributions. If this does not work, then the chair should speak with them in private about their behavior. If meeting chairs try to resolve issues during meetings, additional disruptions might occur when the disruptive attendees escalate their negative behavior in retaliation or withdraw from meeting interactions entirely. When discussing problems, the chair should focus on the behaviors causing problems and not attack the person exhibiting the behaviors. It is also not constructive to mention past arguments between meeting attendees (Kieffer 1988).

Sometimes attendees use disruptive behavior when they want to contribute to a meeting but have not determined the best technique for accomplishing this objective. If a meeting chair recognizes that this is happening, he or she should help the disruptive attendees to participate in the meeting by using strategies such as inviting them to planning meetings, asking them for specific suggestions, or providing them with responsibilities for tasks that result from decisions made during the meeting (Kieffer 1988). Meeting attendees could also share in the responsibility for dealing with disruptive attendees by not allowing unacceptable behavior to go unchecked and by using group censure to neutralize them. If other techniques for neutralizing disruptive behavior are not effective, then meeting attendees have the option of walking out of the meeting when the disruptive behavior prevents them from achieving their objectives, or the disruptive meeting attendee might be excluded from further meetings.

8.6 Conflict Resolution

Conflict resolution is not about avoiding conflicts, but rather it is concerned with managing conflicts without dedicating too much time to the process of solving them. Some of the principal causes of conflicts are listed in Table 8-6.

Conflict results from individual differences, backgrounds, and values, and if it is used constructively, it might help release energy in stressful situations. The negative aspects of conflict include

- Diverting energy away from the tasks that need to be accomplished,
- Destroying morale,
- Polarizing individuals and groups,
- Deepening differences,
- Obstructing cooperative actions,
- Producing irresponsible behavior,
- Creating suspicion and distrust, and
- Decreasing productivity.

Causes of Conflicts

Conflicts arise from four areas:

1. Facts related to the present situation or problem,
2. Methods to be used to achieve goals,
3. Goals related to how people want things to be accomplished, and
4. Values that represent long-term goals and qualities.

Table 8-6. Principal Causes of Conflicts

Causes
Authority issues
Communication failures or inaccuracies
Competition for limited resources
Differences over methods
Frustration and irritability
Lack of cooperation
Personality clashes
Responsibility issues
Substandard performance
Value and goal differences

Conflicts that arise over facts are easier to resolve, and the hardest to resolve are conflicts that arise over value differences.

Conflict Management Strategies

Each individual addresses conflicts by using his or her own inherent strategies, but there are also specific techniques for addressing conflicts. The five main techniques used to address conflicts are:

1. **Withdrawing:** This occurs when an individual retreats from an actual or a potential conflict situation.
2. **Smoothing:** This happens when someone emphasizes areas of agreement and deemphasizes areas where there are differences.
3. **Compromising:** This involves an individual or individuals searching for solutions that bring some degree of satisfaction to the parties in conflict.
4. **Forcing:** This is when someone exerts his or her viewpoint at the potential expense of the viewpoint of someone else, and this could result in open competition.
5. **Confrontation:** This occurs when an individual addresses a disagreement directly and in a problem-solving mode.

When one is trying to reduce the amount of time spent on solving a conflict, it helps to depersonalize the conflict by not allowing judging of the conflicting parties and instead focusing on issues or disagreements about facts. This process keeps the process moving forward, even if it is only in small increments.

When dealing with conflicts, management personnel should realize that each person has his or her own conflict management style. Several common conflict management styles are:

1. **Competitor:** People with this style are power oriented, they prefer win/lose solutions, and they may suppress, intimidate, or coerce people into conflict. This style is effective when quick decisions are required or when important but unpopular decisions must be implemented. This style might be used when a person knows he or she is right and does not have time to listen to all sides of the argument before making a decision.
2. **Avoider:** This style involves low assertiveness and uncooperative behavior. This style works in no-win situations of little importance or for problems that will disappear in the near future.
3. **Accommodator:** This style is used by people who have little concern for personal goals or who realize the conflict is more important to someone else. It is also used when harmony is important, when another person needs to win, or when the parties are bargaining or negotiating several issues.
4. **Compromiser:** This style is most useful when expediency is required above principle and when short-term solutions are required rather than achieving long-term objectives. It is the least time-consuming style.

Table 8-7. Common Sources of Conflicts

Sources
Confusion and uncertainty about change
Favoritism
Interdepartmental or intradepartmental relationship history
Overly competitive environment
Poorly defined tasks
History between two people
Punitive or threatening management style
Severe economic downturns
Unclear or arbitrary standards
Unreasonable levels of pressure and pace

5. **Collaborator:** This style involves depersonalizing conflict and goals. Feelings, attitudes, and opinions are accepted as legitimate concerns, but analyzing facts and potential solutions is also important. The analysis necessary for this style is time consuming, so it is usually used only in important conflicts.

In conflict resolution, it is important to understand the source or sources of conflict because understanding the source of conflict helps identify acceptable solutions. Some common sources of conflict are listed in Table 8-7.

Conflicts do not normally arise from immediate problems; rather, they develop over time. The five stages of escalation to open conflict are as follows:

1. *Anticipation,* which occurs when a change is being made and problems are forecast;
2. *Rumor,* or the realization of conscious but unexpressed differences;
3. *Discussion,* where information is shared, questions are asked, and sides are expressed;
4. *Open dispute,* during which opinions are expressed and lines are drawn; and
5. *Open conflict,* during which people join forces on one side or the other of the argument.

Engineering and Construction Industry Conflicts

If a conflict is addressed in the early stages, before it becomes an open conflict, it might be resolved in less time. In the construction industry, the types of conflicts that frequently occur are **external conflicts** and **internal conflicts**. External conflicts occur between owners and architects/engineers (A/Es), between A/Es and engineers, between A/Es and contractors, and between contractors and material suppliers or subcontractors. Internal conflicts occur at construction jobsites and usually involve disagreements over methods and procedures, who should be performing specific work items, the perception of unequal treatment of workers, the need for overtime

Table 8-8. Categories of External and Internal Conflicts

External Conflict Categories	Internal Conflict Categories
Interdepartmental	Intradepartmental
Interdiscipline	Intradiscipline
Interfunctional	Intrafunctional
Interpersonal	Intrapersonal
Interproject	Intraproject

work, a lack of available resources, and other worker concerns. Table 8-8 lists some of the categories of conflicts.

Each conflict requires time to analyze the situation and the parties involved, and this reduces the time available for other tasks that have to be performed by engineers or project and construction management personnel. If engineers or project and construction managers are more aware of conflict patterns and the typical behaviors of the parties involved in conflicts, they should be able to reduce the amount of time they spend on solving conflicts.

8.7 Planning Process

Planning requires the selection among objectives, alternatives, products, procedures, or programs. It also involves identifying, evaluating, and selecting the courses of action a firm will implement and support. Planning requires establishing objectives by evaluating potential courses of action. Once the objectives are established, then specific plans, policies, and procedures for accomplishing the objectives are developed and implemented by members of the firm. Upper-management planning procedures involve determining the assets and working capital required, cash flow, total capital needs, and sources of capital. During the planning stage, policies and guidelines for operation have to be developed by management personnel.

The following are the steps required to plan activities:

1. Establish objectives.
2. Program tasks to achieve objectives.
3. Schedule tasks.
4. Budget costs.
5. Recognize constraints on freedom to plan (procedures and policies).
6. Devise strategies and analyze alternatives.
7. Create acceptability so that the plan will not be resisted.

Engineers are usually concerned with the material results of their plans and the techniques they use to achieve results, such as flow diagrams, heat balance charts, design criteria, and scheduling programs, all of which are derived from **physical principles**. Managers are more concerned with the **human aspects** of planning and the effects of plans on people. Managers become more involved with minimizing resistance to plans, resolving conflicts, and creating acceptability, and their techniques are derived from **psychological principles**. Engineers start at the top of the list of planning activities and may not spend as much time on the activities listed last. Managers might reverse the list and concentrate on acceptability first, even before they have clearly defined the objectives. Planning and replanning are not tasks to be avoided; rather, engineers and managers should investigate methods for reducing the amount of time they spend on planning and replanning activities.

8.8 Methods for Improving Personal Productivity

This section provides suggestions for improving the individual productivity of engineers and project and construction managers, including reducing productivity roadblocks and distractions.

Work-Space Arrangements

In addition to determining the activities that distract from their assigned work functions, engineers and project and construction management personnel should develop work habits that help improve productivity. The first area where improvements might be implemented is in the arrangement of individual work spaces. The arrangement of desks in work areas should be evaluated to see how it affects the interactions of workers with their coworkers or subordinates. If a desk is facing other workers, interactions will increase, but if it is turned away from other workers, the worker might be able to control interruptions by, for example, not turning around to acknowledge a person talking behind him or her. Similarly, workers with their own offices have the option of facing their desk toward or away from the door to the office. Engineers and construction managers who need to have frequent interaction with coworkers will improve their productivity by locating their desk close to the workers they interact with on a daily basis. If they spend more time analyzing data or making decisions, then their productivity might improve if they work in an office or away from other people in the office.

Some people function well with their desks cluttered, and others need a clean work space before they start a project. The most effective way to prevent a desk from becoming cluttered, and to avoid losing time looking for materials for different projects, is to have an efficient filing system. Workers leave items on their desks to remind them of their importance, and having certain items visible helps some workers remember to work on them. But an efficient filing system should be used to prioritize work items. Priority items should be filed at the front of the filing system and then be moved to the appropriate file once they have been completed by the worker.

Daily Logs

Daily logs are used to keep track of important activities, and they are also used to prioritize activities for each day. A master list of activities should be prepared at the beginning of each week, and then each day a daily priority list is developed by the individuals working on a project. At the end of each day, the activities from the daily log that were not completed are moved to the top of the list for the next day. This prevents activities from being neglected and helps provide a reminder of the importance of activities.

E-mail, Mail, and Telephone Interruptions

One method for increasing productivity is to read and reply to e-mail or mail, not as it comes in, but during designated times. The same system should be used for phone calls. Answering e-mail, mail, and telephone calls throughout the day might disrupt the flow of work. A designated time should be set for returning phone calls, and the best times are usually right before lunch or the end of the day. Returning calls at these times prevents long conversations because both parties will likely be anxious to leave for lunch or leave work on time. People will get to the point faster if they are in a hurry to leave their office. E-mail and mail should be read and answered during these same periods, when a person is less likely to write long replies because they want to leave their office for lunch or to go home.

Office Visitors

Restricting visitors is easier when a person has an office with a door, but there are also techniques for limiting visitors for workers who are assigned to cubicles or who have only a desk. Signs should be posted during the times when the worker does not want to be interrupted by a visitor. Workers who have desks close to one another are able to also screen visitors for each other during certain periods during the day. Workers should also inform coworkers of set periods when they do not want to be disturbed and maintain that schedule on a daily basis. Workers with offices are able to restrict their visitors by posting ***do not disturb*** signs with designated times of when they are available listed on them. If a sign is posted without times, people will learn to ignore the sign and knock on the door with "just one quick question."

When visitors do not leave promptly, a worker might have to rise from his or her desk and escort the visitor out the door or away from the desk. If that technique does not work, the worker should keep walking past the visitor, as if he or she has another meeting to attend somewhere else.

Work Sessions

Having solid blocks of time to complete activities that require accuracy prevents potential mistakes from happening because of an interruption. Not all activities require uninterrupted time, but for those that require it, a *do not disturb* period should be requested and respected by those who work around them. Coworkers, subordinates, and bosses should be trained to respect work sessions, and employees should be allowed to set their time requirements for activities. If engineers and project and construction managers are able to have uninterrupted work sessions during regular work hours, their productivity might increase and the number of hours they devote to overtime should decrease.

8.9 Decision Making

One of the most difficult tasks for engineers and project and construction managers is to make decisions rapidly. Engineers are trained to review all available information before performing any type of analysis, and this technique is mandatory for designing structures or analyzing data. But when a project is under construction, delayed decisions cost firms valuable time, resources, and ultimately, profit. Rather than having all available information before a decision is made, engineers and project and construction managers are forced, because of time constraints, to make decisions with incomplete information. The decisions made under these circumstances may not always be the right decisions, and this causes engineers and project and construction managers stress. One of the main reasons engineers and managers make a decision quickly is because it is better to make a bad decision that keeps construction activities from stopping than to make no decision and disrupt all work.

8.10 Delegating Work

One method for improving the productivity levels of engineers and project and construction managers is to delegate work to coworkers or subordinates. Unfortunately, **delegation** is difficult when only one person is in possession of all the information required to perform an activity. Learning to delegate requires relinquishing the expectation of perfect results and accepting results that meet the minimum criteria for accomplishing an activity. Some of the barriers to delegation are listed in Table 8-9.

Table 8-9. Barriers to Delegation of Work Assignments

Barriers
Demand that everyone "know all of the details"
Disinclination to develop subordinates
Failure to delegate authority commensurate with responsibility
Failure to establish controls and to follow up
Fear of being disliked
Insecurity
Lack of confidence in subordinates
Lack of experience in the job or in delegating
Lack of organizational skill in balancing workloads
Perfectionism, leading to overcontrol
Refusal to allow mistakes
The "I can do it better" fallacy
Uncertainty over tasks and inability to explain

Source: Adapted from Mackenzie (1972), p. 133.

In many instances, the following valid reasons for not delegating tasks to others exist (Mackenzie 1972, p. 133):

- Lack of experience,
- Lack of competence,
- Avoidance of responsibility,
- Overdependence on the boss,
- Disorganization,
- Overload of work,
- Immersion in trivia.

In some situations, the following circumstances might prevent someone from delegating activities (Mackenzie 1972, p. 134):

- One person performance policy;
- No tolerance for mistakes;
- Criticality of decisions;
- Urgency, leaving no time to explain (crisis management);
- Confusion in responsibilities and authority; and
- Understaffing.

8.11 Wasting the Time of Others

Engineers and project and construction managers should realize that they might be wasting the time of others and should be cognizant of the types of activities that waste time. These activities include not providing information when it is required and not communicating instructions clearly. Time is also wasted when engineers and construction managers keep workers waiting for decisions and interrupt their work when it is not necessary.

The methods discussed in the previous sections are used to improve individual productivity levels after a TMS has been conducted to highlight the areas where productivity losses are occurring during each workday. The next section explains how to conduct individual TMSs.

8.12 Conducting Time-Management Studies

One of the typical primary objectives for any E&C project relates to the time required for completion of the project. Engineers and project managers, as well as other project management team members, often become so focused on the project and the execution process that they fail to carefully manage their own time. Successful engineers and project and construction managers develop a system to manage their time based on their current responsibilities. To develop time-management systems, engineers and project and construction managers should have a technique for determining their productivity and the effectiveness of the time they spend at work. Time-management studies provide a technique for quantifying and analyzing the productivity of engineers and project and construction management personnel.

Timing the Study

To conduct a TMS, a method is required to generate random beeps five times per hour. Computers, tablet computers, smart phones, and watches should be set to provide five random beeps per hour. Other creative ways of generating beeps include having a friend call a cell phone five times an hour at random intervals, setting a kitchen timer for different times, and identifying a sound that occurs randomly.

Identifying Critical Activities

The next TMS activity is identifying typical activities that occur during the day and grouping them into categories. Examples of potential activity categories that might be used for the TMS are listed in Table 8-10.

Table 8-10. Typical Activity Categories for Time Studies

Activity Categories
Answering e-mail
Answering questions
Eating
Attending classes
Attending meetings
Project work
Reading e-mail
Surfing the Web for personal information
Surfing the Web for research for work
Talking about non-work-related activities
Talking about work-related activities
Talking to jobsite personnel
Talking on the telephone
Traveling in a car
Sending or receiving personal tweets
Sending or receiving tweets for work
Waiting for other people
Waiting to use a computer or a printer
Walking to a new location
Working on a project or homework assignment
Working on the computer
Writing memos
Writing a report

Each person will have unique activity categories, and a list should be prepared that includes the activities performed on a daily basis by the person conducting the study. Once a list has been prepared, then a capital letter should be assigned to each category to help identify it quickly. For example, *T* can be used for travel, *E* for e-mail, *W* for waiting, *P* for phone call, *D* for driving, and so forth. When starting a TMS, the worker should write the categories and their respective designated letters at the top of the paper or the spreadsheet where the study activities will be recorded for each day. In addition, the heading of the paper or the spreadsheet should include the name of the person conducting the study, along with the date when the activities are being recorded for the study.

Recording Daily Activities

The data-collection forms used to record daily activities should be divided into three columns. Column 1 is where the time is recorded, column 2 is used to record the letter code that represents the activity being recorded, and column 3 is used to record notes pertaining to the activities recorded during the TMS. Enough rows should be included on the forms to record five activities per hour for either an 8-h day or the total number of hours that will be used for the study. An example of a data-collection form that has been filled out for a TMS is shown in Table 8-11.

Table 8-12 provides a blank TMS data-collection form that can be copied and used to record data during TMSs.

Conducting the Study

At the beginning of a TMS, the person performing the study should activate his or her timing system to start approximately 5 min after the study commences. When a beep is heard, the time should be recorded first, then a letter representing the activity that the person is engaging in at that moment is recorded, and then any comments

Table 8-11. Sample Data-Collection Form

Name ____________	**Date** ____________	
	Codes	
T – Traveling	W – Waiting	E – E-mail
F – Eating meals	D – Driving	P – Project work
M – Meetings	C – Class	Ph – Telephone
Com – Computer work	TN – Talking not related to work	TR – Talking related to work
Time	*Code*	*Notes*
8:00 a.m.	D	
8:09 a.m.	M	With boss
8:22 a.m.	M	With boss
8:31 a.m.	E	Reading
8:51 a.m.	P	Commercial structure redesign
9:10 a.m.	Ph	Client
9:23 a.m.	M	

Table 8-12. Blank Data-Collection Form

Name ______________	**Date** ______________	
	Codes	
Time	*Code*	*Notes*

related to the activity are recorded next to the code. This process is repeated until the end of the TMS period.

TMS Data Analysis

When the data-collection process is completed, the data should be summarized in the following manner. First, the number of times that each activity code was recorded is added together and recorded on another piece of paper. Then the total number of observations are calculated and recorded at the bottom of the page. For each activity category, the total number of times an activity occurred is divided by the total number of recorded observations, and this number is multiplied by 100 to determine the percentage of time that each activity occurred during the time study [Eq. (8-1)].

$$\begin{aligned}&\text{percentage of time for each activity}\\&\quad=(\text{total number of observations for activity}/\text{total number of observations})\times 100\end{aligned} \tag{8-1}$$

The percentage of time calculated for each activity provides a summary of the proportion of the day the worker dedicates to each activity, and this summary should be compared with the job responsibilities of the worker to determine whether he or she is spending the appropriate amount of time on each activity. The TMS also highlights the percentage of time spent on activities that are not directly related to work functions or that are distractions. When the percentage of time spent on distractions and activities not directly related to work functions is removed from the calculation, the remaining percentage represents the amount of time that the person performing the study was productive during the day.

8.13 Summary

In addition to analyzing the productivity of workers, determining the productivity rates of engineering and project and construction management personnel is important. Engineers and project and construction managers are responsible for managing workers, but they are also responsible for effectively managing their own time. To determine the productivity of engineering and project and construction management personnel, TMSs should be performed to quantify the percentage of time devoted to productive, unproductive, and non-work-related activities each day.

Engineering and project and construction managers should also be aware of the distractions and roadblocks that prevent them from performing their assigned work tasks. This chapter summarized some distractions and roadblocks and also provided techniques for improving individual productivity.

Project managers spend an average of 40 h per week on meetings, productivity distractions, conflicts, and planning and replanning activities, and techniques that might be used to reduce the amount of time spent on these four activities were discussed in this chapter. This chapter also discussed how to conduct effective meetings and how to address conflicts caused by different conflict management styles.

Time-management studies were introduced in this chapter to provide a method for evaluating the productivity of engineers and project and construction management personnel. Once engineers and project and construction management personnel have an understanding of how they waste time during work, they should be able to start applying some of the suggestions provided in this chapter for improving their productivity.

8.14 Key Terms

accommodator
action items
agenda
avoider
collaborator
competitor
compromiser
compromising
conflict resolution
confrontation
daily logs
delegation
do not disturb
engineering productivity
external conflicts
forcing

human aspects
interdependency
internal conflicts
physical principles
productivity distractions
productivity roadblocks
psychological principles
smoothing
time-management study
withdrawing

8.15 Exercises

8.1 Perform a TMS using a system that randomly provides a beep, ring, or vibration five times during each hour. Use a minimum of one weekday observation. Write a two-page summary that includes statistics and a description of the self-timed usage analysis. The analysis should also include the areas that need improvement.

8.2 Perform a 1-week time analysis. Summarize the results and the statistics for the analysis.

8.3 Use the information gathered in exercises 8.1 and 8.2 to answer the following questions in writing:
- How was time wasted during the TMS?
- What could be done to prevent or reduce wasting time in the future?
- How was the time of others wasted by the person conducting the study?
- Whose time was wasted during the study?
- What could be done to prevent wasting the time of others?
- What activities could be reduced, eliminated, or delegated to someone else?
- What did other people do that wasted the time of the person conducting the study?
- What was accomplished that was directly related to achieving the goals of the person conducting the study?
- Is time being spent pursuing goals that are important? If not, why not, and if so, how?

8.4 At the engineering and project and construction management level, what activities lead to inefficient use of time?

8.5 Of the planning activities listed in the section on the planning process, which one consumes the greatest amount of time and why?

8.6 How would using a log to record daily activities help to improve the productivity of engineers or project and construction managers?

8.7 Why is it so difficult for engineers or project and construction managers to delegate parts of their work to others?

8.8 Select three of the productivity roadblocks listed in Table 8-1 and describe methods for preventing them from occurring during a typical workday.

8.9 Why is it important to use a meeting agenda and to always start meetings on time?

8.10 What is the most efficient method for addressing the problems created by disruptive meeting members?

8.11 Explain which of the conflict management strategies you use and why you use them when confronted with conflict.

References

Kerzner, H. (2005). *Project management*, Wiley, New York.

Kieffer, G. (1988). *The strategy of meetings*, Warner Communications, New York.

Mackenzie, R. (1972). *The time trap: Managing your way out*, Amacon, New York.

Roper, K., and Juneju, P. (2009). "Distractions in the workplace revisited." *Journal of Facilities Management*, (6)2, 91–209.

CHAPTER 9

Computer Applications in Productivity Improvement

This chapter provides background information on the computer hardware, software applications, and simulation models that are used to record and analyze data for productivity improvement studies at construction jobsites. Examples of some of the available computer software programs that include modules used for simulating construction operations are the *General Purpose Simulation System* (*GPSS*) (1960), ***Insight*** (2013), ***Stella Modeling and Simulation Software*** (2013), ***Simscript II*** (2013), and ***MicroCYCLONE***, ***WebCYCLONE***, and ***Stroboscope*** (Halpin 1990a, b; Martinez 1996). The *CYCLONE* and *Stroboscope* simulation software programs were developed by university faculty members rather than by commercial software firms.

This chapter discusses some of the general purpose software applications, and the types of hardware required for their use. Also discussed are the types of networks being used to convey information among project participants. This chapter also provides information on computer applications being used in the engineering and construction (E&C) industry, including software programs, database systems, **management information systems**, radio frequency identification systems, and building information modeling (BIM).

9.1 Engineering and Construction Industry Computer Software

Computers are used in all industries, and examples of the computer software applications used regularly in the E&C industry are listed in Table 9-1.

There are also specialized analytical software programs specific to each engineering, construction, and scientific discipline. Examples of discipline-specific software programs used in the construction industry are listed in Table 9-2.

The computer software applications used by individual firms vary on the basis of a number of factors, with the primary factor being the cost of the software. High-end software programs, such as the project-controls and estimating software ***Timberline*** (2013) by Sage Construction Software, the design and modeling software ***MicroStation*** (2013) by Bentley Systems, the project management and scheduling software *Primavera P6 Professional Project Management* by Oracle, and the three-dimensional modeling software suite by AutoDesk including ***NavisWorks*** (2013), and ***Revit*** (2013), are mainly

Table 9-1. Examples of Computer Software Commonly Used in Engineering and Construction

Computer Software	Computer Software
Accounting	Mathematical modeling
Artificial intelligence	Monitoring systems
Computer aided drafting	Portable document formatting (PDF)
Databases	Presentation graphics
Decision-support systems	Scheduling
Design programs	Simulation modeling
Digital photography	Social networking
Digital video	Spreadsheets
Document management and retrieval	Statistics
Estimating	Structural analysis
File compression	Three-dimensional modeling
File transfer	Web browsers
Financial analysis	Web development
Graphic design	Word processing
Management information systems	

used by medium to large E&C firms. Scaled-down versions of some of these applications, such as *Primavera Contractor* (Oracle) and *AutoCAD* (AutoDesk), perform functions similar to the higher cost software programs but do not contain all of the same features. Other factors that influence the selection of software are ease of use, applicability to the function of a firm, employee knowledge of the use of specific software programs, and client requirements.

9.2 Computer Hardware

The computers selected for use in the E&C industry vary depending on the types of applications required by each firm. The different types of computers include super, mainframe, mini, desktop, laptop, notebook, and tablet. Computers are classified by their memory size including **random-access memory** (RAM) and **hard drive** memory and processing speed. **Super computers** process vast amounts of information, data, and calculations per second. **Mainframe computers** process data rapidly and store large files. **Mini computers** are used as network servers, and they can be linked to mimic the capabilities of mainframe computers. Mini computers are used in distributed database networks, as opposed to centralized data-processing centers, and they are often used as computer-aided design and drafting (CADD) workstations because they are able to process large graphic files.

Desktop computers with monitors are the most widely used computers, but they are limited by the size of their hard drives as to the amount of data they are able to store. Typical desktop computer hard drives store 100 MB to more than a **gigabyte**, which is not adequate if they will be used for storing digital videos, photos, or other large graphic files. Desktop computers could be upgraded to include additional hard

Table 9-2. Examples of Engineering and Construction Computer Software Applications

Discipline-Specific Software
3HTi Pro/Engineer CAD (design)
Autodesk AutoCAD (design)
Autodesk AutoCad Civil 3D (design)
Autodesk Design Review (design)
Autodesk Design Visualization (design)
Autodesk NavisWorks (design and construction)
Autodesk Pro Engineer (design)
Autodesk Revit Architecture (design)
Autodesk Revit Structure (design)
Autodesk Revit Mechanical, Electrical, and Piping (design)
Autodesk Solidworks (design)
Bentley Microstation (design and construction)
Bentley Projectwise Project Team Collaboration (project management)
Bluebeam Review (design and construction document control)
Corecon Technologies V7 (project management)
Graphisosft ArchiCAD (design)
HCSS Heavy Bid Express (estimating)
Innovaya Visual Estimating (estimating)
Innovaya Visual Quantity Take-Off (estimating)
Meridian Systems Prolog (project management)
Microsoft Project Pro (scheduling and project control)
On Center (project management)
Oracle Primavera P6 Professional Project Management (scheduling and project control)
Primavera Suretrack Project Management (scheduling and project control)
Procure Cloud Based Construction (project management)
ProEst Estimating (estimating)
Rib MC^2 iTwo (estimating and project management)
Sage 100 Contractor (formerly Master Builder (project management)
Sage Timberline (estimating)
TurboCAD (design)
Vico Office Suite (construction Management)
Viewpoint V6 (construction management)

drive memory. If users have access to the Internet, they are able to store their files, especially excessively large files, on remote servers or the server of their firm, rather than on the hard drive of their computer.

Notebook computers have capabilities similar to those of desktop computers, but they include built-in screens, instead of external monitors; weigh only a few pounds; and are able to run several hours on the power provided by their internal battery before having to be recharged in a wall outlet. It is not as easy to add memory or other hardware to notebook computers as it is to desktop computers.

Tablet computers and **smart phones** are mainly designed for surfing the Web, sending and receiving e-mail, and texting, but numerous other applications are also available for these devices. Tablets and smart phones have less hard drive memory than desktop and laptop computers, usually 32 or 64 megabytes. Tablet computers are becoming more popular in the construction industry because of the many E&C related applications available and their ability to provide access to documents and drawings stored on a remote server, which allows for the sharing of large files even while at construction jobsites.

Applications for smart phones and tablets are being developed and implemented in the construction industry to allow construction personnel to upload photos, audio, video, and text about projects and receive comments on this material from other personnel. Other applications allow construction workers to receive e-mails when someone has posted something relevant to the settings they have designated on their platforms (*Engineering News Record* 2013).

Tablet computers are being used more frequently at construction jobsites to access project plans and specifications; exchange project-related information with other jobsite and home office personnel; record project documentation; take and save photos or videorecordings of construction operations; and create, send, and receive project-related e-mails.

Computers contain both RAM and **read-only memory** (ROM). Each time a computer is turned on, it accesses the ROM, where permanent instructions are stored that allow the computer to install its **operating system**. Random-access memory is used to temporarily store programs and files. The **central processing unit** (CPU) is where calculations are performed and where data are manipulated and processed while programs are being run on the computer. Hard drives are where the software programs and files are stored. These programs and files are loaded into the RAM when users double-click on icons that represent the programs or files. Figure 9-1 is a diagram of the internal components of a desktop computer, and Figure 9-2 shows the relationship of computer hardware to the input-processing-output cycle. Some computers use flash memory rather than hard drives, and this increases the speed at which the computer is able to perform operations and store data.

For computers to run software, a set of instructions are factory installed into the ROM that allows an operating system to be loaded into the RAM. Once the operating system is installed, it enables the computer to load and run software applications. Operating systems contain computer code that provides instructions to the computer on how to operate.

Newer generations of computers operate using **dual-core processors**, rather than only one processor. Having two processors increases the speed of the computer, but to take advantage of the parallel processors, a computer requires sophisticated software programs that are difficult to write.

Graphical user interfaces, such as those used by Mac OS X and Windows, allow users to provide commands in a user-friendly graphical format rather than in **ASCII**, a code comprised of a series of zeros and ones. Compilers translate ASCII into **machine language**. All letters and numbers have a corresponding ASCII code that contains zeros and ones. Computers view the number 1 as *on* and a 0 as *off*, and they process data based on the series of electrical pulses represented by the zeros and ones of the ASCII code for each letter and number. One character is a **bit**, and eight

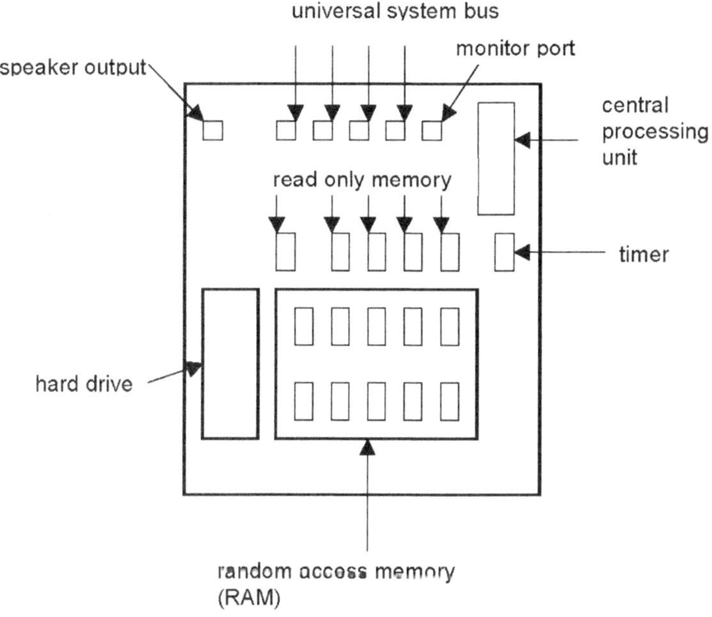

Figure 9-1. Diagram of Interior of a Personal Computer

bits equals one **byte**. Figure 9-3 shows an internal computer binary representation of the letter "A" using ASCII code.

The term *byte* is used to designate the storage capacity of hard drives, **flash drives**, files, and software programs. A **kilobyte** (KB) is a thousand bytes, a **megabyte** (MB) is a thousand kilobytes, a gigabyte (GB) is a thousand megabytes, and a **terabyte** (TB) is a thousand gigabytes. Very large storage capacities include petabytes (PB) (1,000 terabytes), an exabyte (EB) (1,000 terabytes), a zettabyte (ZB) (1,000 exabytes), and a yottabyte (YB) (1,000 zettabytes).

Computer Peripherals

Along with CPUs, RAM, and ROM, computers also require **peripherals** to send, receive, and print data. The basic peripherals required are a keyboard, a mouse or a touchpad, and a printer. In addition, to access the Internet, a computer needs a modem connection, a broadband connection, a high-speed Internet connection, a cellular connection, hot spots, satellite services, access to Wi-Fi, or a high-speed connection card. Other optional peripherals often used along with the standard peripherals to enhance the capabilities of computers are listed in Table 9-3.

Computers normally have one to four built-in **Universal System Bus** (USB) ports. These ports are used for connecting peripherals to the computer. If additional USB ports are needed, external USB extenders are available and when added to the system they provide four to seven or more additional ports. If the USB extenders do not have external power sources, they may not have adequate power to run printers,

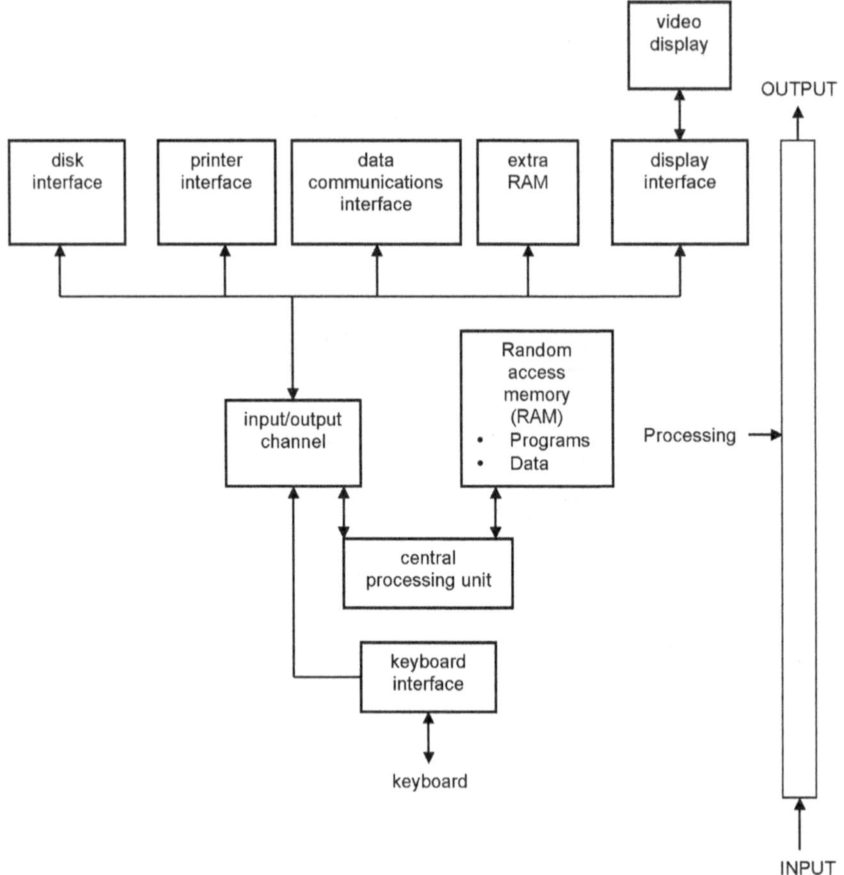

Figure 9-2. Relationship of Computer Hardware to Input-Processing-Output Cycle

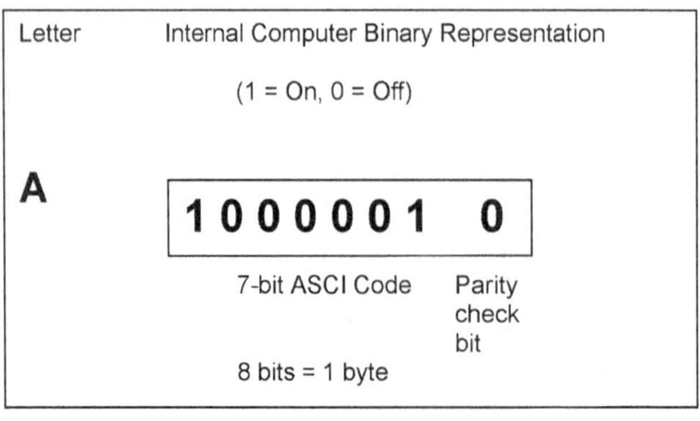

Figure 9-3. Internal Computer Binary Representation of the Letter *A* Using ASCII

Table 9-3. Examples of Optional Computer Peripherals

Peripherals
Hard-wired, wireless, or bluetooth devices, such as keyboard, mouse, speakers, and printers
Digital cameras
DVD drives
External CD drives
External flash-memory cards
External hard drives
Memory card readers
Photo printers
Presentation changers
Scanners
Speakers
Video cameras

scanners, or other optical devices, but they work when used for small peripherals, such as flash drives, keyboards, and mice.

9.3 Computer Networks

Local area networks (LANs) are used to provide members of firms with the ability to link computers for the purpose of sharing data or peripherals. **Wide area network** systems (WANs) provide the ability to share data outside firms and are either private or public. Some WANs operate all over the world. One of the first WANs was the Ethernet system established at universities in the United States with the extension (.edu), and this system is now a nationwide system. There are educational wide area network systems in other countries, but they do not use the .edu extension.

Wide area network systems use phone lines, the Ethernet, cable, fiber-optic lines, satellite, Wi-Fi, or microwave technology. There may be limitations imposed by network providers on data usage or the size of the files that can be sent as attachments. Some Internet service providers also reserve the right to slow down the average upload and download speeds of users that exceed the allowable limit on usage (usually data usage of 500 MB to 1 to 10 GB) to speeds below 1-4 Mbps (megabytes per second). Normal speeds range from 1 to 50 Mbps. If a company has its own WAN, this limitation may not be imposed on users, unless they are trying to prevent users from downloading videos or music, a restriction imposed by many universities. Microwave WANs and satellites require antennas or receivers to receive signals, and service reliability is dependent on the location of the receiver in relation to the towers transmitting and receiving signals or objects that obstruct the signals.

One option for high-speed file sharing is **file transfer protocol** (FTP) sites that allow the transferring of files between computers over the Internet. File transfer protocol sites allow a team of designers, construction managers, contractors, and staff members to access online in-process design drawings, project information, submittals, or other digitized documents. File transfer protocol can be used to transfer documents, manuals, and design drawings, and access to FTP sites is protected by passwords. File transfer protocol sites are set up as Internet sites, and files are uploaded and downloaded directly to the website rather than being sent through an e-mail service provider.

9.4 Computer Applications Used in Construction

The ability to analyze E&C productivity rates is augmented by the use of computer systems. Computers are used for recording and storing data, processing data, performing calculations, organizing databases, **querying** data, and modeling potential alternative workflow scenarios. Data queries allow users to search databases for specific items, such as a name or a title. Some of the ways that computers are used in construction are listed in Table 9-4.

Computers are typically used in construction to track tools, materials, labor, and equipment in order to evaluate the allocation of resources between projects. Computers are also used to schedule activities, labor, material deliveries, and equipment and to help monitor the progress of projects. Financial models are run on computers to monitor the costs associated with each of the elements in the work breakdown structure and to prepare budget status reports. Organization models, including executive information systems, information logs, and procedural monitoring systems that monitor the various phases of projects, are also developed on computers. Computers are used to expedite the processing of environmental models that show the environmental consequences of specific actions and also to help graphically illustrate field conditions and identify problems or compliance concerns.

Table 9-4. Examples of Uses for Computers in Construction

Uses for Computers in Construction
Logging change orders
Creating financial models
Creating organization models
Forecasting
Generating drawings
Generating project management reports
Investigating environmental impacts
Managing and scheduling equipment
Managing and tracking tools and materials
Procuring and expediting materials
Reviewing constructability considerations
Scheduling and monitoring projects
Tracking construction costs
Tracking labor expenditures

Table 9-5. Construction Jobsite Activities That Benefit from Computers

Construction Jobsite Activities
Budgeting and forecasting
Calculations
Cost control
E-mail correspondence
Equipment management
Estimating
Financial planning
Generation of designs using *AutoCAD* or other graphics software programs
Generation of graphics and presentations
Information systems management
Internet research
Inventory management
Material and labor requirement monitoring
Payroll and accounting
Personnel management
Progress planning and tracking
Scheduling
Simulation modeling
Word processing

Computers are used during the design phase for developing constructability models in three-dimensional (3D) CADD software (BIM software) to illustrate construction sequences, test alternative assemblies, and modify designs. Building information models allow users to model modifications to the design or location of elements. Examples of construction jobsite activities that use computer applications are listed in Table 9-5.

Management Information Systems

Management information systems (MIS) are used to monitor work progress at construction jobsites. They allow large quantities of data to be stored in computer databases for easy retrieval. Management information systems also provide a means for controlling and tracking data and information during construction operations. Examples of activities that are monitored through automated MIS are listed in Table 9-6.

Database Programs

Database programs allow users to access and store large amounts of data in either preselected or custom formats. In addition, they provide users with the capability to query data records to locate specific data and to link records to create new records. The main benefit of database programs is the ability to quickly retrieve information that has been stored in database files. Programs such as ***Primavera Project Management***

Table 9-6. Activities Monitored Using Management Information Systems at Construction J Jobsites

Activities	Activities
Accounting	Material inventories
Budgeting	Progress measurement
Change-order logs	Quality control
Cost control	Schedules
Equipment management	Tracking resources
Job tasks	Work orders
Maintenance management	Work quantities

(P6) and ***Microsoft Project*** use databases with a graphical user interface to visually represent data records as activities on a schedule. Database software programs contain standard reporting formats, or users may generate their own custom reports. Many of the software applications used in construction were developed in database platforms.

9.5 Radio Frequency Identification Data

Radio frequency identification data (RFID) programs use small tags—1 in.^{3} or smaller—placed on equipment, materials, tools, or workers to track the tagged item or worker using an *x–y* coordinate system to determine the exact location of the tag. The tags act as a transponder to emit microwaves or ultra high frequency (UHF) radio waves. Radio frequency identification is an automatic identification and data capture (AIDC) technology that includes 1D and 2D bar codes. RFID uses an electronic chip embedded in an item. The information it contains may be read, recorded, or rewritten. Radio frequency identification systems are currently being used to monitor jobsite safety, to locate and track materials, and to monitor job progress. There are additional applications of RFID systems in other industries such as tracking the manufacture of an automobile through an assembly line, tracking pharmaceuticals in warehouses, and tracking lost pets.

9.6 Global Positioning Systems

A **global positioning system** (GPS) allows users to determine their exact location, in terms of longitude and latitude, by using equipment that receives satellite signals and performs triangulation calculations to determine locations. The distance from the satellites to a position is determined by the amount of time it takes radio signals to travel from the ground to the satellites in the triangulation system. Global positioning systems are being used in construction in applications such as electronically locating equipment and structures at construction jobsites, for leveling, and for surveying.

9.7 Building Information Modeling

Many owners and architects are using BIM software on their projects, and some E&C firms are being contractually required to use these 3D modeling technologies. In 2014, the current leaders in BIM technology were AutoDesk, with its *Revit* suite of programs, and *Microstation* by Bentley Systems. Both of these firms provide software that allows the importation of different platforms and formats for design drawings, and 3D models are generated that incorporate the contributions of all of the designers. Other software systems that are **aggregate model viewers** and **conflict resolution tools** are AutoDesk's *NavisWorks*, Bentley's ***ProjectWise Navigator***, ***VICO Contractor***, **Graphisoft's *Virtual Building Explorer***, and ***Tekla Structures***. Figure 9-4 shows a building information model rendering of a brick office building, Figure 9-5 shows a rendering of the inside of the building, and Figure 9-6 provides a section view of the building.

Building information modeling software also allows users to create **four-dimensional** (4D) **schedules** that generate the 3D models in predefined scheduling sequences and **quantity take-offs** that are automatically generated from 3D models. When contractors have the capability to generate simulations of the building process in 3D modeling software, they are able to experiment with different construction sequences for building operations. Being able to test different construction-sequencing scenarios in 3D models allows contractors to initiate and test different work-process flows for construction operations.

Figure 9-4. BIM Rendering of Brick Office Building

Source: Bungert (2009); reproduced with permission.

Figure 9-5. BIM Rendering of Inside of Office Building

Source: Bungert (2009); reproduced with permission.

Figure 9-6. BIM Rendering of Section of Office Building

Source: Bungert (2009); reproduced with permission.

One feature of the BIM software *NavisWorks* by AutoDesk that benefits contractors is **clash detection**, or the **highlighting of interferences** between building elements, and the generation of clash-detection reports. Projects using the clash-detection feature in BIM software experience a significant decline in **change orders** during construction because construction interferences are discovered during the design phase. In addition, construction methods and processes are modeled to determine their viability. With a reduction in the number of change orders issued during construction, the number of claims is also reduced, thus saving owners and contractors money. Even though the initial cost of BIM software seems prohibitive to small and medium-size firms, the cost savings realized through its use are leading to its adoption throughout the E&C industry.

Integrated project delivery (IPD) has led to the development of **collaborative agreements**, which are now available through the **American Institute of Architects**, the **U.S. Army Corps of Engineers**, and other entities and include agreements from **Collaborative Agreements:ConsensusDOCS** LLC, such as ConsensusDOCS 300 Standard Form of Tri-Party Agreement for Collaborative Project Delivery (2007). Agreements that are specific to BIM applications include the following:

1. ConsensusDOCS 301—2008 BIM Addendum,
2. American Institute of Architect's E202—2008 Building Information Model Exhibit and Protocol,
3. Army Corps of Engineers Building Information Modeling Road Map (October 2006),
4. Associated General Contractors Guide to BIM.

Because BIM technology has been available for only a few years, the legal ramifications of its use are not known, and the lawsuits that will eventually set legal precedents are currently being processed. Potential areas for disputes include the following (Salmon 2009):

- Liability issues related to generating 3D models using drawings from several different firms and responsibility for errors,
- Disputes arising from software incompatibility and inaccurate data being uploaded into the BIM software as a result of incompatibility,
- Disputes arising from changes made after drawings are uploaded into the 3D model on the two-dimensional (2D) drawings,
- Disputes from changes resulting from clashes between construction elements designed by members of different firms,
- Disputes resulting from how construction elements are displayed in a 3D model versus how they are shown on individual engineering or shop drawings, and
- Disputes arising when the engineering drawings do not contain all of the changes that were made on the 3D model (although the 3D modeling software is supposed to transpose the 3D model into accurate 2D drawings).

All of these potential disputes might arise on projects that use BIM software programs, and because there are currently no precedent law cases, each new dispute will have to be decided on its own merits until precedents are established by the legal system.

9.8 Construction Operations Simulation Modeling Software

Simulation modeling software allows users to incorporate activities into a model that occur randomly, rather than in set sequences, such as the arrival of trucks that travel through traffic from a dump site to the jobsite. Simulation programs allow users to generate flow diagrams for various procedures, such as paving operations, earthmoving operations, and masonry wall construction. They allow for the incorporation of time and probability distributions of various times and for the development of simulation cycles, which are analyzed to determine optimal cycles. *WebCYCLONE* and *Stroboscope* are two of several simulation programs used to model construction processes that are able to test each proposed adjustment to the work process or flow (Halpin 1990a, b; Martinez 1996). Simulation modeling programs are used to evaluate construction operations such as (Halpin 1990b):

- Asphalt paving operations,
- Concrete-column placing operations,
- Earthmoving operations,
- Loader and truck haul operations,
- Masonry operations,
- Multiple-frame welding operations, and
- Precast concrete operations.

Some of the simulation software programs provide standard templates for construction operations and generate graphic representations of the work processes. Commercially available simulation modeling software programs used to model construction operations are listed in Table 9-7.

The construction operations simulation modeling programs use **graphical elements** to represent construction operations. Typical elements used in these programs are shown in Figure 9-7.

Table 9-7. Examples of Computerized Construction Simulation Software Applications

Simulation Software	Simulation Software
ANSYS	*Revit*
ArchiCAD	*Simscript II*
BAM Construction	*Simulation System*
Construction Simulation INT	*SLAMS*
Construction Simulator	*Stella*
GPSS	*Stroboscope*
Insight	*Symphony.NET*
MicroCYCLONE	*Synchro 4D*
NavisWorks	*VEH SIM*
PROSIDYC	*WebCYCLONE*

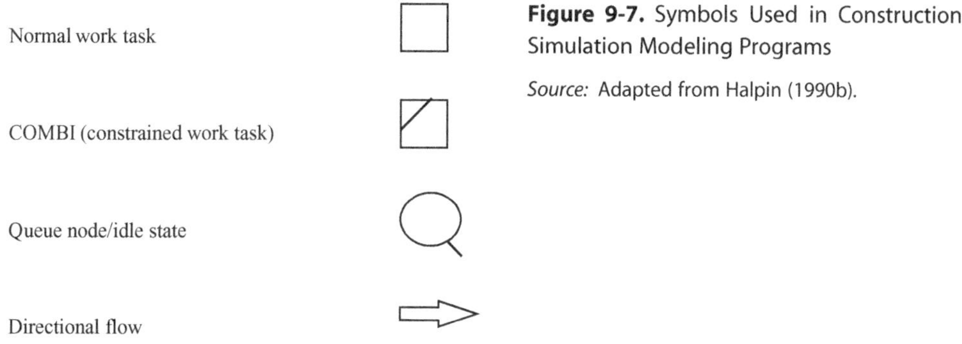

Figure 9-7. Symbols Used in Construction Simulation Modeling Programs

Source: Adapted from Halpin (1990b).

The Normal Element

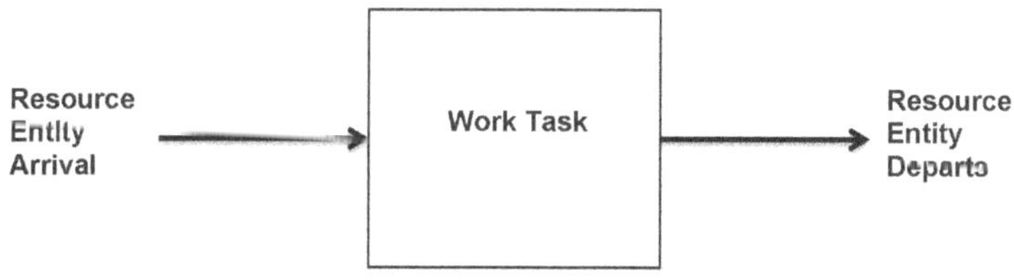

Work Task Time Delay Function

Figure 9-8. Normal Element

Source: Adapted from Halpin (1990b).

The Normal Element

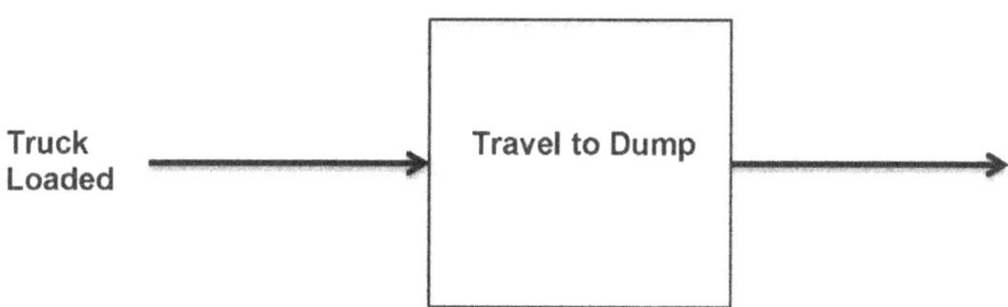

Typical Use of the NORMAL Modeling Element

Figure 9-9. Normal Element with Arrival Resources

Source: Adapted from Halpin (1990b).

Normal elements are construction processes or components of a construction process identified as separate elements (Halpin 1990a). Figure 9-8 shows the normal element with resources arriving and resources departing, and Figure 9-9 shows the normal element with a loaded truck as the arrival resource and the work task as the truck traveling to the dump site. Work tasks each have a time-delay function that is defined by users. This function describes the transit time required for the work task and links this information to the other resources that interact with the work task.

Combi elements include additional precedence requirements, particularly input requirements that have to be completed before a work task is able to start in a cycle. Figure 9-10 shows a combi element with multiple resources arriving, a work task that incorporates the multiple resources, and the resource entity departing after the task (Halpin 1990a). Figure 9-11 shows a combi element with an empty truck arriving and a front-end loader placing dirt into the truck. Once the task has been completed, the loaded truck exits, and the front-end loader is once again available to load another truck. Figure 9-12 shows the same process using simulation symbols.

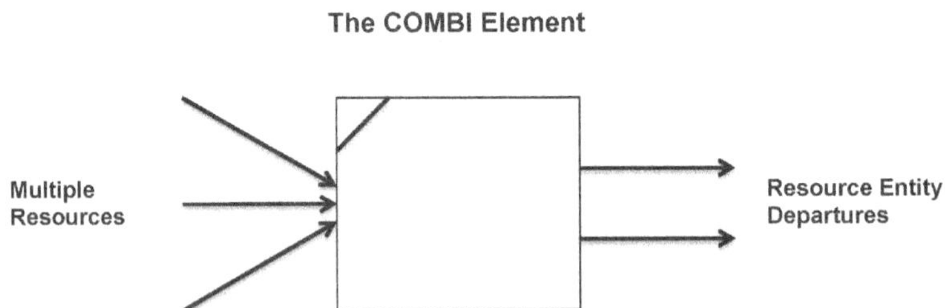

Figure 9-10. Combi Element with Multiple Resources Arriving

Source: Adapted from Halpin (1990b).

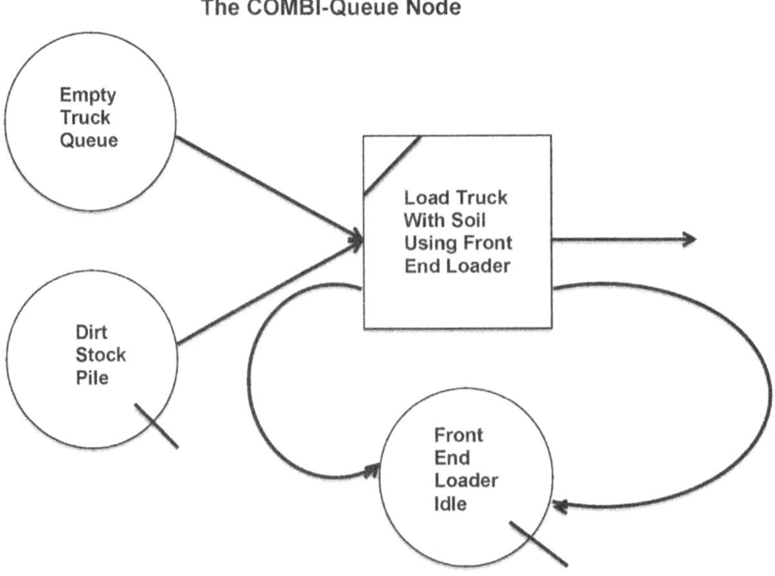

Figure 9-11. Combi Element with Empty Truck Arriving

Source: Adapted from Halpin (1990b).

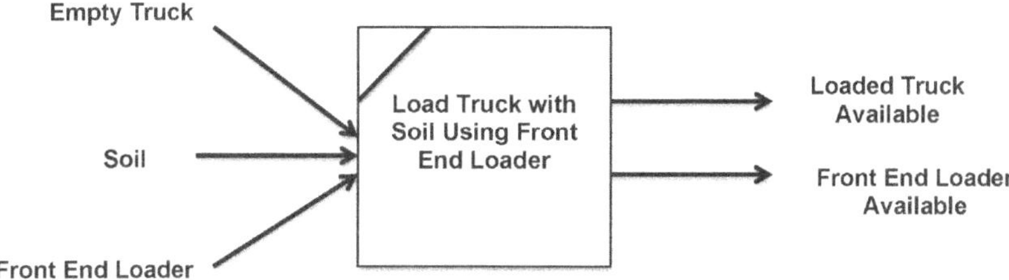

Figure 9-12. Combi-Queue Element Combination

Source: Adapted from Halpin (1990b).

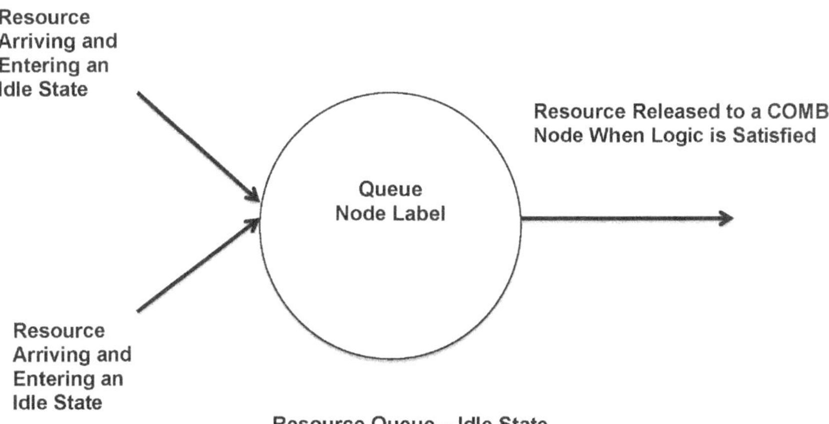

Figure 9-13. Queue Element with Resources Arriving

Source: Adapted from Halpin (1990b).

A **queue element** indicates a period when equipment is delayed or not active, is waiting to move into a queue, or is being constrained by another activity that has to be performed before the next element moves forward. Resources are released one at a time from the **idle state** as soon as all the precedence requirements are met (Halpin 1990b). Figure 9-13 shows the queue element with resources arriving and then waiting until they are released to a combi node when the combi node is available. Figure 9-14 depicts the queue element with a loader completing the loading activity and then becoming idle until another truck is available.

These basic modeling components are used to represent the logic for simulations. One additional symbol is the **cycle-counter element**, which is used to count the

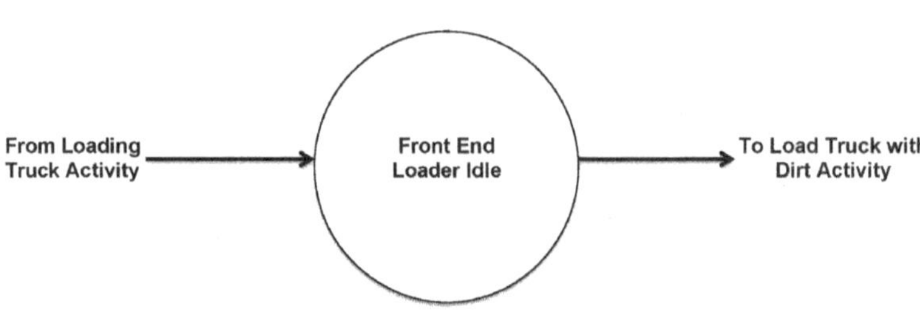

Figure 9-14. Queue Element with Loader Idle

Source: Adapted from Halpin (1990b).

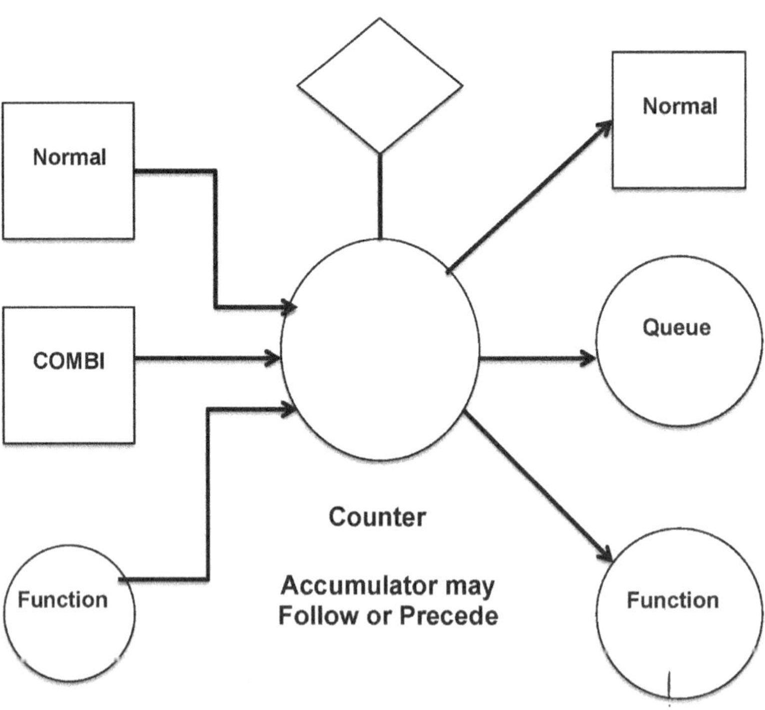

Figure 9-15. Counter Element

Source: Adapted from Halpin (1990b).

number of times a unit passes a particular control point in the network model or the system moves through a cycle. This count is used to measure production and to control the number of times the system cycles before stopping or shutting down (Halpin 1990b). Figure 9-15 shows a counter element with each of the other elements leading into and out of it.

Directional flow indicates the direction in which the activity is traveling or moving in the work processes. Directional flow is represented by an arrow that indicates where the resources are moving to reach the next element. The length of an arrow does not indicate a time segment; it merely represents a connection between activities (Halpin 1990b).

Computer simulation models and a sample computer simulation case study are discussed in Chapter 10.

9.9 Summary

This chapter provided information on computer hardware and software programs used to generate the data incorporated into productivity improvement studies. It explained how hardware and software programs are currently being used at construction jobsites to monitor construction operations. Computer software programs were discussed that could be used during productivity improvement studies to help analyze specific work processes. These include WANs, MIS, database systems, RFID, GPS, and simulation modeling software, such as *WebCYCLONE*, *Stroboscope*, and *GPSS.*

Building information modeling and 3D modeling software programs currently available were introduced in this chapter, along with simulation software programs. A simulation software program is used in a case study presented in Chapter 10 that demonstrates how these programs are used for simulating heavy construction equipment operations.

9.10 Key Terms

aggregate model viewers
American Institute of Architects
ASCII
bit
byte
central processing unit
change orders
clash detection
cloud
collaborative agreements
Collaborative Agreements:ConsensusDocs
combi elements
computer-aided design and drafting
conflict resolution tools
cycle-counter element
database programs
desktop computers
directional flow
dual-core processors
file transfer protocol
flash drives
four-dimensional schedules
gigabyte
global positioning systems
graphical elements
graphical user interfaces
Graphisoft's Virtual Building Explorer
hard drive
highlighting of interferences
idle state
Insight
integrated project delivery
kilobyte

local area networks
machine language
mainframe computers
management information systems
megabyte
MicroCYCLONE
Microsoft Project
Microstation
mini computers
NavisWorks
normal element
notebook computers
operating system
peripherals
Primavera Project Planner
ProjectWise Navigator
quantity take-offs
querying
queue element
radio frequency identification data
random-access memory
read-only memory
Revit
Simscript II
simulation modeling software
smart phone
Stella Modeling and Simulation Software
Stroboscope
super computers
Suretrack
tablet computers
Telka Structures
terabyte
three-dimensional models
Timberline
U.S. Army Corps of Engineers
Universal System Bus
Vico Contractor
WebCYCLONE
wide area network systems

9.11 Exercises

9.1 Which computer software programs could be used during a productivity improvement study, and what could they be used for during the study?

9.2 Which types of construction operations are modeled using currently available computer software programs?

9.3 Are desktop or laptop computers powerful enough to model simulations of construction equipment operations? Why or why not? Provide data to support your conclusions.

9.4 Why might it be necessary to have access to Internet file services such as file transfer protocol to send large data files during construction projects?

9.5 What types of construction operations can be modeled using a program such as *WebCYCLONE*?

9.6 Why are the capabilities of radio frequency identification systems useful during construction operations?

9.7 How are databases and management information systems used in the programs that support construction operations?

9.8 Which type of Internet connection provider is the most reliable and why?

9.9 Why is building information modeling useful to contractors?

9.10 What types of operations can be modeled using simulation modeling software?

References

AutoCAD (Computer software). (2013). San Rafael, CA AutoDesk.

Bungert, T. (2009). Building Information Modeling Renderings, unpublished work. North Dakota State University, Fargo, ND.

Engineering News Record. (2013). "'Facebook of construction' uses social tools to run jobs." 271(13), 30.

General Purpose Simulation Software (*GPSS*) (Computer software). (1960). White Plains, NY, International Business Machines (IBM).

Halpin, D. (1990a). *MicroCYCLONE System Manual,* Purdue University, West Lafayette, IN <http://engineering.purdue.edu/CEM/People/Personal/Halpin/Sim/MicroCYCLONE/Index.html> (March 2014).

Halpin, D. (1990b). *MicroCYCLONE User's Manual,* Purdue University, West Lafayette, IN <https://engineering.purdue.edu/CEM/People/Personal/Halpin/Sim/MicroCYCLONE/Index.html> (March 2014).

Insight (Computer software). (2013). Cary, NC, SAS Institute.

Martinez, J. C. (1996). "Stroboscope: State and Resource Based Simulation of Construction Processes." Ph.D. dissertation, University of Michigan, Ann Arbor MI.

MicroStation (Computer software). (2013). Exton, PA, Bentley Systems.

NavisWorks (Computer software). (2013). San Rafael, CA, AutoDesk.

Primavera Contractor (Computer Software). (2013). Irvine, CA, Oracle.

Primavera P6 Professional Project Management (Computer software). (2013). Redwood Shores, CA, Oracle.

Revit (Computer software). (2013). San Rafael, CA, AutoDesk.

Salmon, J. (2009). "The legal revolution in construction." *Journal of Building Information Modeling,* 3(1), 18–19.

Simscript II (Computer software). (2013). San Diego, CA, CACI Advanced Simulation Lab.

Stella Modeling and Simulation Software (Computer software). (2013). Lebanon, NH, Isee Systems.

Timberline: Sage Construction and Real Estate Software (Computer software). (2013). Irvine, CA.

CHAPTER 10

Computer Simulation Models for Productivity Improvement Studies and Sustainable Heavy Construction Equipment

This chapter elaborates on the concept, introduced in Chapter 9, of using computer simulation modeling software for analyzing construction operations. Construction activities progress at varying speeds, and they are constantly being interrupted and delayed; therefore, modeling construction processes using simulation software programs is beneficial because it provides a method for analyzing the productivity of construction operations. Simulation modeling helps identify potential delays, provides data on balancing equipment and the workforce, assists in the selection of equipment, and potentially helps reduce the amount of time that resources are idle. Computer **simulation models** also provide the ability to analyze specific construction processes.

This chapter discusses the use of computer simulation software for modeling and analyzing construction operations and the application of the queuing theory. A case study is discussed that demonstrates how the simulation modeling software ***General Programming Simulation System*** (*GPSS*) is used to optimize heavy construction equipment fleets. The *GPSS* software program was used for the case study because it is used in many different industries, not just in the construction industry.

Some of the terms relevant to the information discussed in the first part of this chapter require definitions. A **simulator** is "a training device that duplicates artificially the conditions likely to be encountered in some operation" ("Simulator" 2006, p. 1693). A **system** is "a set of facts, principles, rules, etc. classified or arranged in a regular, orderly form so as to show a logical plan linking the various parts" ("System" 2006, p. 1953). Models are used to help describe, design, and analyze systems and to help forecast the outcome of future operations. Some **physical models** include scaled physical objects, such as **iconic models** and mathematical equations and relations, and **abstract models** include graphical representations of data or processes and visual models.

The second part of the chapter discusses sustainable heavy construction equipment and hybrid-electric heavy construction equipment, such as the Caterpillar electric dozer, the Komatsu excavator, the John Deere diesel-electric-hybrid wheel excavator, and the Peterbilt hydraulic-electric truck. It also provides information on tires that help increase the sustainability of heavy construction equipment and the use of biodiesel fuel products to enhance the capabilities of heavy construction equipment. The EPA Tier Four Final Standards are mentioned along with cooled exhaust gas recirculating and selective catalytic reduction systems. Engine repowering, engine upgrading, and diesel-retrofit technologies are introduced to demonstrate alternatives to purchasing hybrid-electric heavy construction equipment.

10.1 Computer Simulation Models

Computer simulation models are used during the design of structures, for procedural analysis, and for performance assessment. Table 10-1 provides examples of where simulation models are used in different industries.

Computer simulation models are also used as **enterprise models** for steel production, hospitals, shipping lines, school districts, facility layout, maintenance scheduling, parking facility design, spacecraft trips, steel mill scheduling, taxi dispatching, and water resources development. Engineers use building information modeling (BIM) software to evaluate and design structures and to test alternative configurations for structural and other elements. Building information modeling is used to test new design concepts, rework structural elements, and provide what-if analysis of alternative scheduling sequences. Building information modeling software programs are used to model construction operations, but they do not provide **simulations** that are time dependent, as is possible with other computer simulation models.

Construction management personnel use computer simulation software to model construction operations, and this helps them make decisions on how to design or alter the design of construction operations. Simulation models are used during construction to

Table 10-1. Industrial Applications of Simulation Models

Applications	Applications
Advertising allocation	Communication system design
Air traffic control and the queuing of airplanes	Consumer behavior prediction
Aircraft maintenance scheduling	Critical path method scheduling
Airport design	Dispatching
Ambulance location and routing	Distribution system design
Assembly-line scheduling	Harbor design
Bank teller scheduling	Industry models
Brand selection	Promotion decisions
Bus scheduling	Railroad operations
Circuit design	Shipyards
Clerical processing system design	Urban traffic system design

Figure 10-1. Earthmoving Operations for Trucks Being Loaded by Loader

Source: Adapted from Halpin (1990b).

- Help explain systems or problems,
- Determine which elements or components are critical to specific construction operations,
- Help synthesize and evaluate alternative solutions, and
- Help forecast future operations.

In construction, computer simulation models are also used to analyze construction operations without having to physically perform different scenarios involving labor or equipment. Figure 10-1 shows a pictorial representation of the construction operation of trucks being loaded by a loader, traveling to a dump site, dumping their load, and returning to the **queue** to be loaded once more by the loader.

To develop a computer simulation model for the heavy construction equipment operation shown in Figure 10-1, the modeling elements shown in Figure 9-7 are used to represent the different construction equipment processes. Figure 10-2 shows the separate components of the same earthmoving operation using simulation modeling software symbols, and Figure 10-3 shows the symbols combined into the total operation (Halpin 1990b, p. 15).

10.2 *WebCYCLONE* Computer Simulation Modeling Software

One computer simulation modeling software program that allows users to model construction operations is *WebCYCLONE.* This program was developed from the computer program *MicroCYCLONE,* which was created at Purdue University by Dr. Daniel Halpin (1990a, b). According to Halpin (1990b), *WebCYCLONE* is composed

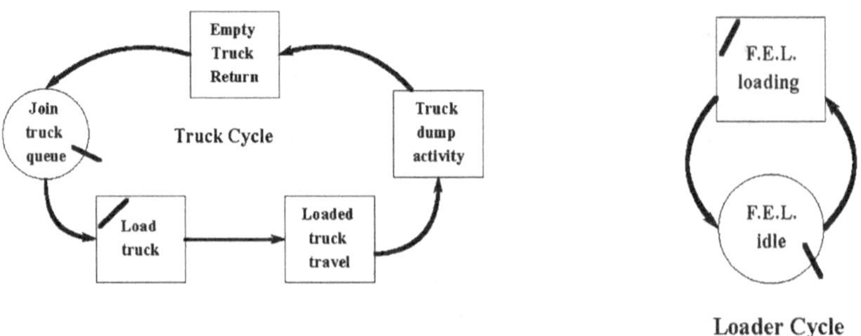

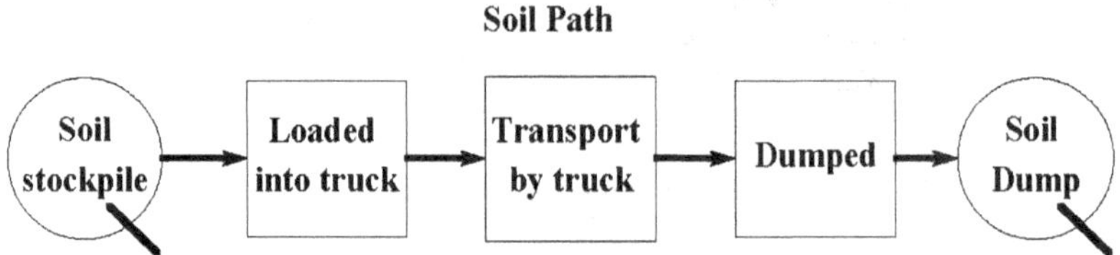

Figure 10-2. Computer Simulation Symbols for Earthmoving Operation

Source: Adapted from Halpin (1990b).

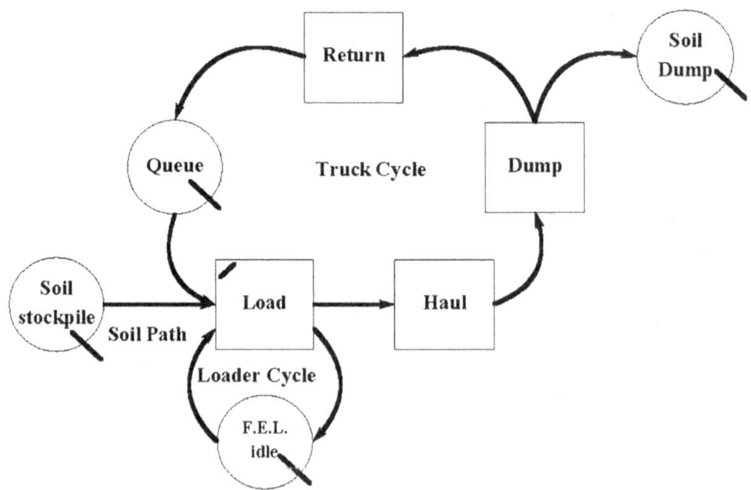

Figure 10-3. Complete Earthmoving Operation

Source: Adapted from Halpin (1990b).

of a series of independent modules, each of which is in control of a particular segment of the overall system. There are four different modules:

1. Data input module
2. Simulation module

3. Report generation module
4. Sensitivity analysis module

WebCYCLONE uses the same elements—combi elements, normal elements, queue functions (idle)—as the elements shown in Figure 9-7. For detailed information on the *WebCYCLONE* program, see the user's manual available online (Halpin 1990b).

10.3 Queuing Theory Models

In construction, computer simulation modeling software is used to assist in solving **queuing theory models**, which are models that attempt to predict the **queuing** (arrival and waiting) time for vehicles waiting in a queue. In construction, queuing theory models apply mainly to heavy construction equipment waiting to be serviced by another vehicle—for example, trucks waiting in line to be loaded by a loader or a scraper waiting for a dozer to push it. The goal of queuing theory models is to optimize the use of heavy construction equipment so that the equipment is not waiting to be serviced and the servicing equipment is not waiting for the other vehicles to arrive for service.

When users work with queuing theory models,

> the mathematical queuing model may take one of several different forms, depending on the characteristics of the problem. The number of arrival units may be finite or infinite. The arrival units may arrive at a deterministic rate or according to some probability distribution—a Poisson process, an exponential process, and so on. In addition, the arrival patterns may either vary with time or be continuous. Service patterns may also be varied. For example, the service rate may be constant, probabilistic and subject to a distribution (such as an exponential distribution), or affected by queue length. There may be more than one server. Queue discipline may also vary from one problem to the next. For example, the first arrival unit may be the first to be served, or the last arrival unit may be served first. So arrival units may be given priority in the queue. (Adrian 2004, p. 399)

Queuing theory problems are developed to include a scenario where the problem is optimized by, for example, determining the number of trucks that are able to be serviced by one loader. If the loader is able to fill the designated number of trucks with soil faster than the trucks are loaded, travel to the dump location, dump the load, and return to the loader location, then the loader will control the operation. Therefore, an optimal number of trucks should be used to keep the loader busy the entire time it is loading trucks. Various combinations of loaders and trucks are tested using computer queuing simulation models. The next section provides an example of a queuing theory computer simulation model.

10.4 Modeling Heavy Construction Equipment Earthmoving Operations

Productivity improvement studies are also used to help increase the production rates of heavy construction equipment. Earthmoving operations are complex because they involve humans, machines, time, resources, and constraints, and using computer software programs to evaluate the interactions required for these operations enhances productivity-improvement studies.

Selecting heavy construction equipment and fleets for specific earthmoving operations and estimating productivity rates and costs are accomplished using one of two methods. The most frequently used method is **deterministic** and involves assigning a specific value for each equipment cycle in the process and then adding up the values for all of the cycles required to complete an activity to find the total operating cost. The total operating time is estimated by adding up all of the cycle times. The second method used for estimating fleet sizes and costs is referencing historical data, production rates provided by equipment manufacturers, and equipment production tables. Some of the factors analyzed to determine optimal fleet sizes are

- The type and quantity of materials to be hauled,
- The type of material to be excavated,
- Haul and return road length,
- The condition of the road surface,
- Grade and rolling resistance,
- Weather,
- Equipment availability, and
- Operator skill.

Other **stochastic variables** (variables whose state is nondeterministic so that the subsequent state of the system is determined probabilistically) that influence cycle time are travel and loading time variability and the irregular interarrival of trucks to loading areas.

10.5 Factors Considered in Equipment Selection

Several factors influence decisions related to heavy construction equipment and operators. Among them are the performance of the equipment alternatives, the type of earth or material, location of the project and soil conditions, length of the haul road, grades and rolling resistance of the road, the condition of the haul road, the volume of material to be moved, job duration, and job cost, including operators and fuel. Other considerations are traffic flow, noise levels, environmental protection, and working hours (Bernold 1986). This information, along with the organization of the work, is used to develop alternative work processes that help lower the cost of operations; however, determining the most efficient combination of equipment to perform specific work is difficult without the assistance of computer software simulation models. Estimating the cost for earthmoving operations is accomplished by

- Using deterministic methods where a fixed value is assigned to each cycle of a unit included in the alternative and the total cycle costs are summed to arrive at the total equipment cost;
- Estimating the overall cost of the operation from past experience, with no concern for factors such as lengths of roads, grade and rolling resistances, types of soil, cycle times, or cost estimates; or
- Using simulation models where all of the factors related to productivity and cost, as well as stochastic variables, are considered to determine the estimated cost.

10.6 Types of Simulation Models Used in Construction

Production times and cost are sometimes estimated **heuristically** (that is, on the basis of previous experience) as well as by applying mathematical techniques because, in the loop system under analysis, the loading task controls the entire process. Using simulation models provides a more efficient method than manual calculations for comparing different alternatives and for selecting the most effective alternative.

The two types of models used to analyze construction processes are

1. Physical models, or **conceptual models**, such as scheduling networks, where activities are represented by lines or nodes that are arranged according to their estimated start and finish times as well as their precedence to other activities (Halpin and Woodhead 1976), and
2. Queuing models, where the units to be processed in a given situation are thought of as either waiting in line, such as trucks waiting to be loaded, moving materials, or traveling empty (Halpin and Woodhead 1976).

Also, models are used to represent **static** (in equilibrium) or **dynamic** (characterized by constant change or activity) situations. One example of a static model is the graphical modeling program *CYCLONE* (*Cyclic Operations Network System*), which is used to model earthmoving processes (Halpin 1990a and 1990b). A conceptual model of earthmoving operations includes loader cycles in combination with truck-hauling cycles. In a queuing model, the loader (server) in an earthmoving operation is represented as keeping a certain position and the trucks move in and out of the system. The arrival of trucks is assumed to be exponential, and the production of the system is determined by the following equation (Halpin and Woodhead 1976):

$$\text{production} = L(1 - P_0)\,\mu\, C = L(\text{PI})\,\mu\, C \tag{10-1}$$

where P_0 = probability that no units are inside the system; μ = processor rate (loads/hour); C = capacity of the unit loaded; L = period of time considered; and PI = productivity index.

The **probability** (P_0), or the measure or estimation of the likelihood of an event occurring, is determined by writing the equations of state for the system and solving for the values of p_i ($i = 0, M$). These values could also be determined by using nomograph charts.

Queuing models allow for the evaluation of the losses caused by irregular arrivals of units, which occur when hauling units are waiting to be loaded by excavators. The advantage of using queuing models is that they allow for the evaluator to explore alternatives using different combinations of heavy construction equipment. To set up a queuing model, all of the factors relevant to earthmoving operations should be included in each of the simulations. The type of materials to be hauled, the loading capacity of the excavator, and the power required to perform the operation are all factors that influence the duration and costs of earthmoving operations (Karafiath 1988).

10.7 Using Simulation Models

Simulation models are used to model complex situations because they mimic the behavior of a system (Carrie 1988). Simulation models generate random numbers (RN) to represent the stochastic variations that occur during construction. Models of earthmoving operations are useful because they generate data by combining the travel times of various types of equipment, excavating and loading times, and dumping times. Computer simulations allow users to concentrate on the logic of the system to be modeled, rather than on computer programming, thus allowing users to develop the models in shorter periods than is required to write original computer code.

The first **discrete simulation languages**, developed in the 1950s, were capable of the following (Carrie 1988):

- Creating random numbers;
- Creating random variables, such as arrival times, interarrival times, loading times, travel times, and dumping times;
- Simulating the movement of time;
- Performing statistical calculations;
- Recording **transactions** that are in models;
- Saving data;
- Detecting errors; and
- Controlling the output.

In the United States, computer simulation models were first used in the steel industry. The first simulation software program was the ***General Simulation Package*** (*GSP*). The first simulation language developed for the steel industry was *GASP*, and this was followed by *SIMSCRIPT*, which permitted event-to-event simulations with discrete logical processes. Other simulation languages are *SLAM*, *SLAM-PC*, *MIDAS*, and *CSML*. The *GPSS* software program, developed for engineering design applications, allows for the solving of nonlinear differential equations.

General Purpose Simulation System

This section discusses the *General Purpose Simulation System (GPSS)* software and how it is used to model heavy construction equipment fleet operations. Other simulation models perform the same functions as *GPSS*, but *GPSS* was selected for this study because of its generic capabilities and its similarity to other commercial simulation modeling programs. This software is also used to simulate production lines in the manufacturing industry and to simulate mining operations.

International Business Machines (IBM) released *GPSS* in 1961. The software allows the use of short statements rather than lengthy commands; two commands in *GPSS* are equal to 10 procedural statements in other programs. The *General Programming Simulation System* is used to model discrete-event simulations, where a model changes only at discrete times but allows for random changes (Schriber 1974).

General Programming Simulation System models include block diagrams and relevant statements. The statements represent specific actions that will be performed in each of the operations modeled. Models have the following elements (Schriber 1974):

- *Block diagrams*: Flowcharts of programs
- *Block statements*: Statements corresponding to the block diagrams
- *Control statements*: Additional types of statements, such as "Start" and "End," to help control the start and end of the simulations
- *Compile directives*: Information on the model used in the compilation

The block diagram is a collection of boxes with different shapes connected in a one-way path. Each block represents a process in the operation—for example, the GENERATE block represents the creation of a transaction in the model, such as trucks arriving. The *General Programming Simulation Software* has 60 block types, each with a specific rule. Some of the blocks, with their corresponding operations and operands, are shown in Figure 10-4.

10.8 Simulation Model Execution

In simulation models, blocks are used to represent resources, information, and decision-making processes. The movements of transactions (units) from one block to another are analogous to the movements of units in the real world. Transaction movements in *GPSS* consist of two phases: (1) the **scan phase** and (2) the **clock update phase**. The scan phase updates the state of the model at the simulated time. The clock update phase advances the clock to the earliest future time at which a transaction has been scheduled to move. The scan and clock update phases are performed alternately by *GPSS* as the overall transaction movement phase proceeds (Schriber 1991).

Each transaction (unit) in a model has a current block, a next block attempted, and additional transactions as part of its life cycle as it moves from block to block. A transaction might stop temporarily at its current block because the next sequential

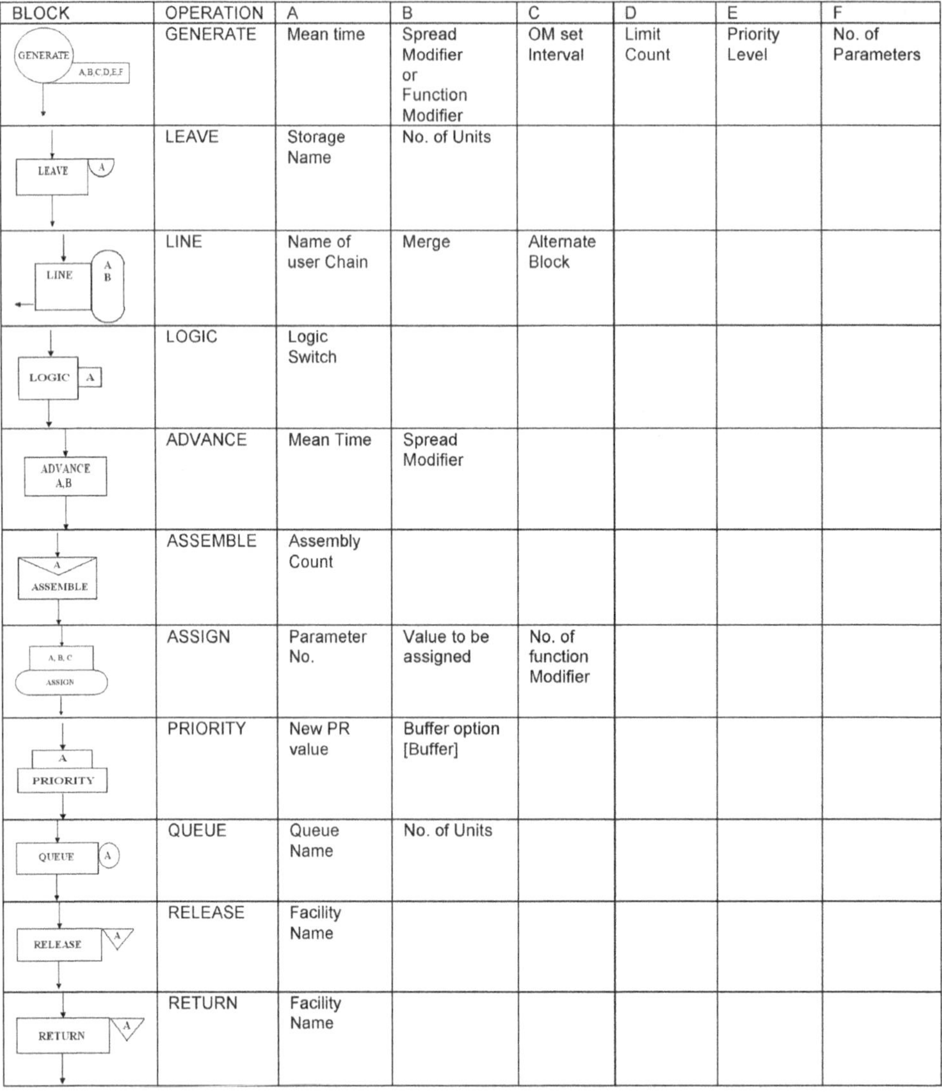

BLOCK	OPERATION	A	B	C	D	E	F
GENERATE A,B,C,D,E,F	GENERATE	Mean time	Spread Modifier or Function Modifier	OM set Interval	Limit Count	Priority Level	No. of Parameters
LEAVE A	LEAVE	Storage Name	No. of Units				
LINE A B	LINE	Name of user Chain	Merge	Alternate Block			
LOGIC A	LOGIC	Logic Switch					
ADVANCE A,B	ADVANCE	Mean Time	Spread Modifier				
A ASSEMBLE	ASSEMBLE	Assembly Count					
A, B, C ASSIGN	ASSIGN	Parameter No.	Value to be assigned	No. of function Modifier			
A PRIORITY	PRIORITY	New PR value	Buffer option [Buffer]				
QUEUE A	QUEUE	Queue Name	No. of Units				
RELEASE A	RELEASE	Facility Name					
RETURN A	RETURN	Facility Name					

Figure 10-4. Samples of *GPSS* Blocks and Their Specifications

Source: Adapated from Schriber (1974), p. 154.

block is engaged with other transactions. After a transaction has executed several blocks, and its cycle is about to end, it will move through the final block, which ends the transaction; this final block is called the TERMINATE block. However, a transaction might circle through the TEST or TRANSFER blocks more than once to represent a certain action, such as a truck going back and forth to load and dump material until a specified time. The segment of the model representing time is composed of a GENERATE block with the time of simulation specified as one unit. A unit of time could be hours, minutes, seconds, or a fraction of a second.

In the time segment model, the TERMINATE block indicates that after the time specified has been reached, the simulation stops. The TERMINATE block has a termination-counter value to indicate the number of times the model should be executed. When the termination counter has decreased to zero, the execution process stops.

The transactions in simulation models are represented by a chain of events. At a given time, a transaction is resident on exactly one of several alternative chains and located at the same time in a block. A *GPSS* chain has a front end and a back end to indicate that the transaction is either at the front of the chain or at the back of the chain, depending on its arrival at the block and its priority specified in the GENERATE block or a PRIORITY block. There are five classifications of chains (Schriber 1991):

1. Current event chain (CEC),
2. Future event chain (FEC),
3. User chain (UC),
4. Interrupt chain (IC), and
5. Link chain (LC).

Analytical Procedures for Simulation Models

Computer simulation programs are used to model the interaction among loader cycles, loaded trucks, and truck-hauling cycles. Figure 10-5 shows a flowchart of the steps for constructing and processing computer simulation models.

The information required to build computer simulation models for heavy construction equipment operations is as follows:

- *Equipment specifications*, including equipment type and size, loading capacities, weights empty, and weights loaded;
- *Road conditions*, including grades for road segments and road-surface type, which are used to determine the grade and rolling resistance and average speeds from the tables in the manufacturer's performance handbook for each segment of the roads;
- *The distance of each segment of the road*, to determine average travel times for each segment;
- *The average loading time* for each truck; and
- *Fixed time*, including spot, maneuver, and reversal of direction.

10.9 Case Study: Heavy Construction Equipment Computer Simulation Modeling Example

(This section discusses a case study provided by Khalid Al-Senani, who is the Performance Advisor for the Ministry of Petroleum and Minerals in Saudi Arabia.)

This case study demonstrates how computer simulation modeling software programs are used in construction to model heavy construction equipment operations.

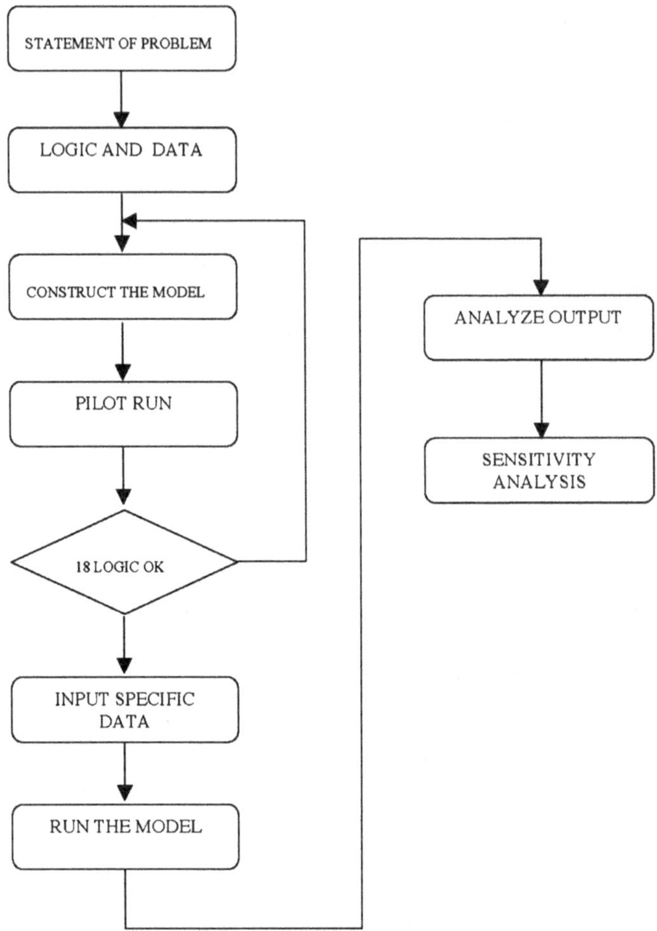

Figure 10-5. Steps for Constructing Computer Simulation Models

Source: Al-Senani (n.d.), Figure 3.2; reproduced with permission.

The off-highway truck type 769C, manufactured by Caterpillar, was used in this heavy construction equipment example to demonstrate how computer simulations are used to determine equipment cycle times. The following specifications for the 769C were obtained from the *Caterpillar Performance Handbook* (Caterpillar 2006):

- Weight (empty): 68,750 lb (31,184.31 kg)
- Weight + weight of the load (gross): 149,000 lb (67,585 kg)
- Capacity (heaped): 30.9 yd^3 (23.62 m^3)

The truck cycle time includes

- Haul time loaded,
- Return time empty,

- Time to load each truck (dependent on which type of loader is used to load the truck),
- Maneuvering time in the load area (6–8 min), and
- Maneuvering and dump time (1.0–1.2 min).

The wheel loader selected for the sample simulation was the 966C, also manufactured by Caterpillar. The specifications for this loader are as follows (Caterpillar 2006):

Hydraulic cycle time (rated load in bucket)

- Raising full bucket: 6.2 s
- Dumping load: 1.6 s
- Lowering bucket (empty): 3.8 s

$$\begin{aligned} \text{total loader cycle time} &= \text{raising full bucket} \\ &\quad + \text{dumping load} + \text{lowering bucket} \\ &= (6.2 + 1.6 + 3.8) \\ &= 11.6\ \text{s} \end{aligned} \tag{10-2}$$

The Caterpillar Performance Handbook (2006) indicates the loader cycle time is from 0.5 to 0.55 min. This time includes loading and dumping, four reversals of direction, a full cycle of hydraulics, and travel time.

The bucket used for this example is a general-purpose bucket with the following specifications:

- Heaped capacity: 4.0 yd^3 (3.06 m^3)
- Overall height (bucket fully extended): 17 ft 8 in. (5.38 m)
- Operation weight: 36,890 lb (16,733.02 kg)
- Cycles to fill one truck: truck capacity (heaped)/loader bucket size = 30.9 yd^3/4.0 yd^3 (23.62 m^3/3.06 m^3) = 7.72 or 8 cycles

Assuming the bucket-fill factor is 100% for loose soil,

$$\begin{aligned} \text{total time to fill one truck} &= \text{number of cycles per truck} \\ &\quad \times \text{number of minutes per cycle} \\ &= (8 \times 0.55) \\ &= 4.4\ \text{min} \end{aligned} \tag{10-3}$$

The time required for each truck is a function of speed, and the speed is governed by

- The power required,
- The power available, and
- The usable power.

The power required is a function of the **grade resistance** (GR) and rolling resistance (RR). **Rolling resistance** is a function of the condition of the road surface and weight on the wheels of the truck. If the condition of the road surface is known, the rolling resistance is determined using the Caterpillar Performance Handbook (2006) or a heavy construction equipment textbook. The **total resistance** is calculated using Eq. (10-4).

$$\text{total resistance} = \text{GR} + \text{RR} \tag{10-4}$$

Rimpull is the force available between the tire and the ground to move the truck and is a function of the available horsepower of the truck.

This case study simulates earthmoving operations with loose soil and different combinations of trucks and loaders. The procedures used for constructing the model were as follows:

1. The number of trucks required for the operation is generated.
2. The truck information is entered into the computer.
3. An identification number is assigned to each truck.
4. Interarrival times are generated for the trucks in the loading area. The interarrival times are a function of the speed and the variables used to calculate the time for each truck to travel to the dump site, dump its load, and return. The interarrival distribution is assumed to be exponential.
5. Each truck enters the queuing line and then waits to be loaded while the loader is loading other trucks. Trucks are loaded on a first-come first-serve (FCFS) basis.
6. The servers (loaders) are represented in the model by a storage symbol. Storage symbols can be applied to more than one server.
7. Trucks are serviced one truck at a time for each loader.
8. Servicing of each truck by a loader is assumed to occur when any of the loaders are available for service and as soon as the previous truck being serviced by the loader leaves.
9. The loading time for each truck is assumed to follow an Erlang distribution (a continuous probability distribution) where $K = 10$ (Gaarslev 1969).
10. After a truck is loaded, the truck volume is recorded and calculated.
11. The time to load each truck is calculated by sampling the Erlang distribution using a random number (RN) generator.
12. Statistics for the queue are determined using the TABLE block to determine the use of the loaders.
13. The loading service times and truck volumes are used to calculate the total volume of the load and the total time to determine the cost and profit.

Assumptions

This section discusses the assumptions used in the computer simulation model. The travel time function distribution was assumed to be exponential, and the travel time for each truck was based on the following information:

- The haul and return road profiles were assumed to be divided into several segments with different grades and types of road surfaces, as shown in Figure 10-6.

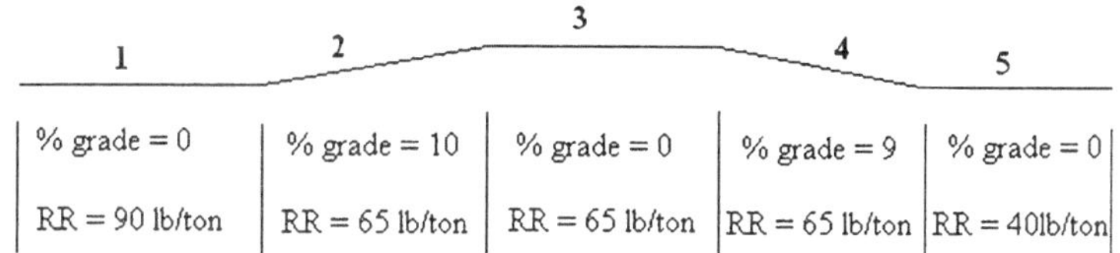

Figure 10-6. Road Profile with Rolling Resistance and Grades

Source: Al-Senani (n.d.), Figure 3.3; reproduced with permission.

Table 10-2. Rimpull and Speed of Loaded Trucks for Each Segment of Road

Section	Distance (feet and meters)	Grade (GR)	Rolling Resistance	Effective Grade	Rimpull (lb and kg)	Speed (mph and kph)
1	3,000 (914.4)	0	90	4.50	7,000 (3,175)	16 (25.75)
2	2,500 (762)	10	65	13.25	9,000 (4,082.3)	8 (12.87)
3	4,400 (1,341.6)	0	65	3.25	5,200 (2,358.7)	29 (46.66)
4	2,300 (701)	−9	65	−5.75	0	42 (67.57)
5	4,600 (1,402)	0	40	2.00	30 (13.6)	43 (69.19)

Source: Al-Senani (n.d.), Table 3.2; reproduced of with permission.

Table 10-3. Rimpull and Speed of Empty Trucks for Each Segment of Road

Section	Distance (feet and meters)	Grade (GR)	Rolling Resistance	Effective Grade	Rimpull (lb and kg)	Speed (mph and kph)
1	3,000 (914.4)	0	90	4.50	3,500 (1,587.6)	42 (67.57)
2	2,500 (762)	10	65	−6.75	0 (0)	42 (67.57)
3	4,400 (1,341.6)	0	65	3.25	2,200 (997.9)	46 (74.01)
4	2,300 (701)	−9	65	12.25	8,100 (3,674)	16 (25.74)
5	4,600 (1,402)	0	40	2.00	900 (408)	49 (78.84)

Source: Al-Senani (n.d.), Table 3.3; reproduced with permission.

- The effective grade, rimpull in pounds, and speed for loaded and empty trucks were calculated using data from the Caterpillar Performance Handbook (2006). See Tables 10-2 to 10-6. The speed functions for loaded and empty trucks were also calculated using the data from Caterpillar (2006).
- Travel times for each segment of the road (haul and return), including acceleration and deceleration, were calculated using data from Caterpillar (2006).

The tables were used to determine the total travel time and the acceleration and deceleration time as follows:

- Total travel time from Table 10-5 = 17.42 min
- Total travel time from Tables 10-2 and 10-3 = 13.97 min
- Acceleration and deceleration time = 17.42 min − 13.97 min = 3.45 min

Table 10-4. Speed Function for Loaded Trucks

Distance (feet and meters)	Speed (mph and kph)	Percentage of Total	Cumulative Percentage
2,500 (762)	8 (12.87)	14.88	14.88
3,000 (914.4)	16 (25.78)	17.86	26.19
4,400 (1,341)	29 (46.68)	26.19	58.93
2,300 (701)	42 (67.59)	13.69	72.62
4,600 (1,402)	43 (69.20)	27.38	100.00

Source: Al-Senani (n.d.), Table 3.4; reproduced with permission.

Table 10-5. Speed Function for Empty Trucks

Distance (feet and meters)	Speed (mph and kph)	Percentage of Total	Cumulative Percentage
2,300 (762)	16 (25.75)	13.69	13.69
5,500 (1,676)	42 (67.59)	32.74	46.43
4,400 (1,341)	46 (74.03)	26.19	72.62
4,600 (1,402)	49 (78.86)	27.38	100.00

Source: Al-Senani (n.d.), Table 3.5; reproduced with permission.

Table 10-6. Travel Time Found by Using Travel Time Chart

Section	Distance (feet and meters)	Effective Grade (GR)	Speed Loaded (mph and kph)	Travel Time (seconds)	Speed Empty (mph and kph)	Travel Time (seconds)
1	3,000 (913)	4.5, 4.5	16 (25.75)	1.80	42 (67.57)	1.2
2	2,500 (762)	13.25, –6.75	8 (12.87)	4.25	42 (67.59)	8.0
3	4,400 (1,341)	3.25, 3.25	29 (46.68)	3.25	46 (74.03)	29.0
4	2,300 (701)	–5.75, 12.5	42 (67.59)	3.00	16 (25.75)	42.0
5	4,600 (1,402)	2.0, 2.0	43 (69.20)	2.00	49 (78.86)	43.0

Source: Al-Senani (n.d.), Table 3.6; reproduced with permission.

The next step was to compare the usable power against the power required to move the fully loaded truck. The usable power was calculated by multiplying the coefficient of traction by the weight on the drive wheels. The distribution of the load for the model 769C is as follows (Caterpillar 2006):

Distribution loaded:

- Front 33% = 49,170 lb (22,303 kg)
- Rear 67% = 99,830 lb (45,282 kg)
- Total weight of the truck loaded = 149,000 lb (67,585 kg), or 74.5 tons (67.59 metric tons)

- Pounds required = 90 lb/ton × 74.5 tons = 6,705 lb (kilograms required = 40.82 kg/0.9072 metric ton × 67.59 metric tons = 3,041.32 kg)

For the haul road segment, the coefficient of traction for loose earth is 0.45.

$$\begin{aligned} \text{usable power} &= \text{coefficient of traction} \\ &\times \text{weight on the drive wheels} \\ &= 0.45 \times 99{,}830\ \text{lb} = 44{,}923\ \text{lb} \\ &(= 0.45 \times 45{,}281.89\ \text{kg} = 20{,}377\ \text{kg}) \end{aligned} \tag{10-5}$$

Because 44,923 lb (20,376.62 kg) is greater than 6,705 lb (3,041.32 kg), the traction is more than adequate for the truck to move when fully loaded.

An average speed was used to calculate the speed of each truck in the system. Travel times were calculated using the average travel time for each section of the haul road, and then acceleration and deceleration times were added to the travel times. Fixed and maneuver times were added to the travel times to establish the truck cycle times, excluding loading times. Equations and the variables to calculate average interarrival time were specified in the model. At the GENERATE block, the interarrival time was assumed to be exponentially distributed.

The average loading time is the time required for a loader to fill one truck, and it is simulated at the ADVANCE block. The function distribution for loading times was assumed to follow an Erlang distribution.

Computer Simulation Model

Block versions of the model are shown in Figure 10-7. The model is divided into three segments: (1) the generation of trucks; (2) the main segment, used to model the entire process; and (3) the timing-control segment, which controls the timing of the simulation. Each block represents a special task of the operation—for example, the ADVANCE block represents the travel time for each truck. Other ADVANCE blocks represent the different segments of the loading time. The model file, which corresponds to the block diagrams, contains the compiler directive, the three model segments, and the run-control statements.

A heavy construction contractor provided data for the hourly costs of loaders and trucks, including maintenance, fuel, operator wages, and other associated costs. In addition, the contractor provided the costs charged for hauling loose soil, including loading and dumping for a distance of approximately 3 mi (4.83 km) between the loading and dumping areas.

Earthmoving operations were simulated for different combinations of vehicles for 3 days of 7 h per day, which is an efficiency factor of 0.875. Each combination was simulated in 10 **non-antithetic** replications (not the direct opposite) and ten **antithetic** replications (direct opposite) to determine the confidence limits. For each replication, there is a rate for arrival times at the loading area, truck waiting times at the queue, and times when dumping is completed by the equipment. The total volume of material hauled and dumped, the cost of operation, and the profit were calculated for each replication during the simulation.

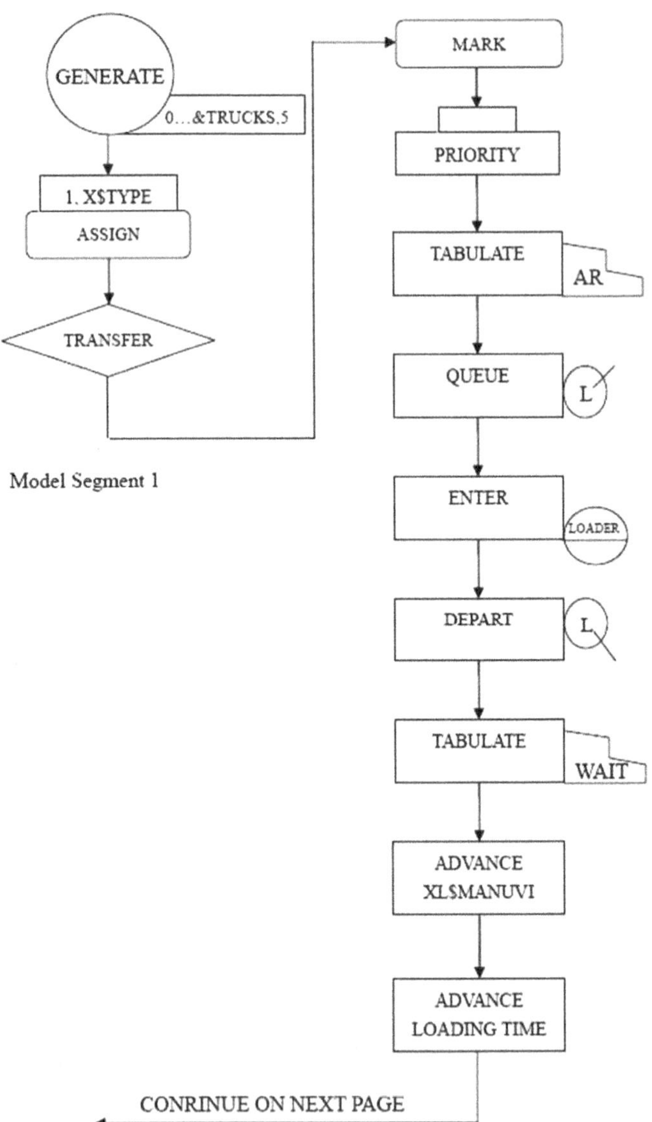

Figure 10-7. Block Diagram of Model

Source: Al-Senani (n.d.), Figure 3.7; reproduced with permission.

Analysis Techniques

A deterministic approach was used to calculate cycle times for the loaders and the trucks in order to determine equipment combinations and total production for the system. Then these numbers were compared with the results of the simulation model to determine its validity. Statistical analysis techniques were used to calculate the confidence limit of the model output. The analysis included determining the number of replications required to represent real-world processes and to simulate a steady-state system. Antithetic techniques were used to increase the statistical precision of the results. Antithetic techniques use pairs of replications where two results are negatively correlated. If one of the results of the pair is extreme and so is the other pair,

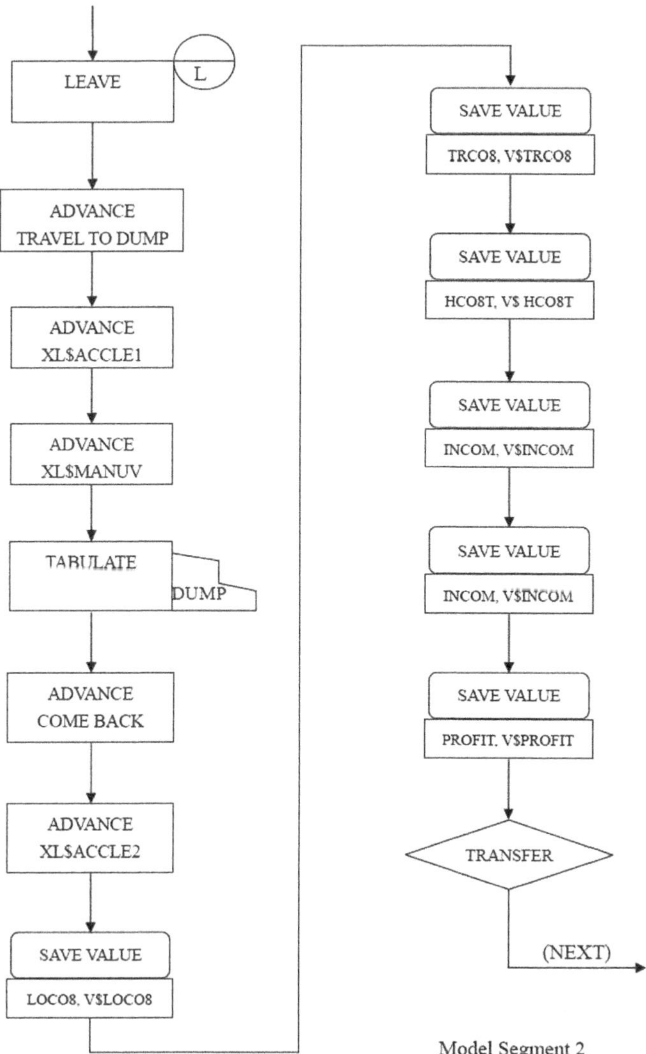

Figure 10-7. *(continued)*

but in the opposite direction, the average of the two results is used in the model (Schriber 1991). **Sensitivity analysis** techniques were applied to the model to determine which of the factors had the greatest effect on production. In addition, the distribution of the loading time was changed from Erlang to uniform in order to observe the effect on the results.

Results

The simulation model was run using the following equipment combinations:

- 6 trucks and 1 loader,
- 6 trucks and 2 loaders,
- 7 trucks and 1 loader,

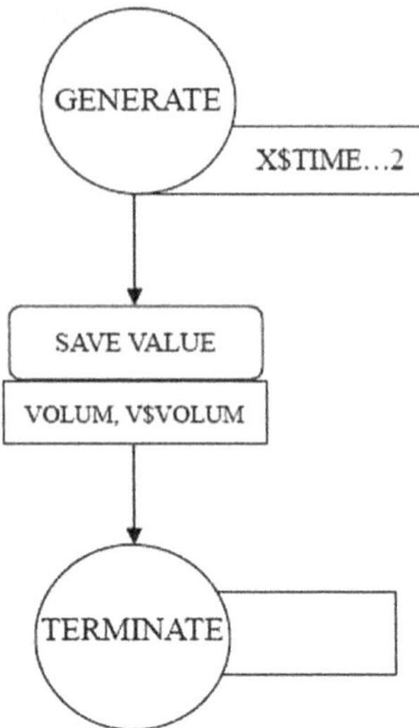

Model Segment 3

Figure 10-7. *(continued)*

- 7 trucks and 2 loaders,
- 8 trucks and 1 loader,
- 8 trucks and 2 loaders,
- 9 trucks and 1 loader,
- 9 trucks and 2 loaders,
- 10 trucks and 1 loader, and
- 10 trucks and 2 loaders.

Figures 10-8 to 10-17 summarize the output that was obtained from the computer simulations. The figures represent the output for each equipment combination for the fifth antithetic replication. Tables 10-7 to 10-16 summarize the output of the simulation model and show each of the non-antithetic and antithetic replications with their respective output. The outputs of the model are the average waiting times for the trucks in the loading area; the total volume of loose soil loaded, hauled, and dumped; and the profit for the operation for each combination.

The following is additional information used in the models:

- Maneuver time = 1.2 min
- Mean loading time = 8 min × 0.57 = 4.56 min
- Truck cycle time (assuming no delays) = 24.1 min

$$\begin{aligned} \text{total cycle time} &= \text{maneuver time} + \text{loading time} \\ &\quad + \text{truck cycle time} \\ &= 1.2 + 4.56 + 24.1 \\ &= 29.86 \text{ min} \end{aligned} \tag{10-6}$$

Text continued on p.310

Average-Utility-During

Storage	Total Time	Avail Time	Unavailable Time	Entries	Average Time/Unit	Current Status	Percent Available	Capacity	Average Contents	Current Contents	Maximum Contents
Loading	0.789			196	5.071	Avail	100.0%	1.0	0.789	1.0	1.0

Queue	Maximum Contents	Average Contents	Total Entries	Zero Entries	Percent Zeros	Average Time/Unit	SAverage Time/Unit	QTable Number	Current Contents
Line	5	0.632	196	60	30.6	4.064	5.857		0

TABLE - ARRIVAL

Entries in Table	Mean Argument	Standard Deviation	Sum of Arguments	
21.000	9.3333	2.2657	197.0000	Non-Weighted

Upper Limit	Observed Frequency	Percent of Total	Cumulative Percentage	Cumulative Remainder	Multiple of Mean	Deviation from Mean
10.0000	14.0000	66.67	66.67	33.33	1.0714	0.2942
20.0000	7.0000	33.33	100.00	0.00	2.1429	4.7079

TABLE - WAITING

Entries in Table	Mean Argument	Standard Deviation	Sum of Arguments	
196	4.5477	4.6450	891.3412	Non-Weighted

Upper Limit	Observed Frequency	Percent of Total	Cumulative Percentage	Cumulative Remainder	Multiple of Mean	Deviation from Mean
1.00	62.00	31.63	31.63	60.37	0.2199	-0.7638
3.00	28.00	14.29	45.92	54.08	0.6597	-0.3332
5.00	32.00	16.33	62.24	37.76	1.0995	0.0974
7.00	29.00	14.80	77.04	22.96	1.5393	0.5280
9.00	16.00	8.16	85.20	14.80	1.9790	0.9585

Figure 10-8. 6 Trucks and 1 Loader Combination

Source: Al-Senani (n.d.), Figure 4.1; reproduced with permission.

continued

11.00	17.00	8.67	93.88	6.12	2.4188	1.3891
13.00	3.00	1.53	95.41	4.59	2.8586	1.8197
15.00	3.00	1.53	96.94	3.06	3.2984	2.2503

Average Value of Overflow is 27.00

TABLE – DUMP

Entries in Table	Mean Argument	Standard Deviation	Sum of Arguments	
190	29.3977	79.7302	5680.56	Non-Weighted

Upper Limit	Observed Frequency	Percent of Total	Cumulative Percentage	Cumulative Remainder	Multiple of Mean	Deviation from Mean
8.00	24.00	12.63	12.63	87.37	0.2676	-0.2746
10.00	14.00	7.37	10.00	80.00	0.3345	-0.2496
12.00	18.00	9.47	29.47	70.53	0.4014	-0.2245
14.00	20.00	10.53	40.00	60.00	0.4483	-.01994
16.00	22.00	11.58	51.58	48.42	0.4683	-0.1743
18.00	17.00	8.95	60.53	39.47	0.5352	-0.1492

Average Value of Overflow is 82.2655

Non Zero Full Word Save Values: (Name:Value)

LoadT: 31 Time: 1260 Volume: 5890 Income: 32395 Locos: 95

Trcos: 480 Mcost: 575 Tcost: 12075 Profit: 20320

The above is Antithetic Replication Number 5

Number of Trucks: 6

Number of Loaders: 1

Figure 10-8. *(continued)*

Average-Utility-During

Storage	Total Time	Avail Time	Unavailable Time	Entries	Average Time/Unit	Current Status	Percent Available	Capacity	Average Contents	Current Contents	Maximum Contents
Loading	0.433			214	5.103	Avail	100.0%	2.0	0.867	0.0	2.0

Queue	Maximum Contents	Average Contents	Total Entries	Zero Entries	Percent Zeros	Average Time/Unit	SAverage Time/Unit	QTable Number	Current Contents
Line	4	0.076	214	177	82.7	0.461	2.667		0

TABLE - ARRIVAL

Entries in Table	Mean Argument	Standard Deviation	Sum of Arguments	
21.000	10.1905	2.8744	214.0000	Non-Weighted

Upper Limit	Observed Frequency	Percent of Total	Cumulative Percentage	Cumulative Remainder	Multiple of Mean	Deviation from Mean
10.0000	12.0000	57.24	57.14	42.86	0.9813	-0.0663
20.0000	9.0000	42.86	100.00	0.00	1.9626	3.4128

TABLE - WAITING

Entries in Table	Mean Argument	Standard Deviation	Sum of Arguments	
214	0.9510	1.5352	203.5060	Non-Weighted

Upper Limit	Observed Frequency	Percent of Total	Cumulative Percentage	Cumulative Remainder	Multiple of Mean	Deviation from Mean
1.00	183.00	85.51	85.51	14.49	1.0516	0.0319
3.00	20.00	9.33	94.86	5.14	3.1547	1.3347
5.00	4.00	1.87	96.73	3.27	5.2578	2.6374
7.00	5.00	2.34	99.07	0.93	7.3609	3.9402

- - -

Figure 10-9. 6 Trucks and 2 Loaders Combination

Source: Al-Senani (n.d.), Figure 4.2; reproduced with permission.

continued

- - -

13.00	1.00	0.47	99.53	0.47	13.6703	7.8484
15.00	1.00	0.47	100.00	0.00	13.7734	9.1511

TABLE – DUMP

Entries in Table	Mean Argument	Standard Deviation	Sum of Arguments	
211	26.3547	75.2124	5560.8355	Non-Weighted

Upper Limit	Observed Frequency	Percent of Total	Cumulative Percentage	Cumulative Remainder	Multiple of Mean	Deviation from Mean
8.00	47.00	22.27	22.27	77.73	0.3036	-0.2440
10.00	28.00	13.27	35.55	64.45	0.3794	-0.2174
12.00	22.00	10.43	45.97	54.93	0.4553	-0.1909
14.00	19.00	9.00	54.98	45.02	0.5312	-0.1643
16.00	19.00	9.00	63.98	36.02	0.6071	-0.1377
18.00	6.00	2.84	66.82	33.18	0.6830	-0.1111
20.00	13.00	6.16	72.99	27.01	0.7589	-0.0845

Average Value of Overflow is 84.9989

Non Zero Full Word Save Values: (Name:Value)

LoadT: 31 Time: 1260 Volume: 6541 Income: 35305 Locos: 190

Trcos: 480 Mcost: 670 Tcost: 14070 Profit: 21735

The above is Antithetic Replication Number 5

Number of Trucks: 6

Number of Loaders: 2

Figure 10-9. *(continued)*

Average-Utility-During

Storage	Total Time	Avail Time	Unavailable Time	Entries	Average Time/Unit	Current Status	Percent Available	Capacity	Average Contents	Current Contents	Maximum Contents
Loading	0.925			229	5.088	Avail	100.0%	2.0	0.925	1.0	1.0

Queue	Maximum Contents	Average Contents	Total Entries	Zero Entries	Percent Zeros	Average Time/Unit	SAverage Time/Unit	QTable Number	Current Contents
Line	6	1.441	229	37	16.2	7.931	9.459		0

TABLE - ARRIVAL

Entries in Table	Mean Argument	Standard Deviation	Sum of Arguments	
21.000	10.9048	2.4679	222.0000	Non-Weighted

Upper Limit	Observed Frequency	Percent of Total	Cumulative Percentage	Cumulative Remainder	Multiple of Mean	Deviation from Mean
10.0000	10.0000	47.62	47.62	52.38	0.9170	-0.3666
20.0000	11.0000	52.38	100.00	0.00	1.8341	3.6854

TABLE - WAITING

Entries in Table	Mean Argument	Standard Deviation	Sum of Arguments	
229	8.4148	7.0059	1926.9909	Non-Weighted

Upper Limit	Observed Frequency	Percent of Total	Cumulative Percentage	Cumulative Remainder	Multiple of Mean	Deviation from Mean
1.00	41.00	17.90	17.90	82.10	0.1188	-1.0584
3.00	24.00	10.48	28.38	71.62	0.3565	-0.7729
5.00	22.00	9.61	37.00	62.01	0.5943	-0.4874
7.00	24.00	10.48	48.47	51.53	0.8319	-0.2019
9.00	27.00	11.79	60.26	39.74	1.0695	0.0835

Figure 10-10. 7 Trucks and 1 Loader Combination

Source: Al-Senani (n.d.), Figure 4.3; reproduced with permission.

continued

11.00	16.00	6.99	67.25	32.75	1.3072	0.3690
13.00	19.00	8.30	75.55	24.45	1.5449	0.6545
15.00	15.00	6.55	82.10	17.90	1.7826	0.9299
17.00	14.00	6.11	34.67	11.79	2.0202	1.2254

Average Value of Overflow is 28.6156

TABLE – DUMP

Entries in Table	Mean Argument	Standard Deviation	Sum of Arguments	
225	28.2102	36.5227	6347.2874	Non-Weighted

Upper Limit	Observed Frequency	Percent of Total	Cumulative Percentage	Cumulative Remainder	Multiple of Mean	Deviation from Mean
8.00	14.00	6.22	22.27	77.73	0.3036	-0.5534
10.00	11.00	4.89	35.55	64.45	0.3794	-0.4986
12.00	21.00	9.33	45.97	54.93	0.4553	-0.4458
14.00	18.00	8.00	54.98	45.02	0.5312	-0.3891
16.00	14.00	6.22	63.98	36.02	0.6071	-0.3343

Average Value of Overflow is 49.1863

Non Zero Full Word Save Values: (Name:Value)

LoadT: 31 Time: 1260 Volume: 8975 Income: 38362 Locos: 95

Trcos: 560 Mcost: 655 Tcost: 13755 Profit: 24807

The above is Antithetic Replication Number 5

Number of Trucks: 7

Number of Loaders: 1

Figure 10-10. *(continued)*

Average-Utility-During

Storage	Total Time	Avail Time	Unavailable Time	Entries	Average Time/Unit	Current Status	Percent Available	Capacity	Average Contents	Current Contents	Maximum Contents
Loading	0.509			252	5.091	Avail	100.0%	2.0	1.015	1.0	2.0

Queue	Maximum Contents	Average Contents	Total Entries	Zero Entries	Percent Zeros	Average Time/Unit	SAverage Time/Unit	QTable Number	Current Contents
Line	5	0.178	252	180	71.4	0.890	3.114		0

TABLE - ARRIVAL

Entries in Table	Mean Argument	Standard Deviation	Sum of Arguments	
21.000	10.9048	2.4679	222.0000	Non-Weighted

Upper Limit	Observed Frequency	Percent of Total	Cumulative Percentage	Cumulative Remainder	Multiple of Mean	Deviation from Mean
10.0000	9.0000	42.86	42.86	57.14	0.8333	-0.4984
20.0000	11.0000	52.38	95.24	4.76	1.6667	1.9938
30.000	1.000	4.76	100.00	0.00	2.5000	4.4860

TABLE - WAITING

Entries in Table	Mean Argument	Standard Deviation	Sum of Arguments	
252	1.3420	2.0048	338.1881	Non-Weighted

Upper Limit	Observed Frequency	Percent of Total	Cumulative Percentage	Cumulative Remainder	Multiple of Mean	Deviation from Mean
1.00	188.00	74.60	74.60	25.40	0.7451	-0.1671
3.00	25.00	9.92	84.52	15.48	2.2354	0.8100
5.00	29.00	11.51	96.03	3.97	3.7257	1.7872
7.00	6.00	2.38	98.41	1.59	5.2160	2.7643

Figure 10-11. 7 Trucks and 2 Loaders Combination

Source: Al-Senani (n.d.), Figure 4.4; reproduced with permission.

continued

9.00	1.00	0.40	98.81	1.19	6.7063	3.7414

TABLE – DUMP

Entries in Table	Mean Argument	Standard Deviation	Sum of Arguments	
247	23.6891	52.0620	5851.2145	Non-Weighted

Upper Limit	Observed Frequency	Percent of Total	Cumulative Percentage	Cumulative Remainder	Multiple of Mean	Deviation from Mean
8.00	61.00	24.70	24.70	75.30	0.3377	-0.3014
10.00	36.00	14.57	39.27	60.73	0.4221	-0.2629
12.00	33.00	13.36	52.63	47.37	0.5066	-0.2245
14.00	13.00	5.26	37.89	42.11	0.5910	-0.1861
16.00	21.00	3.50	36.40	33.60	0.6754	-0.1477

Average Value of Overflow is 77.7114

Non Zero Full Word Save Values: (Name:Value)

LoadT: 31 Time: 1260 Volume: 7657 Income: 42113 Locos: 190

Trcos: 560 Mcost: 750 Tcost: 15750 Profit: 26363

The above is Antithetic Replication Number 5

Number of Trucks: 7

Number of Loaders: 2

Figure 10-11. *(continued)*

Average-Utility-During

Storage	Total Time	Avail Time	Unavailable Time	Entries	Average Time/Unit	Current Status	Percent Available	Capacity	Average Contents	Current Contents	Maximum Contents
Loading	0.955			235	5.119	Avail	100.0%	1.0	0.955	1.0	1.0

Queue	Maximum Contents	Average Contents	Total Entries	Zero Entries	Percent Zeros	Average Time/Unit	SAverage Time/Unit	QTable Number	Current Contents
Line	7	1.938	237	21	8.9	10.303	11.305		0

TABLE - ARRIVAL

Entries in Table	Mean Argument	Standard Deviation	Sum of Arguments	
21.000	11.2857	11.802	237.0000	Non-Weighted

Upper Limit	Observed Frequency	Percent of Total	Cumulative Percentage	Cumulative Remainder	Multiple of Mean	Deviation from Mean
10.0000	6.0000	28.57	28.57	71.43	0.8861	-0.7062
20.0000	15.0000	71.53	100.00	0.00	1.7722	4.7867

TABLE - WAITING

Entries in Table	Mean Argument	Standard Deviation	Sum of Arguments	
235	10.8461	8.4693	2548.8367	Non-Weighted

Upper Limit	Observed Frequency	Percent of Total	Cumulative Percentage	Cumulative Remainder	Multiple of Mean	Deviation from Mean
1.00	26.00	11.06	11.06	88.74	0.0922	-1.1626
3.00	17.00	7.23	18.03	81.70	0.2766	-0.9264
5.00	28.00	11.91	30.21	69.79	0.4610	-0.6905
7.00	25.00	10.64	40.85	59.15	0.6454	-0.4541
9.00	22.00	9.63	50.01	49.79	1.8298	0.2180

Figure 10-12. 8 Trucks and 1 Loader Combination

Source: Al-Senani (n.d.), Figure 4.5; reproduced with permission.

continued

11.00	16.00	6.81	57.02	42.98	1.0142	0.0182

Average Value of Overflow is 28.6156

TABLE – DUMP

Entries in Table	Mean Argument	Standard Deviation	Sum of Arguments	
232	32.5462	41.0736	7550.7235	Non-Weighted

Upper Limit	Observed Frequency	Percent of Total	Cumulative Percentage	Cumulative Remainder	Multiple of Mean	Deviation from Mean
8.00	6.00	2.59	2.59	97.41	0.2458	-0.5976
10.00	17.00	7.33	9.91	90.09	0.3073	-0.5489
12.00	16.00	6.90	16.81	83.19	0.3587	-0.5002
14.00	14.00	6.03	22.84	77.16	0.4302	-0.4515
16.00	11.00	6.74	27.59	72.41	0.4916	-0.4028

Average Value of Overflow is 49.1863

Non Zero Full Word Save Values: (Name:Value)

LoadT: 31 Time: 1260 Volume: 7192 Income: 39385 Locos: 95

Trcos: 640 Mcost: 755 Tcost: 15435 Profit: 23950

The above is Antithetic Replication Number 5

Number of Trucks: 8

Number of Loaders: 1

Figure 10-12. *(continued)*

Average-Utility-During

Storage	Total Time	Avail Time	Unavailable Time	Entries	Average Time/Unit	Current Status	Percent Available	Capacity	Average Contents	Current Contents	Maximum Contents
Loading	0.558			275	5.115	Avail	100.0%	2.0	1.116	0.0	2.0

Queue	Maximum Contents	Average Contents	Total Entries	Zero Entries	Percent Zeros	Average Time/Unit	SAverage Time/Unit	QTable Number	Current Contents
Line	6	0.228	275	179	65.1	1.045	2.995		0

TABLE - ARRIVAL

Entries in Table	Mean Argument	Standard Deviation	Sum of Arguments	
21.000	13.0952	4.5266	275.0000	Non-Weighted

Upper Limit	Observed Frequency	Percent of Total	Cumulative Percentage	Cumulative Remainder	Multiple of Mean	Deviation from Mean
10.0000	5.0000	23.81	25.81	76.19	0.7636	-0.6838
20.0000	15.0000	71.43	95.24	4.75	1.5273	1.5254
30.0000	1.0000	4.76	100.00	0.00	2.2909	3.7345

TABLE - WAITING

Entries in Table	Mean Argument	Standard Deviation	Sum of Arguments	
275	1.5216	2.2787	418.4469	Non-Weighted

Upper Limit	Observed Frequency	Percent of Total	Cumulative Percentage	Cumulative Remainder	Multiple of Mean	Deviation from Mean
1.00	192.00	69.82	69.82	30.18	0.6572	-0.2289
3.00	36.00	13.09	82.91	17.09	1.9716	0.6588
5.00	32.00	11.64	94.55	5.45	3.2860	1.5264
7.00	9.00	3.27	97.82	2.18	4.6003	2.4041

Figure 10-13. 8 Trucks and 2 Loaders Combination

Source: Al-Senani (n.d.), Figure 4.6; reproduced with permission.

continued

9.00	1.00	0.36	98.18	1.82	5.1947	3.2818
11.00	1.00	0.36	98.55	1.45	7.2291	4.1595
13.00	3.00	0.36	98.91	1.09	8.4535	5.0371

TABLE – DUMP

Entries in Table	Mean Argument	Standard Deviation	Sum of Arguments	
190	29.3977	79.7302	5680.56	Non-Weighted

Upper Limit	Observed Frequency	Percent of Total	Cumulative Percentage	Cumulative Remainder	Multiple of Mean	Deviation from Mean
8.00	53.00	19.56	19.56	80.44	0.3440	-0.3676
10.00	39.00	14.39	33.95	66.05	0.4300	-0.3194
12.00	32.00	11.81	45.76	54.24	0.5160	-0.2714
14.00	20.00	7.53	53.14	48.86	0.6020	-0.2230
16.00	30.00	11.07	64.21	35.79	0.6880	-0.1748

Average Value of Overflow is 82.2655

Non Zero Full Word Save Values: (Name:Value)

LoadT: 31 Time: 1260 Volume: 6401 Income: 16035 Locos: 190

Trcos: 640 Mcost: 830 Tcost: 17430 Profit: 28605

The above is Antithetic Replication Number 5

Number of Trucks: 8

Number of Loaders: 2

Figure 10-13. *(continued)*

Average-Utility-During

Storage	Total Time	Avail Time	Unavailable Time	Entries	Average Time/Unit	Current Status	Percent Available	Capacity	Average Contents	Current Contents	Maximum Contents
Loading	0.986			244	5.090	Avail	100.0%	1.0	0.986	1.0	1.0

Queue	Maximum Contents	Average Contents	Total Entries	Zero Entries	Percent Zeros	Average Time/Unit	SAverage Time/Unit	QTable Number	Current Contents
Line	8	2.682	247	8	3.2	13.683	14.141		0

TABLE - ARRIVAL

Entries in Table	Mean Argument	Standard Deviation	Sum of Arguments	
21.000	11.7619	2.5081	247.0000	Non-Weighted

Upper Limit	Observed Frequency	Percent of Total	Cumulative Percentage	Cumulative Remainder	Multiple of Mean	Deviation from Mean
10.0000	6.0000	28.57	28.57	71.43	0.8502	-0.7-25
20.0000	15.0000	71.43	100.00	0.00	1.7004	3.2854

TABLE - WAITING

Entries in Table	Mean Argument	Standard Deviation	Sum of Arguments	
244	14.2233	8.6776	3470.4924	Non-Weighted

Upper Limit	Observed Frequency	Percent of Total	Cumulative Percentage	Cumulative Remainder	Multiple of Mean	Deviation from Mean
1.00	11.00	4.51	4.51	95.49	0.0705	-1.5239
3.00	10.00	4.10	8.61	91.39	0.2109	-1.2934
5.00	18.00	7.36	15.98	84.02	0.3515	-1.0629
7.00	22.00	9.02	25.00	75.00	0.4921	-0.8324
9.00	21.00	8.61	33.61	66.39	0.6328	-0.6019

Figure 10-14. 9 Trucks and 1 Loader Combination

Source: Al-Senani (n.d.), Figure 4.7; reproduced with permission.

continued

11.00	15.00	6.15	39.75	60.25	0.7734	-0.3715
13.00	15.00	6.15	45.90	54.10	0.9140	-0.1410

TABLE – DUMP

Entries in Table	Mean Argument	Standard Deviation	Sum of Arguments	
240	38.2882	44.2007	8949.1786	Non-Weighted

Upper Limit	Observed Frequency	Percent of Total	Cumulative Percentage	Cumulative Remainder	Multiple of Mean	Deviation from Mean
8.00	5.00	2.08	2.08	97.92	0.2145	-0.6623
10.00	4.00	1.67	3.75	96.26	0.2682	-0.6171
12.00	8.00	3.33	7.08	92.92	0.3218	-0.5719
14.00	7.00	2.92	10.00	90.00	0.3755	-0.5266

Average Value of Overflow is 52.1966

Non Zero Full Word Save Values: (Name:Value)

LoadT: 31 Time: 1260 Volume: 7440 Income: 40920 Locos: 95

Trcos: 720 Mcost: 815 Tcost: 17115 Profit: 23805

The above is Antithetic Replication Number 5

Number of Trucks: 9

Number of Loaders: 1

Figure 10-14. *(continued)*

Average-Utility-During

Storage	Total Time	Avail Time	Unavailable Time	Entries	Average Time/Unit	Current Status	Percent Available	Capacity	Average Contents	Current Contents	Maximum Contents
Loading	0.606			298	5.124	Avail	100.0%	2.0	1.1212	1.0	2.0

Queue	Maximum Contents	Average Contents	Total Entries	Zero Entries	Percent Zeros	Average Time/Unit	SAverage Time/Unit	QTable Number	Current Contents
Line	7	0.289	298	192	64.4	1.222	3.436		0

TABLE - ARRIVAL

Entries in Table	Mean Argument	Standard Deviation	Sum of Arguments	
21.000	14.1905	4.7183	298.0000	Non-Weighted

Upper Limit	Observed Frequency	Percent of Total	Cumulative Percentage	Cumulative Remainder	Multiple of Mean	Deviation from Mean
10.0000	4.0000	19.05	19.05	80.85	0.7047	-0.8881
20.0000	14.0000	66.67	85.71	14.29	1.4094	1.2313
30.0000	3.0000	14.29	100.00	0.00	2.1141	3.3507

TABLE - WAITING

Entries in Table	Mean Argument	Standard Deviation	Sum of Arguments	
298	1.7144	2.6271	510/8847	Non-Weighted

Upper Limit	Observed Frequency	Percent of Total	Cumulative Percentage	Cumulative Remainder	Multiple of Mean	Deviation from Mean
1.00	198.00	66.44	66.44	33.56	0.5833	-0.2719
3.00	42.00	14.09	80.54	19.46	1.7499	0.4894
5.00	37.00	12.42	92.95	7.05	2.9165	1.2507
7.00	10.00	3.36	96.31	3.69	4.0831	2.0129

Figure 10-15. 9 Trucks and 2 Loaders Combination

Source: Al-Senani (n.d.), Figure 4.8; reproduced with permission.

continued

9.00	5.00	1.68	97.99	2.01	5.2497	2.7753

TABLE – DUMP

Entries in Table	Mean Argument	Standard Deviation	Sum of Arguments	
292	23.6011	39.6144	5210.8847	Non-Weighted

Upper Limit	Observed Frequency	Percent of Total	Cumulative Percentage	Cumulative Remainder	Multiple of Mean	Deviation from Mean
8.00	59.00	20.21	20.21	79.79	0.3390	-0.3918
10.00	37.00	12.87	32.88	67.12	0.4237	-0.3416
12.00	48.00	16.44	49.31	50.68	0.5085	-0.2914
14.00	21.00	7.19	56.51	43.49	0.5932	-0.2411
16.00	17.00	5.82	62.33	37.67	0.6779	-0.1909
18.00	20.00	6.85	69.18	30.82	0.7627	-0.1407

Average Value of Overflow is 67.0711

Non Zero Full Word Save Values: (Name:Value)

LoadT: 31 Time: 1260 Volume: 9052 Income: 49613 Locos: 190

Trcos: 720 Mcost: 910 Tcost: 19110 Profit: 30505

The above is Antithetic Replication Number 5

Number of Trucks: 9

Number of Loaders: 2

Figure 10-15. *(continued)*

Average-Utility-During

Storage	Total Time	Avail Time	Unavailable Time	Entries	Average Time/Unit	Current Status	Percent Available	Capacity	Average Contents	Current Contents	Maximum Contents
Loading	0.994			246	5.089	Avail	100.0%	1.0	0.994	1.0	2.0

Queue	Maximum Contents	Average Contents	Total Entries	Zero Entries	Percent Zeros	Average Time/Unit	SAverage Time/Unit	QTable Number	Current Contents
Line	9	3.501	246	4	1.6	17.786	18.078		2

TABLE - ARRIVAL

Entries in Table	Mean Argument	Standard Deviation	Sum of Arguments	
21.000	11.8095	2.8393	248.0000	Non-Weighted

Upper Limit	Observed Frequency	Percent of Total	Cumulative Percentage	Cumulative Remainder	Multiple of Mean	Deviation from Mean
10.0000	7.0000	33.33	33.33	67.67	0.8468	-0.6373
20.0000	14.0000	66.67	100.00	0.00	1.6935	2.8846

TABLE - WAITING

Entries in Table	Mean Argument	Standard Deviation	Sum of Arguments	
246	18.4085	9.5195	4528.4880	Non-Weighted

Upper Limit	Observed Frequency	Percent of Total	Cumulative Percentage	Cumulative Remainder	Multiple of Mean	Deviation from Mean
1.00	4.00	1.63	1.63	98.37	0.0543	-1.8287
3.00	6.00	2.44	4.07	95.93	0.1630	-1.6186
5.00	5.00	2.03	6.10	93.90	0.2716	-1.4085
7.00	14.00	5.69	11.79	88.21	0.3805	-1.1984

Figure 10-16. 10 Trucks and 1 Loader Combination

Source: Al-Senani (n.d.), Figure 4.9; reproduced with permission.

continued

9.00	9.00	3.66	15.45	84.55	0.4589	-0.9883
11.00	27.00	10.98	26.42	73.58	0.5976	-0.7782

TABLE – DUMP

Entries in Table	Mean Argument	Standard Deviation	Sum of Arguments	
238	41.3425	43.9036	9839.5147	Non-Weighted

Upper Limit	Observed Frequency	Percent of Total	Cumulative Percentage	Cumulative Remainder	Multiple of Mean	Deviation from Mean
8.00	3.00	1.26	1.26	98.74	0.1935	-0.7594

12.00	3.00	1.26	2.52	98.48	0.2903	-0.6684
14.00	3.00	1.26	3.78	98.22	0.3386	-0.6228

Average Value of Overflow is 51.0998

Non Zero Full Word Save Values: (Name:Value)

LoadT: 31 Time: 1260 Volume: 7378 Income: 40579 Locos: 95

Trcos: 800 Mcost: 895 Tcost: 18795 Profit: 21784

The above is Antithetic Replication Number 5

Number of Trucks: 10

Number of Loaders: 1

Figure 10-16. *(continued)*

Average-Utility-During

Storage	Total Time	Avail Time	Unavailable Time	Entries	Average Time/Unit	Current Status	Percent Available	Capacity	Average Contents	Current Contents	Maximum Contents
Loading	0.626			308	5.122	Avail	100.0%	2.0	1.252	1.0	2.0

Queue	Maximum Contents	Average Contents	Total Entries	Zero Entries	Percent Zeros	Average Time/Unit	SAverage Time/Unit	QTable Number	Current Contents
Line	8	0.318	308	187	60.7	1.302	3.314		0

TABLE - ARRIVAL

Entries in Table	Mean Argument	Standard Deviation	Sum of Arguments	
21.000	14.7549	2.9686	540.5041	Non-Weighted

Upper Limit	Observed Frequency	Percent of Total	Cumulative Percentage	Cumulative Remainder	Multiple of Mean	Deviation from Mean
10.0000	2.0000	9.82	9.82	90.48	0.6818	-1.2326
20.0000	18.0000	85.71	92.24	4.76	1.3636	1.4087
30.00	1.0000	4.76	100.00	0.00	2.0455	4.0501

TABLE - WAITING

Entries in Table	Mean Argument	Standard Deviation	Sum of Arguments	
308	1.7549	2.9686	540.5041	Non-Weighted

Upper Limit	Observed Frequency	Percent of Total	Cumulative Percentage	Cumulative Remainder	Multiple of Mean	Deviation from Mean
1.00	202.00	68.58	65.58	36.42	0.5698	-0.2543
3.00	56.00	18.18	83.77	16.23	1.7095	0.4194
5.00	25.00	8.12	91.88	8.12	2.8492	1.0931
7.00	13.00	4.22	96.10	3.90	3.9889	1.7668

Figure 10-17. 10 Trucks and 2 Loaders Combination

Source: Al-Senani (n.d.), Figure 4.10; reproduced with permission.

continued

9.00	1.00	0.32	96.43	3.57	5.1285	2.4406
11.00	3.00	0.97	97.40	2.60	6.2682	3.1443

TABLE – DUMP

Entries in Table	Mean Argument	Standard Deviation	Sum of Arguments	
302	22.0176	37.6782	6921.1095	Non-Weighted

Upper Limit	Observed Frequency	Percent of Total	Cumulative Percentage	Cumulative Remainder	Multiple of Mean	Deviation from Mean
8.00	64.00	21.19	21.19	78.81	0.3491	-0.3959
10.00	50.00	16.56	37.75	62.25	0.4363	-0.3428
12.00	37.00	12.25	50.00	50.00	0.5236	-0.2898
14.00	18.00	5.96	55.96	44.04	0.6109	-0.2567
16.00	21.00	6.95	62.91	37.09	0.6982	-0.1836

Average Value of Overflow is 61.8970

Non Zero Full Word Save Values: (Name:Value)

LoadT: 31 Time: 1260 Volume: 9362 Income: 51941 Locos: 190

Trcos: 800 Mcost: 990 Tcost: 20790 Profit: 30701

The above is Antithetic Replication Number 5

Number of Trucks: 10

Number of Loaders: 2

Figure 10-17. *(continued)*

Table 10-7. Results for 6 Trucks and 1 Loader Combination

Replication Pair	Non-antithetic			Antithetic		
	Average Waiting (minutes)	Volume (yd^3 and m^3)	Profit (dollars)	Average Waiting (minutes)	Volume (yd^3 and m^3)	Profit (dollars)
1	6.1307	4,526 (3,460)	12,647	4.3234	6,231 (4,764)	22,195
2	5.4725	5,921 (4,527)	20,490	2.7367	4,247 (3,247)	10,942
3	5.3418	6,169 (4,726)	21,854	3.3772	4,433 (3,389)	12,306
4	4.5973	5,115 (3,911)	16,057	3.1798	5,053 (3,863)	15,716
5	3.0204	4,464 (3,414)	12,477	4.5477	5,890 (4,503)	20,320
6	2.4593	4,402 (3,366)	12,136	7.2319	6,820 (5,214)	25,435
7	5.6648	5,890 (3,403)	20,149	3.5647	5,487 (4,195)	17,933
8	4.3523	5,332 (4,077)	17,251	4.8477	5,642 (4,314)	18,956
9	4.8331	5,146 (3,924)	16,057	3.8080	4,588 (3,508)	13,159
10	3.6877	4,216 (3,223)	11,113	4.2820	5,053 (3,863)	15,716

Source: Al-Senani (n.d.), Table 4.1; reproduced with permission.
Note: Average waiting = 4.3729 min; average volume = 5,231.25 yd^3 (3,200 m^3); average profit = $16,645.45.

Table 10-8. Results for 6 Trucks and 2 Loaders Combination

Replication Pair	Non-antithetic			Antithetic		
	Average Waiting (minutes)	Volume (yd^3 and m^3)	Profit (dollars)	Average Waiting (minutes)	Volume (yd^3 and m^3)	Profit (dollars)
1	0.9642	5,549 (4,242)	16,449	0.8846	7,099 (5,427)	24,633
2	0.9326	6,541 (5,001)	21,905	0.9571	4,154 (3,176)	8,777
3	0.9999	7,533 (5,759)	27,361	0.7282	5,270 (4,029)	14,403
4	0.9618	5,828 (4,456)	17,643	0.8185	5,177 (3,948)	14,403
5	0.8375	4,247 (3,247)	8,947	0.9510	6,541 (5,001)	21,735
6	0.7923	5,456 (4,171)	15,597	1.2736	8,091 (6,186)	30,430
7	0.9865	6,758 (5,166)	22,758	1.0258	6,665 (5,095)	22,417
8	0.9116	6,169 (4,716)	19,859	0.9283	5,890 (4,503)	17,984
9	0.9560	6,726 (5,142)	22,928	0.8565	4,526 (4,360)	10,823
10	0.8263	5,270 (4,029)	14,915	1.0301	5,053 (3,863)	13,721

Source: Al-Senani (n.d.), Table 4.2; reproduced with permission.
Note: Average waiting = 0.9312 min; average volume = 5,927.15 yd^3 (4,531.60 m^3); average profit = $18,384.40.

Table 10-9. Results for 7 Trucks and 1 Loader Combination

Replication Pair	Non-antithetic			Antithetic		
	Average Waiting (minutes)	Volume (yd^3 and m^3)	Profit (dollars)	Average Waiting (minutes)	Volume (yd^3 and m^3)	Profit (dollars)
1	6.4284	6,014 (4,598)	33,077	6.3686	6,820 (5,214)	23,584
2	7.9437	6,665 (5,096)	22,732	3.7050	4,619 (3,531)	11,649
3	7.5544	6,975 (5,333)	24,607	6.3686	6,045 (4,622)	19,151
4	6.4918	5,797 (4,431)	18,128	5.0603	6,107 (4,669)	24,607
5	3.9704	5,766 (4,408)	17,958	8.4148	6,975 (5,332)	24,607
6	3.5350	5,642 (4,314)	17,105	11.1643	7,223 (5,522)	25,971
7	7.9672	6,634 (5,072)	22,732	6.2480	6,293 (4,811)	20,686
8	6.5534	6,603 (5,048)	22,561	6.5908	5,828 (4,379)	18,299
9	6.2425	5,735 (4,385)	17,787	5.9233	5,177 (3,948)	14,718
10	4.4938	5,053 (3,863)	14,036	7.1934	6,448 (4,930)	21,709

Source: Al-Senani (n.d.), Table 4.3; reproduced with permission.
Note: Average waiting = 6.4109 min; average volume = 6,120.95 yd^3 (4,679.47 m^3); average profit = $20,546.50.

Table 10-10. Results for 7 Trucks and 2 Loaders Combination

Replication Pair	Non-antithetic			Antithetic		
	Average Waiting (minutes)	Volume (yd^3 and m^3)	Profit (dollars)	Average Waiting (minutes)	Volume (yd^3 and m^3)	Profit (dollars)
1	1.2115	5,828 (4,456)	16,304	1.2262	8,215 (6,281)	29,432
2	1.2599	7,099 (5,428)	23,294	1.0703	4,681 (3,579)	9,825
3	0.9693	8,556 (6,514)	31,308	1.0061	6,355 (4,759)	19,202
4	1.0870	6,789 (5,190)	21,589	0.9613	6,758 (5,167)	21,419
5	0.9335	5,642 (3,549)	15,281	1.3420	7,657 (5,854)	26,363
6	1.0392	6,293 (4,811)	18,861	1.5074	9,300 (7,110)	35,059
7	1.3892	8,060 (6,162)	28,409	1.1405	6,510 (4,977)	19,884
8	1.1709	7,073 (5,410)	22,783	1.1157	6,696 (5,119)	20,907
9	1.1193	6,913 (5,285)	21,930	1.0181	5,704 (4,361)	15,281
10	1.1231	5,673 (4,337)	15,451	1.0132	6,696 (5,119)	21,078

Source: Al-Senani (n.d.), Table 4.4; reproduced with permission.
Note: Average waiting = 1.1352 min; average volume = 6,824.9 yd^3 (5,217.64 m^3); average profit = $21,683.00.

Table 10-11. Results for 8 Trucks and 1 Loader Combination

Replication Pair	Non-antithetic			Antithetic		
	Averge Waiting (minutes)	Volume (yd^3 and m^3)	Profit (dollars)	Average Waiting (minutes)	Volume (yd^3 and m^3)	Profit (dollars)
1	9.4506	6,541 (5,001)	20,540	7.8720	7,006 (5,356)	23,098
2	11.1792	6,975 (5,333)	22,927	5.4128	5,208 (3,982)	13,038
3	8.9690	7,378 (5,641)	25,144	7.2122	6,231 (4,764)	18,835
4	8.3084	6,231 (4,764)	18,665	5.7805	6,200 (4,740)	18,494
5	5.8515	6,138 (4,693)	18,153	10.8461	7,192 (5,498)	23,950
6	4.9554	6,262 (4,788)	19,006	13.9704	7,192 (5,422)	24,121
7	10.7322	7,068 (5,404)	23,268	7.2405	6,386 (4,882)	19,176
8	8.7882	6,944 (5,309)	22,757	8.6779	6,200 (4,740)	18,494
9	8.9930	6,355 (4,858)	19,517	7.6720	5,952 (4,321)	17,130
10	5.9143	5,921 (4,527)	16,960	11.2255	7,068 (5,404)	23,268

Source: Al-Senani (n.d.), Table 4.5; reproduced with permission.
Note: Average waiting = 8.0690 min, average volume = 6,522.40 yd^3 (4,986.48 m^3); average profit = \$20,327.05.

Table 10-12. Results for 8 Trucks and 2 Loaders Combination

Replication Pair	Non-antithetic			Antithetic		
	Average Waiting (minutes)	Volume (yd^3 and m^3)	Profit (dollars)	Average Waiting (minutes)	Volume (yd^3 and m^3)	Profit (dollars)
1	1.3712	6,820 (5,214)	20,080	1.2980	8,711 (6,660)	30,480
2	1.4093	9,021 (6,897)	32,015	1.2182	5,890 (4,503)	14,624
3	1.5032	9,610 (7,347)	35,084	1.1218	7,161 (5,485)	21,955
4	1.3623	7,998 (6,115)	26,559	1.1612	6,975 (5,333)	20,932
5	1.0573	6,355 (4,858)	17,522	1.5216	8,401 (7,187)	28,605
6	0.9699	7,068 (5,404)	21,273	1.9353	9,641 (7,310)	35,425
7	1.5284	8,804 (6,731)	30,310	1.3877	8,494 (6,494)	29,116
8	1.3578	7,750 (5,925)	24,854	1.1600	7,223 (5,522)	21,955
9	1.2486	7,254 (5,546)	22,467	1.3181	6,696 (5,119)	19,398
10	1.2804	7,316 (5,593)	22,467	1.3284	8,181 (6,255)	27,241

Source: Al-Senani (n.d.), Table 4.6; reproduced with permission.
Note: Average waiting = 1.3269 min; average volume = 7,768.6 yd^3 (5,929.09 m^3); average profit = \$22,590.01.

Table 10-13. Results for 9 Trucks and 1 Loader Combination

Replication Pair	Non-antithetic			Antithetic		
	Average Waiting (minutes)	Volume (yd^3 and m^3)	Profit (dollars)	Average Waiting (minutes)	Volume (yd^3 and m^3)	Profit (dollars)
1	10.7263	6,448 (4,930)	18,008	10.9660	7,502 (5,735)	24,146
2	12.5699	7,006 (5,356)	21,247	6.8246	6,076 (4,645)	16,132
3	12.6179	7,595 (5,806)	24,657	9.4677	6,758 (5,167)	19,542
4	11.7238	7,161 (5,475)	21,759	8.5170	6,758 (5,166)	20,054
5	6.7743	6,572 (5,024)	19,031	14.2233	7,440 (5,688)	23,805
6	6.2295	6,913 (5,285)	20,736	17.9718	7,378 (5,641)	23,293
7	13.2736	7,192 (5,498)	22,441	11.0068	7,037 (5,800)	21,418
8	12.3614	7,378 (5,656)	23,464	12.9146	7,161 (5,479)	22,270
9	11.3326	6,727 (5,143)	19,201	9.2946	6,355 (4,859)	17,667
10	8.0882	6,479 (4,953)	18,519	14.1549	7,192 (5,422)	22,441

Source: Al-Senani (n.d.), Table 4.7; reproduced with permission.
Note: Average waiting = 11.05189 min; average volume = 7,294.30 yd^3 (5,576.49 m^3); average profit = $22,224.40.

Table 10-14. Results for 9 Trucks and 2 Loaders Combination

Replication Pair	Non-antithetic			Antithetic		
	Average Waiting (minutes)	Volume (yd^3 and m^3)	Profit (dollars)	Average Waiting (minutes)	Volume (yd^3 and m^3)	Profit (dollars)
1	1.5625	7,719 (5,901)	23,174	1.4418	8,804 (6,731)	29,312
2	1.8922	9,300 (7,110)	32,040	1.3645	6,882 (5,261)	18,857
3	1.8070	10,571 (8,082)	39,030	1.5076	8,401 (6,423)	27,095
4	1.8251	8,308 (6,352)	26,584	1.5656	8,494 (6,494)	27,607
5	1.3386	7,595 (5,806)	22,662	1.7144	9,052 (6,921)	30,505
6	1.3112	8,494 (6,494)	27,095	2.1839	10,726 (8,200)	39,712
7	1.8388	9,238 (7,063)	31,358	1.6994	8,835 (6,755)	29,312
8	1.7492	9,207 (7,039)	31,187	1.5300	8,742 (6,683)	28,971
9	1.5819	8,308 (6,352)	26,072	2.4289	6,665 (5,095)	17,377
10	1.4011	8,370 (6,399)	26,754	1.8983	9,114 (6,968)	30,846

Source: Al-Senani (n.d.), Table 4.8; reproduced with permission.
Note: Average waiting = 1.6821 min; average volume = 8,641.25 yd^3 (6,606.24 m^3); average profit = $28,720.80.

Table 10-15. Results for 10 Trucks and 1 Loader Combination

Replication Pair	Non-antithetic			Antithetic		
	Average Waiting (minutes)	Volume (yd³ and m³)	Profit (dollars)	Average Waiting (minutes)	Volume (yd³ and m³)	Profit (dollars)
1	14.5642	6,820 (5,214)	18,715	13.3841	7,750 (5,925)	23,830
2	16.3855	7,254 (5,546)	20,931	9.9706	6,944 (5,309)	19,226
3	15.8522	7,688 (5,878)	22,977	9.5237	6,634 (5,072)	17,692
4	13.7054	7,223 (5,522)	20,590	12.7773	7,130 (5,451)	20,249
5	8.8623	7,037 (5,380)	19,567	18.4085	7,378 (5,641)	21,784
6	10.3108	7,595 (5,806)	22,977	22.8895	7,409 (5,664)	21,954
7	17.0890	7,440 (5,688)	22,125	14.9970	7,316 (5,591)	21,272
8	15.5337	7,505 (5,758)	22,466	16.1731	7,316 (5,591)	21,272
9	13.8724	7,130 (5,451)	20,420	11.9584	6,820 (5,214)	18,715
10	11.8609	7,130 (5,451)	20,249	18.1878	7,223 (5,522)	20,761

Source: Al-Senani (n.d.), Table 4.9; reproduced with permission.
Note: Average waiting = 14.3153 min, average volume = 6,880.45 yd³ (5,260.10 m³); average profit = $20,888.60.

Table 10-16. Results for 10 Trucks and 2 Loaders Combination

Replication Pair	Non-antithetic			Antithetic		
	Average Waiting (minutes)	Volume (yd³ and m³)	Profit (dollars)	Average Waiting (minutes)	Volume (yd³ and m³)	Profit (dollars)
1	2.0648	9,641 (7,371)	32,235	1.8100	9,610 (7,347)	32,065
2	2.2836	10,416 (7,963)	36,498	1.6904	8,370 (6,399)	25,245
3	2.3204	11,160 (8,532)	40,590	1.5246	8,587 (6,565)	26,268
4	2.1381	9,610 (7,347)	31,553	1.7657	8,339 (6,375)	25,074
5	1.4983	7,874 (6,035)	22,517	1.7549	9,362 (7,158)	30,701
6	1.2723	8,773 (6,707)	27,461	2.6443	11,532 (8,816)	42,465
7	2.2973	10,664 (8,138)	37,862	2.1000	9,300 (7,110)	30,189
8	2.1203	9,951 (7,608)	30,189	2.1911	9,021 (6,897)	28,655
9	1.9004	10,012 (7,554)	34,111	1.6384	7,564 (5,783)	20,641
10	1.6627	8,897 (6,802)	28,143	1.9436	9,083 (6,944)	28,996

Source: Al-Senani (n.d.), Table 4.10; reproduced with permission.
Note: Average waiting = 1.9311 min; average volume = 9,388.35 yd³ (7,139.16 m³); average profit = $30,572.90.

For 10 trucks and 2 loaders, the productivity of the system theoretically would be as follows:

Truck Productivity

$$\begin{aligned}&\text{Truck cycles per hour} = 60/24.1 \text{ min}\\&= 2.48 \text{ cycles/h}\\&\text{For } T \text{ number of trucks,}\\&\text{truck production} = \text{number of cycles per hour}\\&\times \text{number of minutes per cycle} \times T\\&= 2.48 \times 29.86 \times T = 74.05T \text{ yd}^3 (56.62 \text{ m}^3)\end{aligned} \quad (10\text{-}7)$$

Loader Productivity

$$\begin{aligned}&\text{Loader cycles per hour} = 60/0.75 = 80 \text{ cycles/h}\\&L = \text{number of loaders,}\\&\text{loader production per loader}\\&= \text{number of cycles per hour} \times \text{bucket capacity}\\&= 80 \times 4 \text{ yd}^3\\&= 320 \text{ loose yd}^3/\text{h}\\&(80 \times 3.06 \text{ m}^3 = 244.66 \text{ loose m}^3/\text{h})\end{aligned} \quad (10\text{-}8)$$

For the system to be balanced, the theoretical maximum loader production needs to equal truck production.

$$\begin{aligned}&\text{balance point} = \text{loader production/truck production}\\&= 320 \text{ loose yd}^3/74.05 \text{ loose yd}^3 \ (244.66 \text{ loose m}^3/56.62 \text{ loose m}^3)\\&= 4.32 \text{ trucks/loader}\end{aligned} \quad (10\text{-}9)$$

With two loaders the total number of trucks that can be serviced is

$$= 4.32 \times 2 = 8.64 \text{ or } 9 \text{ trucks}$$

Whereas the deterministic method indicates that the most efficient combination would be 9 trucks and 2 loaders, the simulation model showed the most efficient combination to be 10 trucks and 2 loaders.

$$\begin{aligned}&\text{productivity of the system by the deterministic method}\\&= \text{production of loaders per day} \times \text{number}\\&\text{of loaders} \times \text{number of loads}\\&= 320 \text{ loose yd}^3/\text{day} \times 2 \times 21\\&= 13{,}440 \text{ yd}^3\\&(.244.66 \text{ loose m}^3/\text{day} \times 2 \times 21 = 10{,}275.72 \text{ m}^3)\end{aligned} \quad (10\text{-}10)$$

The productivity of the system determined by the model was 9,388.35 yd^3 (7,171.86 m^3).

Confidence Limit

Confidence limits for the results obtained in the 10 trucks and 2 loaders combination simulation were calculated to determine the reliability of the results. The small sample confidence interval for the mean was found as follows:

$$\overline{X} \pm t\alpha/2(S/\sqrt{n}) \qquad (10\text{-}11)$$

where $\overline{X}$ = mean; $t\alpha/2$ = t statistics; S (standard deviation) = $\sqrt{\sum(\overline{x}-x)^2/(n-1)}$; n = number of samples.

To use the above equation, it was assumed that the values for the results obtained from the model were normally distributed. The t statistics are from statistical tables and are based on confidence levels of 95% (any value from 80% to 95% confidence level could be assumed), and the sample size was 10 (the number of replication pairs).

Sample means, standard deviations, and confidence intervals were t statistics for a 95% confidence level and 10 samples = 2.26.

1. Average time waiting at the truck queue
   - Mean = 1.9311 min
   - Standard deviation (S) = 0.1699
   - 95% confidence interval = [1.8927, 1.970]
2. Volume of soil
   - Mean = 9,388.35 yd^3 (7,177.86 m^3)
   - Standard deviation (S) = 529.226 yd^3 (404.62 m^3)
   - 95% confidence interval = [9,010.126 yd^3, 9,766.574 yd^3], or [6,888.74 m^3, 7,467.03 m^3]
3. Profit
   - Mean = $30,572.90
   - Standard deviation (S) = $2,859.29
   - 95% confidence interval = [$28,529.43, $32,616.36]

The standard deviation for the results was from 0.6% to 0.9% of the mean, which indicated that the reliability of the model was high.

Sensitivity Analysis

Sensitivity analysis was used to determine the deciding factor in the selection of equipment. The distribution for truck travel was changed from exponential to uniform. The choice of equipment combinations was the same, but production was overestimated, and the results were close to the deterministic values. The distribution for the loading time was changed from an Erlang distribution to a constant value (5 min), and the actual results obtained during construction were close to the results predicted by the simulation model.

10.10 Sustainable Heavy Construction Equipment

This section discusses some of the sustainable technologies available in the heavy construction equipment industry that are considered when selecting heavy construction equipment and calculating its productivity. Sustainable technologies include tires, **engine repowering**, engine upgrades, cooled exhaust gas recirculation, selective catalytic reduction, remanufacturing and rebuilding engines, and hybrid-electric heavy construction equipment. This section also introduces the EPA Tier Four Final Standards for heavy construction equipment.

Tires

Sustainable technologies are being incorporated into manufacturing practices for producing components of heavy construction equipment. For instance, manufacturers of truck tires have developed a technology for retreading old tires. A retreaded tire sometimes lasts for up to 80% of the mileage of a new tire (Michelin 2014). New tires contain 20% to 30% natural rubber; therefore, retreading tires saves natural resources and decreases the number of tires that will be disposed of at the end of their useful life. When heavy construction equipment with retreaded tires is evaluated to determine its productivity, the analysis needs to include the 20% reduction in the life of the tires.

In 1992 Michelin launched a third generation of energy-saving tires called **Energy Saver Green tires**. The tread technology in these tires improves traction and makes the tires self-cleaning. The ability of the tire to expel the soil that collects between the tread blocks helps improve their gripping potential and reduces rolling resistance and fuel consumption. The use of these tires reduces fuel consumption by up to 3% (which is equivalent to 1 gal. per 62.14 mi, or 1 L per 100 km, of fuel consumption) compared with traditional tires used on three-axle trailers (Michelin 2014). In addition, lower rolling resistance translates into a reduction in CO_2 emissions. The reduction in fuel consumption and rolling resistance should be included in productivity studies involving equipment with Energy Saver Green tires.

Emissions

Reducing heavy construction equipment emissions is another method for introducing sustainability practices into the construction industry. Diesel engines, for example, emit less hydrocarbon (HC), carbon monoxide (CO), and other toxic air pollutants than gasoline engines. Using diesel engines on heavy construction equipment also increases fuel economy; however, diesel engines have the disadvantage of emitting significant amounts of particulate matter (PM) and nitrogen oxide (NO_x).

A report released by the EPA in April 2008 summarized energy consumption and greenhouse gas (GHG) emissions data pertaining to the operation of off-road heavy construction and mining equipment. The report indicated that in 2006 heavy construction equipment consumed 5,968 million gal. (4,968.96 million imperial gallons, or 22,603.8 million L) of diesel fuel, which is equivalent to 827.8 trillion Btu (730,012 trillion joules) at 138,700 Btu/gal., and 688 million gal. (2,605.8 million L or 572.83 million imperial gal.) of gasoline, which is equivalent to 86.04 trillion Btu at 138,700 Btu/gal. of diesel fuel. In 2006 in the United States, the construction industry used 913.85 trillion Btu, and this represented 1.2% of the total U.S. energy consumption. The EPA has estimated that 47% of mobile source diesel particulate matter emissions nationwide comes from off-road diesel engines, and 25% of mobile source NO_x comes from off-road diesel engines (EPA 2013).

In May 2004, the EPA introduced a diesel engine pollution control measure called the Clean Air Nonroad Diesel Rule to help reduce the pollution caused by diesel-powered equipment used in the agriculture, construction, and mining industries. This off-road diesel program was implemented in 2008 and resulted in a reduction in annual PM emissions of 129,000 tons (117,028.8 metric tons) and nitrogen oxide (NO_x) emissions of 738,000 tons (669,513.6 metric tons) (EPA 2009a).

Engine manufacturers have developed new engines with advanced emission-control technologies, such as diesel-retrofit systems. **Diesel-retrofit technologies** (DRT) are devices that can be attached to the engines of heavy construction equipment to help remove pollutants, such as PM and NO_x emissions, from the engine exhaust systems. Retrofit equipment is being installed on school buses, long-haul trucks, heavy construction equipment, and mining equipment.

The two most common diesel-retrofit technologies are **diesel oxidation catalysts** (DOCs) and **diesel particulate filters** (DPFs). Both of these help reduce PM, CO, and HC emissions. Diesel oxidation catalyst devices employ chemicals that react with the exhaust stream gases to convert them into inert or less harmful products. A DOC might reduce the concentration of PM by 20%, CO by 40%, and HC by 50% in a diesel engine exhaust system. Diesel oxidation catalysts are also called catalytic converters. Diesel oxidation catalysts are used not only with conventional diesel fuel but also with biodiesel and other alternative diesel fuels (Wescott 2005).

Diesel particulate devices use filters to reduce PM in exhaust systems. They might be used on their own but are best used in conjunction with an **ultra-low-sulfur diesel** (ULSD) **fuel** because of the damage caused by the sulfur in off-road diesel vehicles. Ultra-low-sulfur diesel fuel is an environmentally friendly fuel that contains less than 15% sulfur. The use of DPFs and ULSD could reduce PM, HC, and CO emissions by 60% to 90%.

Ainslie et al. (1999) wrote a report for the Society of Automotive Engineers (SAE) about a testing program conducted to study the emissions and duty cycles from five heavy-duty construction vehicles. The authors confirmed that retrofitting exhaust emission control technologies used on off-road construction equipment is feasible and that emission reductions could be achieved by retrofitting vehicles. For instance, a Caterpillar wheeled loader equipped with a catalyzed DPF reduced emissions by 97% PM and improved gaseous control. A backhoe equipped with active DPF showed PM reductions of 81%. According to the EPA, reductions in NO_x and PM emissions from off-road diesel engines provide public health benefits (EPA 2009b). The EPA estimates that by 2030, controlling these emissions should annually prevent 12,000 premature deaths, 8,900 hospitalizations, and 1 million workdays lost (EPA 2009b).

Biodiesel Fuel

Although productivity improvement studies typically focus on increasing production rates, they might also include recommendations for the use of environmentally sustainable technologies, such as those listed in this section. If the options also reduce fuel consumption, then there is even more incentive for recommending their implementation on projects.

Biodiesel fuel is a renewable plant- or animal-based diesel fuel substitute composed of mono-alkyl esters of long-chain fatty acids derived from vegetable oils or animal fats (National Biodiesel Board 2014). Biofuels are usually blended with gasoline or diesel fuel at low levels, such as 20% (B20) or less, but they can also be used at 100% (B100).

The ASTM specification for B100 diesel fuel (ASTM 2011) covers biodiesel fuel blend stock, B100, in Grades S15 and S500 for use as a blend component with middle distillate fuels. This specification describes the required properties of biodiesel fuels. The use of biodiesel fuel in a conventional diesel engine results in a substantial reduction in unburned HC, CO, and PM, decreases the solid carbon fraction of PM, and helps to eliminate sulfur, while the HC fraction remains the same or is increased. According to the EPA, B20 biodiesel fuel decreases PM by about 10% but increases NO_x by approximately 2% (EPA 2013). According to a manufacturer of emissions-controlled devices, retrofitted DOCs and DPFs can operate effectively on vehicles using a biodiesel-blended fuel up to B20. The EGR engines manufactured by John Deere operate efficiently with traditional low-sulfur diesel (LSD) fuels as well as B5 to B20 biodiesel fuel blends.

In *A Comprehensive Analysis of Biodiesel Impacts on Exhaust Emissions*, the EPA (2002) concluded that even though the use of biodiesel fuels reduces the emissions of PM, HC, and CO, it also increases the emission of NO_x. Specific NO_x increases depend on the fuel blend used, equipment type, and operating patterns of the equipment or vehicle. Using B20 fuel results in a NO_x emissions increase of approximately 2%, and with B100 in heavy-duty highway engines, the increase was approximately 10%.

The cost of corn and other biofuel feedstock is causing the price of biofuels to increase along with the price of the foods derived from these farm products. The European Commission has written a report on indirect land use change (ILUC) that says carbon emissions are increasing as croplands are converted for ethanol or biodiesel fuel production in response to the increased global demand for biofuels.

Biodiesel fuels might be included in productivity improvement recommendations for reducing the operating cost of heavy construction equipment, as well as for reducing emissions of PM, HC, and CO.

In response to the U.S. EPA Interim Tier 4 (IT4)/Stage III B emissions regulations for diesel engines 174 hp and above, equipment manufacturer Caterpillar indicated that its C18 ACERT industrial engines are designed to use B20 biofuel (Caterpillar 2013). ACERT industrial engines are compliant with the U.S. EPA Tier 3 emissions regulations governing off-road machines, which took effect January 1, 2005, for engines of 300 to 750 hp. The fuel system allows for multiple injections during each combustion cycle. Small amounts of fuel are injected at precise times to achieve the combined goals of fuel economy and lower emissions. An advanced air system provides more cool air in the combustion chamber. A waste-gate turbocharger provides excellent low-end response. In addition, cross-flow cylinder heads provide a direct path of air to the engine (Caterpillar 2013). The Caterpillar 349E hydraulic excavator has an Interim IT4, 396-net-hp C13 ACERT engine. This excavator is able to operate on either ultra low, sulfur diesel ULSD or B20 fuel or a combination of diesel fuel and 20% biodiesel fuel (Caterpillar 2013).

Environmental Protection Agency Tier Four Final Standards

In 2014, the EPA fully implemented its ***Tier Four Final Standards*** that were part of the phase in of the multi-tiered emissions reduction process that started in 1996 to reduce PM and NO_x emissions by 90%. As a result of the new Tier Four requirements, manufacturers have created **cooled exhaust gas recirculation** (CEGR) and **selective catalytic reduction** (SCR) systems for heavy construction equipment. Selective catalytic reduction systems use higher combustion temperatures and urea-based diesel exhaust fluid (DEF) after-treatment. Cooled exhaust gas recirculation systems mix the exhaust gases with fresh air before recirculating it. A diesel particulate filter (DPF) is still required to filter soot. Meeting Tier Four requirements might add an additional cost of 25% to the equipment. In addition, the engines are more sensitive to water, dust, and extreme temperatures. They may require low-ash engine oil, ultra-low sulfur diesel, and an extra fuel filtration system (*Engineering News Record* 2014a).

In addition to emission reductions, these technologies also reduce fuel consumption. For instance, according to the heavy construction equipment manufacturer John Deere, extensive testing of its key products featuring cooled exhaust gas recirculation (EGR) platform engines for NO_x control—including the 350D excavator, 700J crawler dozer, and the new 772G motor grader—showed a 10% or more increase in material moved per unit of fuel used, as compared with the equipment manufactured by their competitors (John Deere 2011).

Specific pieces of equipment have been designed to meet Tier Four Standards such as the Kelly Tractor Company IMT Tier 4 A150 hydraulic drill rig. It is powered by a Caterpillar C7.1 ACERT engine that delivers 118,000 ft/lb (53,523.62 m/kg) of rotary torque using 217 hp (*Engineering News Record* 2014b). The Manitowoc lattice-boom crawler crane also is available with either Tier 3 or Tier 4 Cummins engine.

Engine Repowering and Engine Upgrades

Another method for reducing emissions in heavy construction equipment is engine repowering, and it involves replacing an existing engine with a new engine that meets lower emission standards than the original engine. Engine repowering involves the use of on-road engines to replace existing off-road engines. Depending on the type and year of manufacture of the on-road engine, the replacement engine could reduce PM emissions by 90% and NO_x by 70%, compared with the off-road engine. Repowering could lead to credits in retrofit requirements in environmental regulations (Caterpillar 2010).

Engine upgrading occurs when emissions-reducing parts are added to an existing engine during an engine rebuild. This involves installing an upgrade kit to bring the old construction equipment up to current codes. According to the EPA (2007), upgrading an engine during a rebuild allows companies to modernize equipment at a lower marginal cost.

The emission upgrade design manufactured by Caterpillar (Model 3306 diesel engines with mechanical direct fuel injection for off-road applications, compatible with model years 1988–1995) has been verified by the EPA to reduce PM emissions by 22%, NO_x by 37%, HC by 71%, and CO by 13%. According to the EPA (2007) report *Cleaner Diesels: Low Cost Ways to Reduce Emissions from Construction Equipment*, the cost-effectiveness of repowering a piece of equipment depends on the make and model of the machine and the availability of funds to defray the costs. For Sukut Equipment Inc., repowering single-engine scrapers costs up to $120,000 (Sacramento Metropolitan Air Quality Management District 2011). Therefore, the cost of repowering should be compared with the cost of buying new equipment before a decision is made to repower.

Remanufacturing and Rebuilding Equipment

Caterpillar offers the **CAT Remanufacturing Service**, a one-for-one exchange in which end-of-life products are returned for remanufactured products. The service reduces waste and minimizes the need for raw materials to produce brand-new heavy construction equipment (Caterpillar 2011). Caterpillar defines remanufacturing as the process of using a manufacturing and quality-control system to refurbish worn-out equipment. The equipment is rebuilt to operate like a new machine at a fraction of the cost of a new product. The core is completely disassembled into its constituent parts, down to the level of individual nuts and bolts. The parts are cleaned using environmentally friendly processes and then inspected for remanufacturability using detailed Caterpillar remanufacturing criteria. Fewer resources are consumed to remanufacture a component than to build a completely new one. Remanufacturing is more environmentally friendly than recycling in that remanufacturing dramatically lowers the use of new resources (Caterpillar 2011). According to its 2008 sustainability report, *The Big Picture*, Caterpillar processes nearly 3 billion lb (1.36 billion kg) of remanufactured products per year and uses close to 70% recycled materials in the manufacture of its engines, transmissions, hydraulic locomotives, and railcars (Caterpillar 2008).

Other Technological Advances in Heavy Construction Equipment

Advancements in technology have made it possible to optimize construction processes for efficiency. One example is **intelligent compaction** (IC), a technology used to measure, monitor, and evaluate the stiffness of the layers of soil, aggregate bases, and asphalt materials during road construction. The IC system employs modern vibratory rollers equipped with an in situ measurement system and feedback control. Often, global positioning system (GPS)–based mapping is included, along with software that automates the documentation of results. The ability to continuously measure stiffness, both during the compaction process to aid in optimum compaction and as an acceptance or design tool on the in-place material, improves highway engineering. The possible benefits are immediate identification of weak areas that need to be recompacted and the avoidance of harmful overcompaction, both of which save time and money.

Another new technology that enhances the sustainability of a process in the construction sector is the Caterpillar AccuGrade grade-control system. The AccuGrade system increases productivity by up to 40%. It is factory integrated, is sensor independent, and features a suite of products, including cross-slope, sonic, laser, and GPS technology (Caterpillar 2011). By combining digital design data, in-cab operator guidance features, and automatic blade controls, the AccuGrade grade-control system enhances grading accuracy and helps eliminate the need for survey stakes.

10.11 Hybrid-Electric Heavy Construction Equipment

Hybrid-electric heavy construction equipment may reduce the consumption of gasoline but may not always be a sustainable alternative. If hybrid vehicles use electricity to recharge and the electricity in the recharging system is generated by burning coal, then the GHG emissions could be comparable to using gasoline or diesel fuel. If the electricity comes from a mix of renewable and traditional energy sources and coal constitutes only half of them, then hybrid vehicles reduce emissions of GHGs by 50% (Begley 2008).

The EPA requirements for reduced emissions and the need to develop fuel-efficient engines have resulted in heavy construction equipment manufacturers developing hybrid-electric vehicles. Hybrid-electric vehicles use more than one power source. Hybrid-electric systems for construction equipment reduce fuel consumption and CO_2 emissions when the electric motor that turns the upper structure of the hybrid hydraulic excavator converts kinematic energy—regenerated when the turning of the upper structure slows down—into electric energy. This electric energy is then stored in the capacitor and reused for the next turning of the upper structure. The power-generating motor also reuses it as extra energy to accelerate the engine revolution speed.

Volvo Hybrid Wheel Loader

Volvo Construction Equipment (Volvo CE) created the L220F hybrid wheel loader, which offers more power, better performance, and a 10% reduction in fuel consumption (Volvo Construction Equipment 2011). The Volvo hybrid system includes an **integrated starter generator** (ISG), mounted between the engine and the transmission, that is coupled to an advanced battery. An ISG is an electronically controlled electrical unit used in place of the conventional starter motor and generator used in internal combustion engines. It can be used as a starter, a booster electric motor, a generator, and an electric propulsion unit. When a wheel loader is being used, it idles for up to 40% of the time. The ISG allows the diesel engine to shut off when the machine is stationary and to restart almost instantly by rapidly spinning the engine up to optimum working speed using the high-powered battery. The ISG also mitigates the problem of low torque at low engine speeds by automatically offering a massive electric torque boost. The ISG's 50-kW electric motor offers torque up to 516 lb/ft (700 Nm) from standstill (Van Hampton et al. 2008).

Caterpillar D7E Electric Dozer

In 2008, Caterpillar released a D7E electric bulldozer. According to the *Engineering News Record* (*Engineering News Record* 2008, p. 1),

> A 9L ACERT diesel drives a generator, whose wiring harness has effectively replaced the driveshaft. It runs a power inverter wired to two WC **liquid-cooled electric motors** mated to an axle containing two **double-reduction gear sets**. A third plenary set in between, powered hydraulically, controls differential steering. Transmission is continuously variable, eliminating the need for extra valving and gearing. The engines are beltless, and the entire machines weights about 3,000 lb [1,360.77 kg] less than the current D7R. By taking out 60% of the moving parts and lightening the load, Cat is able to cut down on parasitic energy losses to get a 20% fuel economy improvement.

Figure 10-18 shows the Caterpillar D7E dozer. In 2013, Caterpillar released another hybrid construction vehicle, the 336H excavator, which operates under similar principles as the D7E bulldozer.

Figure 10-18. Caterpillar D7E Electric Dozer

Source: Open Source Photo.

Komatsu PC200LC-8 Hybrid Excavator

In 2008, Komatsu, the second-largest producer of heavy construction equipment in the world, released its PC200-8 hybrid excavator and HB215LC-1 excavator (Komatsu 2008). The HB215LC-1 excavator has three main components in its hybrid drive system. Its design incorporates an electric swing motor, an ultracapacitor, and a generator. Electricity is stored in the ultracapacitor, which sends energy to the electric swing or to the generator/motor to power the engine. One key feature is that the electric swing motor generates and stores electricity during swing braking that is then reused by the capacitor. Figure 10-19 shows the Komatsu PC200LC-8 hybrid excavator.

Regenerating its own energy is what allows the hybrid excavator to be more efficient and increase fuel savings. The 20-ton (18.144-metric ton) hybrid excavator also reduces emissions, and the average fuel savings are 25% when compared with traditional heavy construction equipment that is the same size. The emissions reduction is equivalent to using 14 hybrid vehicles. In an average work year, the HB215LC-1 hybrid excavator reduces fuel consumption by 1,500 gal. (1,248.9 imperial gal. and 5,681.25 L) of diesel fuel or 6,300 gal. (5,245.38 imperial gal. and 23,861.25 L) of crude oil and will produce 25% less CO_2 than a standard excavator without hybrid technology (Komatsu 2008).

Komatsu is marketing the 22-ton (19.96-metric ton) PC200LC-8 hybrid excavator in Asia and the United States. The excavator is 25–40% more fuel efficient than the diesel-powered version and emits 22 lb (9.98 kg) less of CO_2 per hour of operation. Unfortunately, in 2013, the new Komatsu hybrid excavator cost 50% more than the diesel version of the same model.

The Komatsu hybrid excavator uses a "diesel engine, an electric-swing motor, a generator, a capacitor and pumps. As the swinging superstructure slows down, kinetic energy converts to electricity, which is sent through an inverter and then is captured

Figure 10-19. Komatsu PC200LC-8 Hybrid Excavator

Source: Open Source Photo.

by a capacitor.... The generator/motor is located behind the engine and the hydraulic pumps. It can charge the capacitor during periods of downtime, and it can receive power from the capacitor for engine assist, determined by the power controller" (*Engineering News Record* 2010, pp. 12–13). The hybrid excavator is rated at 138 hp and has a four-cylinder 4.5-L engine. The traditional version of the excavator is rated at 148 hp and has a 6.7-L six-cylinder Komatsu turbo-diesel engine.

John Deere Diesel-Electric 644K and 944K Hybrid Wheel Loaders

John Deere manufactures a diesel-electric construction 644K hybrid wheel loader and a more-efficient 944K loader. The new John Deere technology uses internal combustion engines with electric motors. Some of the existing loaders used for quarrying operations consume up to $200,000 per year in fuel if they are operated two or three shifts every day. The new 644K is projected to save between 15% and 20% per year, and the 944K should save 25% to 30% per year in fuel costs. The 644K uses existing technology, and the 944K uses all new technology. According to John Deere (2011, pp. 26–27):

> Deere's 944K hybrid starts with a 13.5 L diesel engine that produces about 500 hp. It connects to two generators, each of which powers two motors, one at each wheel. Sandwiched between the engine and generators is a pump drive, a simple gearbox that grabs power from the engine's flywheel to drive the hydraulic pumps for the bucket and steering. The pumps run 20% faster than engine speed; the generators run at three times engine speed.
>
> The generators send AC power to an inverter assembly, which converts the power to DC current to run accessories, then switches it back to AC to run the four outboard electric wheel motors. Overall the system runs at 700 volts. A computer can sense when the wheels are slipping and adjust the power to boost traction.
>
> Though it has no traditional energy storage, the 944K captures some regenerative braking when the machine is slowing down by sending power back to the generators to drive the hydraulics. Unused energy is "cooked off" in brake resistors.

Figure 10-20 shows the John Deere 644K hybrid-diesel-electric wheel loader.

Peterbilt Hydraulic-Hybrid Truck

A hydraulic-hybrid truck was developed by the EPA in conjunction with Cleveland-based **Eaton Hybrid Power Systems**, **Parker Hannifin**, and **Peterbilt** (Dumaine 2010, p. 14):

> The energy from deceleration is stored in a pressurized tank called an accumulator, which is full of hydraulic fluid and nitrogen. When the truck starts

Figure 10-20. John Deere 644K Hybrid-Diesel-Electric Wheel Loader

Source: Open Source Photo.

Figure 10-21. Peterbilt Model 320 Hybrid Class 8 Refuse Truck

Source: Open Source Photo.

> moving pressure released from the tank drives the wheels, saving the diesel engine from having to kick in. The system is great for stop and go driving. Annual fuel savings should reach 1,000 gallons [832.6 imperial gallons and 3,787.5 liters] of diesel per truck per year, about a 30% improvement over traditional haulers. Greenhouse gas emissions are reduced 20% or more.

Figure 10-21 shows the Peterbilt Model 320 hybrid Class 8 refuse truck.

Research Comparing Traditional Diesel to Hybrid Heavy Construction Equipment

A two-year study of hybrid construction equipment was conducted at the University of California—Riverside and completed in 2013. According to the results of the research, hybrid construction equipment does save fuel, with the results varying by the type of construction equipment. For the Caterpillar D7E bulldozer, the fuel savings averaged 14%, and the CO_2 emissions were also reduced by 14%. The Komatsu HB215LC-1 hybrid excavator reduced both fuel and CO_2 emissions by 16%. Unfortunately, the research also indicated that for the Caterpillar bulldozer the NO_x emissions increased by 13% for the hybrid machine and for the Komatsu excavator they increased by 1%, when compared with the emissions of a Caterpillar D6T and Komatsu PC200, respectively. Even though the NO_x emissions were higher for both types of hybrid construction equipment, they did not exceed federal limits. Additional research will be conducted as new models of construction equipment hybrids become available in the industry (*Engineering News Record* 2013).

10.12 Summary

This chapter explained computer simulation modeling software programs. The *GPSS* computer simulation model was used to demonstrate how simulation software programs are used to model heavy construction operations and determine optimal fleet configurations. Other simulation software modeling programs can be used in the same manner as *GPSS* to model heavy construction operations. The case study presented in this chapter demonstrated that different combinations of heavy equipment might be tested using a computer simulation model in order to determine the most cost-effective heavy construction equipment combinations.

This chapter discussed sustainable alternatives for heavy construction equipment and hybrid-electric heavy construction equipment, such as the Caterpillar electric dozer, the Komatsu excavator, the John Deere diesel-electric-hybrid wheel excavator, and the Peterbilt hydraulic-electric truck. It also presented information on tires that help increase the sustainability of construction equipment and the use of biodiesel fuel products to enhance the capabilities of heavy construction equipment. The EPA Tier Four Final Standards were introduced along with information on cooled exhaust gas recirculating and selective catalytic reduction systems. Engine repowering, engine upgrading, and diesel-retrofit technologies were also covered in this chapter to demonstrate alternatives to purchasing hybrid-electric heavy construction equipment.

10.13 Key Terms

abstract models
antithetic
CAT Remanufacturing Service
clock update phase
conceptual models
cooled exhaust gas recirculation

deterministic
diesel oxidation catalysts
diesel particulate filters
diesel-retrofit technologies
discrete simulation languages
double-reduction gear sets
dynamic
Eaton Hybrid Power Systems
Energy Saver Green tires
engine repowering
engine upgrading
enterprise models
General Programming Simulation System
General Simulation Package
grade resistance
heuristically
iconic models
integrated starter generator
intelligent compaction
International Business Machines
liquid-cooled electric motors
non-antithetic
Parker Hannifin
Peterbilt
physical models
probability
queue
queuing
queuing models
queuing theory models
rimpull
rolling resistance
scan phase
selective catalytic reduction
sensitivity analysis
simulation models
simulations
simulator
static
stochastic variables
system
Tier Four Final Standards
total resistance
transactions
ultra-low-sulfur diesel fuel

10.14 Exercises

10.1 Using the symbols for the simulation modeling elements in Figure 9-7, develop a flow-process diagram simulation model for any type of construction operation. Define the scope of the model and its key components, and discuss the labor and equipment required for the operation.

10.2 Using the simulation elements in Figure 9-7, develop a flow-process diagram simulation model for a scraper and tractor pusher construction operation using three scrapers and two tractor pushers.

10.3 What are the types of models used to represent data or visual items?

10.4 What are the factors that should be considered when one is selecting equipment for construction operations?

10.5 What are queuing theory models, and how are they used in construction operations?

10.6 Explain what computer simulation models are and how they can be used in construction.

10.7 Explain how transaction units are used in computer simulation models.

10.8 What was the computer simulation model in the case study in this chapter testing, and what were the results of the simulation?

10.9 Explain how the deterministic method is used to compute the total operating costs for construction equipment.

10.10 How is production calculated for construction equipment? Provide a numerical example of the production formula.

10.11 Describe how using Energy Saver Green tires is sustainable.

10.12 Explain whether the efficiency achieved by using hybrid-electric heavy construction equipment justifies the increase in its purchase price over traditional diesel-engine heavy construction equipment.

10.13 Explain engine repowering. Why would it be used in heavy construction equipment?

10.14 What is diesel-retrofit technology, and how does it benefit the environment?

10.15 Describe how hybrid-electric heavy construction equipment is different from traditional diesel-powered equipment.

10.16 Explain why it is important to reduce the toxic emissions from heavy construction equipment.

10.17 What is a major disadvantage to using biodiesel fuel in heavy construction equipment?

10.18 Discuss the different types of emissions generated by the use of diesel engines.

10.19 Explain what diesel-retrofit technologies are used to accomplish when installed on heavy construction equipment.

10.20 Explain why using biodiesel fuel is advantageous as compared to using conventional diesel fuel.

10.21 Discuss why the John Deere diesel-electric hybrid wheel loaders could be viable alternatives to traditional diesel wheel loaders.

References

Adrian, J. (2004). *Construction productivity: Measurement and improvement*, Stipes, Champaign, IL.

Ainslie, B., Rideout, G., Cooper, C., and McKinnon, D. (1999). "The impact of retrofit exhaust control technologies on emissions from heavy-duty diesel construction equipment." *SAE Technical Paper 1999-01-0110*, Society of Automotive Engineers, Warrendale, PA.

Al-Senani, K. (n.d.). "Efficient equipment selection for earthmoving operations using GPSS simulation language." Master's report, University of Colorado, Boulder, Co.

ASTM. (2011). "Standard specification for biodiesel fuel blend stock (B100) for middle distillate fuels." *D6751-11A*, West Conshohocken, PA.

Begley, S. (2008). "Sounds good, but..." *Newsweek*, 151(15), 20.

Bernold, L. E. (1986). "Low level artificial intelligence and computer simulation to plan and control earthmoving operations." *Earthmoving and heavy equipment*, G. D. Oberlender, ed., American Society of Civil Engineers, New York, 156–165.

Carrie, A. (1988). *Simulation of manufacturing systems*, Wiley, New York.

Caterpillar. (2006). *Caterpillar performance handbook*, Peoria, IL.

Caterpillar. (2008). "The Big Picture, Annual Sustainability Report," Peoria, IL. <http://www.caterpillar.com/en/company/sustainability/sustainability-report.html>. (March 2014).

Caterpillar. (2010). "Independent construction Caterpillar 633D scraper tier 2 engine repower fact sheet." Peoria, IL.

Caterpillar. (2011). "AccuGrade grade control system." <http:www.cat.com/technology/earth-moving-solutions/accugrade-grade-control-system> (May 30, 2011).

Caterpillar. (2013). "ACERT Technology." <http://www.cat.com/en_US/articles/solutions/acert-technology.html> (Dec. 2013).

Dumaine, B. (2010). "The big rigs go hybrid." *Time*, September 20, 14.

Engineering News Record. (2008). "Electric dozer moves like a cat but looks like a dog." 269(8), 11.

Engineering News Record. (2010). "Komatsu's hybrid excavator takes first stab at U.S. market." 262(6), 12.

Engineering News Record. (2013). "Big hybrids need more greening." 271(13), 8–9.

Engineering News Record. (2014a). "Diesel emissions rules add maintenance costs." *Contractor Business Quarterly*, February, 28.

Engineering News Record. (2014b). "Tier 4 drill rig debuts at ConExpo." March, 31.

EPA. (2002). "A comprehensive analysis of biodiesel impacts on exhaust emissions." *EPA420-P-020001*, Washington, DC. <http://www.epa.gov/oms/models/analysis/biodsl/p02001.pdf> (May 30, 2011).

EPA. (2007). *Cleaner diesels: Low cost ways to reduce emissions from construction equipment*, Washington, DC. <http://www.epa.gov/sectors/pdf/emission_0307.pdf> (May 28, 2011).

EPA. (2009a). *Potential for reducing greenhouse gas emissions in the construction sector*, Washington, DC. <http://www.epa.gov/sectors/pdf/construction-sector-report.pdf> (May 2011).

EPA. (2009b). "Verified technologies list." National Clean Diesel Campaign, Washington, DC. <http://epa.gov/cleandiesel/verification/verif-list.htm> (May 2011).

EPA. (2013). *Inventory of US greenhouse gas emissions and sinks: 1990-2012.* EPA 430-R-11-005. <http://www.epa.gov/climatechange/ghgemissions/usinventoryreport.html> (April 2014).

Gaarslev, A. (1969). "Stochastic models to estimate the production of material handling in the construction industry." *Technical Rep. No. 111*, Department of Civil Engineering, Stanford University, Palo Alto, CA.

Halpin, D. (1990a). *MicroCYCLONE System Manual*, Purdue University, West Lafayette, IN. <https://engineering.purdue.edu/CEM/People/Personal/Halpin/Sim/MicroCYCLONE/Index.html> (March 2014).

Halpin, D. (1990b). *WebCYCLONE User's Manual*, Purdue University, West Lafayette, IN. <https://engineering.purdue.edu/CEM/People/Personal/Halpin/Sim/MicroCYCLONE/Index.html> (March 2014).

Halpin, D., and Woodhead, R. W. (1976). *Design of construction and process operations*, Wiley, New York.

John Deere. (2011). "IT4 emissions solutions." <http://www.deere.com/en_US/ProductCatalog/FR/media/pdf/8r_series/it4_45549_2011.pdf> (May 2011).

Karafiath, L. (1988). "Rolling resistance of off-road vehicles." *J. Constr. Eng. Manage.*, 114(3), 458–471.

Komatsu. (2008). “Komatsu introduces the world’s first hydraulic excavator: Hybrid evolution plan for construction equipment.” <http://www.komatsu.com/CompanyInfo/press/2008051315113604588.html> (May 31, 2011).

Michelin. (2014). “The future of tire technology.” <http://www.michelintruck.com/michelintruck_en_us/goGreen/smartway-landing.jsp> (March 2014).

National Biodiesel Board. (2014). “Biodiesel—Commonly asked questions.” <http://www.biodiesel.org/docs/ffs-basics/commonly-asked-questions.pdf?sfvrsn=6> (March 2014).

Sacramento Metropolitan Air Quality Management District. (2011). “Projects funded with MIT fees.” <http://www.airquality.org/ceqa/ProjectsfundedwithMitFees.pdf> (May 2011).

Schriber, T. (1974). *Simulation using GPSS,* Wiley, New York.

Schriber, T. (1991). *An introduction to simulation using GPSS/H,* Wiley, New York.

“Simulator.” (2006). *Webster’s New Universal Unabridged Dictionary,* Simon and Schuster, New York.

“System.” (2006). *Webster’s New Universal Unabridged Dictionary,* Simon and Schuster, New York.

Van Hampton, T., Illia, T., Tuchman, J. L., and Krizan, W. G. (2008). “LeTourneau’s electric legacy haunts Las Vega mega-show.” *Engineering News Record,* 260(6), 37. <http://enr.construction.com/news/finance/archives/080319a-2.asp> (May 28, 2011).

Volvo Construction Equipment. (2011). “LF220F Rugged Loader.” <http://www.volvoce.com/constructionequipment/na/en-us/products/wheelloaders/wheelloaders/l220f/pages/specifications.asp> (April 2014).

Wescott, R. (2005). *Cleaning the air: Cost effectiveness of diesel retrofit versus current CMAQ projects,* Emissions Control Technology Assessment, Washington, DC.

CHAPTER 11

Global Productivity Issues

In the global engineering and construction (E&C) industry, an essential component of preparing competitive bid estimates is determining labor costs, and this requires accurate productivity rates. The cost of construction has been dramatically increasing throughout the world in the 21st century because of the volatility of the cost of construction materials and fluctuating wage rates attributable to increases in employee benefits. As with domestic projects, the goal of global construction projects is to complete them on time and under budget, and productivity improvement methods should be used to help reduce both time and cost factors.

In spite of their significance, some of the concepts of labor productivity are elusive. Labor productivity rates are defined in a number of ways, ranging from the value of gross output per worker (referred to as work hour) to the careful measurement of the physical output of labor taking into account other factors that affect production. External factors that affect the physical output of labor vary from location to location; therefore, the conditions under which labor productivity values are obtained also need to be known.

Several analysis methods are available for measuring global labor productivity and for deconstructing processes into distinct activities. Global labor productivity data are obtained through the various work-measurement techniques discussed in Chapter 6, such as work sampling, time-series techniques, crew balance analysis, and process analysis, all of which were developed to analyze construction processes in industrialized nations. However, these methods might have limited use in developing countries because of external influences beyond the control of project managers. In addition to using the work-measurement techniques introduced in Chapter 6, productivity improvement studies in developing countries should be adapted to local conditions and include an analysis of nutrition, education level, motivation, and incentives.

Work-measurement techniques yield data that might be interpreted in several ways depending on the particular needs of the person or agency interpreting the data, and they are used by contractors or governments to measure the impact of a variety of parameters on labor productivity. The quality of labor is crucial in cross-country comparisons of labor productivity, and several dimensions of labor quality, including education, training, experience, health, and nutritional status, should be examined during investigations of the primary sources of labor productivity differences.

In this chapter, statistical methods are provided for assessing the impact of several factors on productivity rates in developing and emerging countries, as contrasted with industrialized nations. A case study is presented that analyzes productivity rates from several construction projects in Nigeria and the United States. Global productivity indices that could be used to analyze productivity rates and prepare bid estimates are also included in this chapter. Labor productivity factors for developing countries are discussed—including labor quality, education, nutrition, motivation, financial incentives, management, cultural motivation, and process and labor issues. The last section of the chapter provides a case study that describes a productivity improvement investigation conducted in Saudi Arabia.

11.1 Global Labor Productivity Variations

Labor productivity factors were compared in a report by the **International Labor Organization** with a norm of 1.0 for the Washington, D.C., area (ILO 1969 p. 299). The International Labor Organization (ILO) report is useful to construction industry personnel when they are planning global construction projects. Forty-seven labor productivity factors were developed using data from the ILO survey of global contractors and other research projects.

Productivity rates vary throughout the world, and Table 11-1 provides a summary of construction labor productivity rates for various countries as compared with Washington, D.C., which was assigned a productivity rate of 1.0. Table 11-1 lists **base labor productivity rates** for different regions of the world, along with low and high rates. The data in Table 11-1 demonstrate that productivity factors vary substantially throughout the world. The table was developed by the author from a variety of different global sources that evaluated productivity in individual countries. Base labor productivity is defined as the productivity that would be expected on a standard project, and it includes projects requiring *x* number of work hours (companies set their own minimum number of hours) of direct labor at a standard location. Some of the other elements of base productivity are

- Area construction activity at a normal level;
- Normal execution basis;
- Normal area weather;
- Local work rules and practices;
- Normal worker attitudes;
- Normal area workweek;
- Average contractor performance;
- Normal material availability and deliveries for project location;
- Sufficient areas for staging, lay down, and work; and
- Ratio of national to expatriate labor based on historical standards and the number of work hours.

Table 11-1. Global Base Labor Productivity Rates

Country	Rate (range)	Country	Rate (range)
Argentina	2.30 (1.30–2.60)	Italy	1.45 (1.10–1.48)
Australia	1.20 (0.96–1.45)	Jamaica	2.00 (1.49–3.05)
Austria	1.60 (1.67–2.10)	Japan	1.90 (1.00–2.00)
Belgium	1.45 (1.30–2.60)	Korea, South	1.23 (1.00–1.67)
Brazil	2.80 (1.23–3.10)	Kuwait	2.00 (1.32–4.55)
Canada, East	1.14 (1.08–1.17)	Lebanon	1.25 (1.00–1.67)
Canada, West	1.07 (1.02–1.11)	Mexico	2.00 (1.54–3.15)
Ceylon	3.50	Netherlands	1.80 (1.60–3.30)
Chile	2.07 (2.00–2.09)	Nicaragua	2.67
China	4.00 (2.60–4.50)	Norway	1.23
Costa Rica	1.71 (1.25–3.00)	Oman	1.67 (1.32–4.55)
Denmark	1.28 (1.24–1.28)	Pakistan	2.64 (1.67–7.14)
Dominican Republic	2.00 (1.49–3.03)	Panama	1.71 (1.43–3.03)
Egypt	2.04 (1.30–5.00)	Philippines	1.80
El Salvador	2.00 (1.50–3.00)	Portugal	3.00
England	1.52 (1.11–2.08)	Puerto Rico	1.80
Finland	1.28 (1.24–1.28)	Saudi Arabia	1.90 (1.27–4.50)
France	1.33 (0.89–1.54)	Singapore	4.00
Germany	1.28 (1.00–1.33)	South Africa	1.58
Ghana	3.50	Spain	2.95
Greece	1.43 (1.00–1.59)	Sweden	1.13 (1.10–1.20)
Guatemala	2.13 (1.50–3.05)	Switzerland	1.05 (1.0–1.10)
Haiti	2.33 (1.49–2.94)	Taiwan	1.94 (1.69–3.33)
Honduras	2.22 (1.50–3.03)	Turkey	2.32 (2.22–2.44)
India	4.50 (2.50–10.0)	United States (Washington, DC)	1.00
Iran	4.00		
Iraq	3.50	Venezuela	1.22
Israel	1.23 (0.83–2.44)		

Source: Adapted from various sources.

Base labor productivities are then modified using a **job size/activity correction curve**, as well as **workweek correction curves**. Base productivities could also be modified for other criteria, such as execution basis; abnormal weather; travel time to site; temporary lay-down; work or staging areas; site accessibility; work rules (craft restrictions, breaks, etc.); worker attitudes; and contract performance. Construction firms involved in global contracting usually develop their own in-house base labor productivity factors for locations where they frequently build projects.

One example of how the data in Table 11-1 is used is the following: The labor factor for Costa Rica is 1.71, and this indicates that 71% more work hours are required to complete a project in Costa Rica, as compared with Washington, D.C.

The data in Table 11-1 indicate that labor productivity rates in the construction industry in the United States are higher than rates in many countries, and this suggests that workers in the United States are more productive than workers in other countries. The countries that have comparable rates to the United States are Sweden and Switzerland, with 1.13 and 1.05, and Canada, with 1.07 to 1.14. The next tier of countries includes Australia (1.20); Venezuela (1.22); Norway, Israel, and South Korea (1.23); Lebanon (1.25); and Denmark, Germany, and Finland (1.28).

Global labor productivity rates indicate that construction labor productivity rates in newly emerging countries are lower than the rates in industrialized nations. Construction operations in newly emerging countries are affected by the nonpayment of wages, work interruptions attributable to a lack of proper materials or tools, inadequate management, and various other socioeconomic factors.

11.2 Global Labor Productivity Factors

This section discusses labor productivity variations and other factors that affect the efficiency of labor in countries throughout the world.

Labor Quality

Labor productivity differentials might be explained on the basis of factors such as education, health, nutrition status, age, and education. Studying variations in labor productivity rates throughout the world provides insight into the effects of education and health improvements, not only on the productivity of construction workers, but also on the productivity of workers in other industries as well (U.S. Government 1977).

Education Levels

Education is a central determinant of productivity, as it affects the choice of occupation as well as productivity within an occupation. Higher levels of education both increase productivity and aid in the adoption of new technology. Institutional vocational training programs do not increase productivity as much as on-the-job training programs, and formal education increases the ability of a worker to gain from on-the-job training.

One study conducted in the United States on designer construction knowledge indicated that it is beneficial for civil engineering students to take construction courses in their educational programs and for design engineers to visit jobsites more often than just to check conformance to the plans and specifications (Yates and Battersby 2003). The study also indicated that the quality of education, the type of formal education, and the subjects covered in formal educational programs are all important to the ability of graduates to apply what they have learned in a manner that helps to improve productivity.

Nutrition Levels

The effects of nutrition on the performance of individual workers is difficult to quantify, as nutrition interacts with several other factors, such as disease, education, and motivation, in a complex way. One example that illustrates how nutrition affects worker productivity is a road construction project built as part of the Pan-American Highway, which connects the United States to Mexico. Productivity increased by up to 500% when workers were housed in camps and provided with all of their meals.

The most clear-cut evidence of nutritional impact on worker productivity relates to specific nutrients. One study found a correlation in cross section between performance and iron intake in construction workers (International Bank for Reconstruction and Development 1974). Two other studies of road construction workers in India and Kenya indicated a correlation between productivity and iron levels in the blood (U.S. Government 1975, 1981). Results for caloric intake are less clear than they are for iron intake. Some studies have indicated that productivity is not significantly affected by current calorie intake but rather by indicators of past calorie intake, such as height, weight, and arm circumference.

The effects of reduced nutrition are evident in countries where workers observe the Muslim holiday of Ramadan. For one month, workers fast during daylight hours and eat only after the sun has set. The strength of construction laborers is impaired while they are fasting during the day; therefore, some jobsites in Muslin countries operate for only 4 h a day or shut down completely during the month of Ramadan.

Motivation

The job content of workers, perceived status, and increasing sense of participation all significantly affect productivity rates. In some situations, poor worker motivation causes low productivity. Frederick Herzberg and Russell (1953) developed a theory of attitudes toward work that categorizes characteristics associated with any job as related to either motivation or **hygiene factors**. Motivation factors give workers satisfaction at work and include achievement, a sense of responsibility, and pleasure from the work itself. Hygiene factors include salary, working conditions, company policy, and administration, and these are principal sources of dissatisfaction (Herzberg and Russell 1953).

If rewards are primarily intrinsic to (belonging to the essential nature of) the task performed, the provision of extrinsic (coming from the outside) incentives may have a negative effect on performance. If a task is viewed as requiring highly skilled labor and if the emphasis is on producing quantities of it, quality may deteriorate as a result. Group motivation influences productivity, and the size of the work group is important, as it affects the strength of social relations. Productivity may vary inversely with group size, and workers often set output restrictions for certain piece-rate jobs.

Financial Incentives

The effects of financial incentives on increasing worker productivity were demonstrated on a road construction project in India. Managers on this job discovered that by changing from time rates to piece rates, or task rates, worker productivity rates improved by 50% to 75% (U.S. Government 1975). International Labor Organization studies on road construction projects in Nigeria and Tanzania confirmed that when workers were paid in a manner that made them feel like they were working for themselves, productivity improved on projects (Koehn and Caplan 1987). In India, changing from a day rate to a finish-and-go rate system raised productivity 45% to 53%, and changing to a paid task rate increased productivity by 60%.

Management

Managers affect labor productivity in a variety of ways because they are responsible for setting general goals for firms, as well as for supervising day-to-day tasks. Managers make a wide range of allocation and technical decisions, such as choosing construction materials, the methods used for construction, and the order in which tasks will be accomplished at jobsites. Unspecified management factors have been cited as being responsible for labor productivity differences even when similar techniques are being used, especially in global comparisons where there are differences in management styles.

Productivity in India and Indonesia for earthwork tasks is as much as four times the standard figures for those countries if there are good management, high incentives (piecework), and longer working days. Conversely, in Chad, a country with no tradition of labor-based construction, in the first months of a project, under conditions of low incentives (daily paid work), fair management, and a 7-h workday, worker productivity rates were one-third the standard figures (Coukis 1983). In some situations, lower levels of productivity in Africa are the result of poor management rather than the inherent lack of ability of the workers. Poor working conditions and a lack of financial incentives also reduce the motivation of workers in Africa.

11.3 Factors That Affect Productivity on Global Projects

While working on global projects, E&C personnel have to manage workers from nations throughout the world who speak different languages. Even within one country, several languages may be spoken, and citizens practice different religions. Having a basic understanding of the culture of a host country and knowing something about local religious beliefs is helpful when designing methods for improving productivity on construction projects in foreign countries.

Cultural Motivations

In societies based on the Protestant (Quaker) work ethic, such as the United States, people are taught that hard work will be rewarded by raises and promotions, and individuals, rather than groups, are praised for their performance. In Islamic cultures, people accept their position in life, and they do not live to work, rather they work to live. Because their status in society is determined by birth, their clan, their tribal affiliation, or their place of birth, work is not a means for them to elevate their status. Social networks are important to workers, and they may only socialize or work with people from their own region or clan.

South American and Asian cultures are **paternalistic societies** where companies are like families. In addition to supervising work, managers are parent figures, and they are admired at work and treated with reverence by their employees. The group is more important than the individual, and groups, not individuals, are rewarded when objectives are met.

Motivation techniques vary among Protestant, Islamic, and paternalistic societies because workers in each of these types of societies are motivated by different objectives. If workers are taught to be content with whatever job they have, rather than to attempt to find a better job, as is the case in some Muslin cultures, they are not motivated by anticipated promotions. Productivity rates are influenced by how motivated workers are to perform a task, and their motivation may be tied to the work ethic of a particular region of the world. Workers in Islamic countries might not understand the importance of working rapidly because they expect to be rewarded in their afterlife, not their present life.

In rural areas of the world, construction workers may not be familiar with the type of facility they are working on if the project was designed in a foreign country. They may also not understand why the facility is essential to their society because they have lived without certain structures their entire lives. Workers may not comprehend how a facility will personally benefit them when it is complete; therefore, they do not understand the urgency of meeting deadlines during construction. If workers are being compensated on a daily rate, it is to their benefit to work slowly so that they will be paid for additional work.

Work ethics are also affected by whether workers have worked in a communist society, a dictatorship, or a society influenced by communism or a dictator in the past. If the objective of projects that workers worked on previously was to employ large numbers of people, the workers may have difficulty transitioning to working on profit-driven projects and adopting methods designed to increase productivity. Also, if workers were not previously compensated for overtime, then they may be reluctant to work past normal working hours without additional compensation.

Process Issues

This section discusses some of the process issues that affect productivity rates on construction projects.

The location of a project affects productivity rates. In rural areas, productivity declines because of problems associated with securing and transporting construction materials and equipment to jobsites. In some countries, such as parts of Africa, Afghanistan, China, and Pakistan, materials are still transported to jobsites by water and by mule, horse, elephant, camel, water buffalo, or other beasts of burden.

Trucks might be available to transport construction materials, but they might be unreliable because of a lack of trained mechanics or spare parts. Some of the less-developed regions of the world have advanced rail systems and port facilities, but once materials arrive at train terminals or ports, no transportation systems are available to move materials to rural jobsites.

It is also challenging to transport materials to jobsites located in crowded cities, as lay-down yards are scarce and materials might have to be brought to jobsites on a daily basis. Urban jobsites have restricted access because of the congestion caused by existing structures in the surrounding environment. Trucks may be allowed on city streets only during certain hours of the night or day.

Materials may not be delivered to jobsites as scheduled because of delays caused by bureaucratic governments. Without the assistance of **business agents** (**fixers**) and the payment of unofficial **processing fees**, or **gratuities**, officials in the customs office of a host country might delay foreign material deliveries. Business agents are familiar with how local systems operate and are able to assist with expediting government administrative paperwork and obtaining the release of materials from customs.

Construction firms may need to employ a business agent before they are able to hire local labor to work on construction projects. In some regions of the United States and some of the northern European countries, business agents are individuals who work directly for labor unions. They manage union affairs, help settle union disputes, and operate hiring halls where union workers wait to be hired by construction firms to work on projects. Business agents might be union or government officials or unofficial agents who are familiar with local laborers and are able to locate workers for projects. In eastern European countries and countries with weak or changing governments, unofficial agents or members of the local **mafia**, or illegal criminal organizations, control materials and the labor supply. In tribal societies, the clan or tribal chief may be the only person who determines who will work on projects.

If a project is being built in a clan or tribal society, workers might slow down their productivity so that additional workers, such as their family or clan members, will be hired to work on the project. Tribal construction workers may refuse to perform work if they are required to work with people who are not from their clan or tribe. Workers may also refuse to work for a supervisor or take directions from someone who is not from their tribe or clan. Supervisors who are not from the local clan might have to explain work processes to a clan member who will in turn explain the work to laborers.

In some countries, workers may worship their tools or have to share their tools. Workers might have to share their tools with workers from different trades, which slows down construction as workers wait for their tools to be returned to them. In developing countries, workers might lack appropriate tools to work on certain types of construction and use tools that are available locally to perform work. For example, they might use iron pile rods to dig for electrical and plumbing conduit below grade level during excavation or wood 2 × 4s instead of hoes or shovels for spreading concrete as it is being placed in formwork.

Material Issues

Materials in developing countries might be inferior to those produced for construction projects in industrialized nations. If materials called for in the specifications are not available locally or are illegal to import, substitutions have to be made at the jobsite, which delays projects or completely stops work. Local contractors sometimes order materials and have them delivered to jobsites without the knowledge of project managers.

Labor Issues

In parts of the Middle East, such as Saudi Arabia, construction labor is difficult to locate because certain classes of people will not perform manual labor; therefore, workers come from other countries such as Pakistan, the Philippines, Sri Lanka, Thailand, Egypt, and South Korea. Members of Japanese, American, Russian, and European firms are involved in supervising construction projects in many Middle Eastern countries.

In countries with high unemployment rates, workers are desperate to keep their jobs and may try to perform work beyond their capabilities. Managers should be cognizant of this problem and not allow workers to perform work that is beyond their capabilities because this could lead to accidents, injuries, or deaths.

Issues of **face**, which involve public embarrassment, directly influence productivity at jobsites in countries where it is part of the culture, such as Asian countries. If a worker loses face, he or she will not return to work, and if he or she is part of a team, this affects the entire team. Team members might reduce their output in protest over one of their team members losing face.

Laborers take smoke (Asian countries), tea (Britain or former British colonies, such as India), kava (Fiji Islands), qat (Yemen), cinnamon cigarettes (Indonesia), coffee (the United States), or other types of breaks during the workday. Productivity might be positively or negatively affected by what is ingested by workers during their breaks (stimulants versus depressants). Workers are not as productive right before, during, or after holidays, as they are involved in preparations required for holidays after regular work hours. In India, the start of construction could coincide with a religious festival, and certain rituals (***poojas***) are performed during groundbreaking ceremonies at jobsites.

11.4 Global Productivity Variables

Although productivity rates vary throughout the world, they are all related to the following variables (Koehn and Brown 1986, p. 299):

- *Management*: proper planning, realistic scheduling, adequate coordination, and suitable control
- *Labor*: union agreements, restrictive work practices, absenteeism, turnover, delays, availability of labor, the skill of the workers, and use of equipment
- *Government*: regulations, social characteristics, environmental rules, climate and political ramifications
- *Contracts*: fixed-price, unit-cost, and cost-plus fixed-fee
- *Owner characteristics*
- *Financing*

In addition, individual national productivity rates are dependent on an array of diverse factors, such as national characteristics, climate conditions, local labor practices, and whether the workers accept the use of mechanized equipment. To improve productivity, the impact of each of these variables on labor productivity should be assessed using statistical methods, and special attention should then be given to the parameters that adversely affect productivity.

Statistical methods are available that measure the impact of one variable, the **dependent variable**, on another variable, the **independent variable**. These methods allow users not only to predict the value of the dependent variable on the basis of information about an independent variable but also to measure the strength of the relationship between these variables. One measure of the strength of the relationship between two variables x and y is the **coefficient of linear correlation** (r), or simply the **correlation coefficient**. Given n pair of observations $(x_p y_i)$, the sample correlation coefficient r could be computed as follows (Yates and Guhathakurta 1993):

$$r = S_{xy} / S_{xx} S_{yy} \tag{11-1}$$

where

$$S_{xy} = \Sigma xy - (\Sigma x)(\Sigma y) / n \tag{11-2}$$

$$S_{xx} = \Sigma x^2 - (\Sigma x)^2 / n \tag{11-3}$$

$$S_{yy} = \Sigma y^2 - (\Sigma y)^2 / n \tag{11-4}$$

To find the proportion r^2 of the total variability of the y values accounted for by the independent variable x, the following equation is used:

$$r^2 = S^2_{xy} S_{xx} S_{yy} \tag{11-5}$$

Similarly, $1 - r^2$ represents the proportion of the total variability of the y value not accounted for by the x variable. Eqs. (11-1) to (11-5) can be used if, and only if, there is a linear relationship between x and y. Other models become necessary when the relationship between x and y is not linear, that is, when y increases or decreases with x but not in a linear fashion. This approach is the rank correlation coefficient, which measures the monotonic (having the property of either never increasing or never decreasing as the values of the independent variable or the subscripts of the terms increase) relationship between y and x, that is, y increases or decreases with x even when the relationship between y and x is nonlinear.

Another statistical method that is used to analyze variables is the **rank-order correlation coefficient**, which is calculated by first ranking all x values and y values separately and then calculating the ordinary correlation coefficient for the ranks. The relationship based on the ranks is called the Spearman's rank-order correlation coefficient r_s.

11.5 Global Comparisons of Labor Productivity

Data obtained through work-measurement techniques yield valuable information for future planning and productivity forecasting. Several factors affect labor productivity, and these factors are broadly categorized as general, organizational, technical, and social. The extent to which these factors affect labor productivity varies from country to country. For example, **socioeconomic factors** may have a greater impact in newly emerging countries than in industrialized nations (Moavenzadeh 1978). In global construction, these types of comparisons are useful for planning resources and management methods and selecting appropriate technology.

Although the extent to which each of these factors affects productivity varies from country to country, a method is required to make international comparisons in order to establish the *relative* importance of the variables for different countries, and a **rank-agreement factor** allows such a comparison. The rank-agreement factor is a method whereby factors are ranked according to their impact on productivity for each country, and these rankings allow a cross comparison of the relative importance of variables for different countries. For any two countries, the rank of the *i*th item for Country A is R_i^1, and for Country B it is R_1^2. Then, the absolute difference, D_i, between any rankings of the *i*th item for the countries would be as follows (Yates and Guhathakurta 1993, p. 16):

$$D_i = |R_i^1 - R_i^2| \tag{11-6}$$

where $i = 1, 2\ldots, N$, and there are N items.

Define $D_{\max} = \sum_{i=1}^{N} |R_{i1} - R_{j2}|$, where $j = N - i + 1$

The formula for the rank-agreement factor is

$$\mathrm{RA} = \sum_{i=1}^{N} \frac{|R_{i1} - R_{i2}|}{N} \tag{11-7}$$

with a maximum RA: $\mathrm{RA}_{\max} = \sum_{i=1}^{N} \frac{|R_{i1} - R_{j2}|}{N}$

The formula for the percentage disagreement between countries is

$$\mathrm{PD} = 100^{i=1} 1 \frac{\sum |R_{i1} - R_{i2}|}{\mathrm{N}} \tag{11-8}$$

$$\sum |R_{i1} - R_{j2}|_{i=1}$$

The formula for percentage agreement, PA, is

$$\mathrm{PA} = 100 - \mathrm{PD} \tag{11-9}$$

A larger percentage agreement would indicate that the ranking of a particular factor under consideration has global significance and is not peculiar to any one of the countries under consideration. A larger percentage disagreement would indicate that the rankings for a particular factor are peculiar only to the countries under consideration. These data have many potential uses in planning for global construction, as they might help identify factors that have a greater relative effect on the productivity of labor in foreign countries, as compared with labor productivity rates in the native country.

11.6 Case Study: Worker Productivity in Nigeria and the United States

This section describes the results of a research investigation that explored productivity issues in Nigeria in comparison with the United States. The areas identified by workers (**artisans**) in Nigeria as factors that influence their productivity rates were as follows (Olomolaiye et al. 1987, p. 320):

- Lack of materials,
- Lack of tools,
- Repeat work instruction delays,
- Inspection delays,
- Absenteeism,
- Supervisors' incompetence, and
- Changing crew members.

Some of these areas and their impact on productivity rates in both Nigeria and the United States are discussed in the following sections.

Lack of Materials

Transporting construction materials was ranked as being the most frequent difficulty encountered in each of five case studies in the United States, with an estimated average loss of from 6.40 to 8.40 h per week per worker. The time lost on Nigerian sites ranged from 2.5 to 5.5 h per week per worker. The reasons given for higher productivity losses in the United States were as follows (Borcherding et al. 1980):

- Lack of cranes, or trucks, or both for transporting construction materials;
- Not enough laborers to retrieve material orders from warehouses;
- Excessive paperwork required for requisitioning materials;
- The nonexistence of certain items at jobsites;
- Improper materials delivered to work areas; and
- A lack of proper preplanning by either supervisors or general supervisors.

In Nigeria, workers identified outright shortages of certain materials as the main problem. This was traced to incessant cash-flow problems of the contractor, which resulted in suppliers not being paid for materials previously delivered to jobsites; therefore, the suppliers were not willing to fulfill new orders for materials. Workers also identified lack of proper planning and on-site transportation difficulties as the second- and third-most frequent sources of problems, and delays were the fourth problem source (Olomolaiye et al. 1987).

Lack of Proper Tools

At U.S. sites, an average of 3.41 to 5.08 h per worker was lost each week due to a lack of tools. In the United States, craftsmen indicated poor quality, improper maintenance, and insufficient tools as the three most frequent causes of problems. **Pilferage** (stealing) was also a significant problem. Pilfering from other crews was condoned and sometimes encouraged by some supervisory personnel. Some craftsmen worked with broken tools, and there were insufficient quantities of power-driven tools. The indifferent attitude toward tools contributed to related problems, according to warehouse officials (Borcherding et al. 1980).

In Nigeria, poor-quality tools and poor maintenance of tools and equipment were cited as major problems. At some sites, steel-bending machines were not available; consequently, **steel fixers** (ironworkers) were forced to bend steel reinforcing materials by hand, which resulted in productivity losses. Concrete operations would stop for an entire day when concrete vibrators were not working properly, and **joiners** took turns using power saws because some of the saws were not working (Olomolaiye et al. 1987).

Repeat Work

The amount of time spent on rework was from 4.92 to 7.73 h per week per worker in the United States and 1 to 7 h per week per worker in Nigeria. In the United States, most of the work that had to be redone was caused by change orders, substandard workmanship, poor instructions that resulted from the misinterpretation of drawings and specifications, and other regulations imposed by inspectors or supervisors.

In the United States, a majority of the rework was attributable to poor-quality engineering designs. Incomplete drawings, coupled with errors and inconsistencies, were a continuous dilemma for supervisors and tradesmen. In Nigeria, workers blamed most of the rework on design changes. On the other hand, steel fixers claimed that repeat work was mainly caused by poor or inadequate instructions from supervisors.

Inspection Delays

The effects of inspection delays varied between both of the sites investigated and by who was surveyed as to their contribution to lost time and degree of effect. The estimated average weekly loss of hours attributable to inspection delays was from 2.06 to 4.06 h per worker in the United States and 1 to 3 h per worker in Nigeria. A considerable number of workers, both in the United States and Nigeria, indicated that some of the inspection delays were caused by supervisor incompetence. The workers thought that a high proportion of inspection delays could have been avoided if their supervisors had provided clearer instructions (Olomolaiye et al. 1987).

Thus, a comparison of the two countries indicates that worker productivity was affected by similar problems in both countries. The number of hours lost per worker per week was higher in the United States for each of the factors affecting productivity.

Because the amount of time required to build a square meter of finished structure varied from 6.44 to 16.78 worker days in Nigeria, compared to 1.53 worker days in the United States, other sources of productivity differences had to account for the lower productivity rates in Nigeria.

11.7 Productivity Improvement Global Case Study: Saudi Arabia

This section provides information about a global productivity improvement case study conducted by researchers in Saudi Arabia. The research was summarized in *Improving the GCC Construction Productivity (Part I)* (Zakieh 2007b) and *The Essence of Improving Construction Productivity in Practice* (Zakieh 2007a), which were written by Dr. Rashad Zakieh and published by Saudi Aramco in Jedah, Saudi Arabia.

Construction projects in the Middle East have been averaging approximately $200 billion a year. If productivity could be improved on these projects by 20%, this would result in annual savings of over $20 billion. With this in mind, Zakieh conducted a research investigation to determine the factors affecting productivity rates in the Middle East with the goal of determining the most efficient methods for improving productivity (Zakieh 2007b).

The Saudi Arabian research investigation measured productivity levels on several projects by examining productivity indices of earned values for designated project activities.

The formula for a **productivity index** used in this case study was

$$\text{productivity index} = \text{actual production per work hour} / \text{estimated production per work hour} \quad (11\text{-}10)$$

A productivity index of 1.25 indicates that productivity is 25% better than the estimated productivity. A productivity index of 0.75 means that productivity is only 75% of what it was expected to be at that point in time.

For the Saudi Arabian productivity improvement study, the following factors were listed as affecting productivity (Zakieh 2007b, p. 2):

- Interruptions;
- Inadequate selection, training, and motivation of the workforce;
- Working overtime;
- Unplanned increases in the labor force.

Zakieh (2007b, p. 3) used the following methodology:

- Six sites were studied that ranged in value from $70 million to over $3 billion.
- The number of crews ranged from one to three for each activity.
- Major causes of interruptions, and the quantity of work hours lost due to interruptions, were analyzed over a four- to six-week period.

- The study determined daily productivity variations for the same crew, and productivity variations among crews, doing similar work under the same conditions.
- Productivity improvements were implemented to determine whether they would improve productivity rates.

The investigation found that the major causes of interruptions included waiting for information such as design information or instructions, and waiting for materials, inspections, permits, or machinery when machinery broke down. Interruptions consumed 10–40% of the work hours, resulting in an average productivity rate of 64%. Zakieh (2007b) also indicated that a work stoppage of 1 h caused a loss of productivity of 25% for the time that a crew remained at work after the work stoppage. For the piping and groundwork operations, the interruptions that led to productivity interruptions of 36% and 23%, respectively, are listed in Table 11-2.

The following are examples of interruptions that occurred during steel-framing operations (Zakieh 2007a, p. 23):

- Taking the crane off of the pipe rack and moving it to the storage yard required 30 to 40 minutes.
- Requesting and unloading materials from a trailer required 1 to 4 hours because sometimes there was no trailer available,
- Bringing the crane to the pipe rack required 30 to 40 minutes,
- Uploading materials from the trailer required 20 to 30 minutes.

Table 11-3 summarizes the significance of the interruptions experienced during the research investigation.

Productivity during the Saudi Arabian study varied by 40% on a daily basis for some crews and over 25% among crews that were performing similar activities under the same circumstances. Table 11-4 summarizes the productivity variations within a crew and between crews for bricklayers.

Table 11-2. Interruptions That Affected Piping and Groundwork

Piping Work	Total Interruptions 36%	Groundwork	Total Interruptions 23%
Interruption	Percentage of 36%	Interruption	Percentage of 23%
Inspection	11.9	Material and equipment	17.0
Materials	9.8	Machinery breakdown	2.5
Design	7.6	Inspection	2.1
Permits	3.6	Permits	0.9
Weather	1.8	Sequencing	0.5
Sequencing	1.4		

Source: Adapted from Zakieh (2007a), Figure 3.

Table 11-3. Significance of Interruptions to Construction Work

Duration of Interruption (hours)	Productivity Loss (percentage)	Duration of Interruption (hours)	Productivity Loss (percentage)
0.0	0	1.6	33
0.2	5	1.8	35
0.4	10	2.0	38
0.6	15	2.2	40
0.8	20	2.4	40
1.0	25	2.6	40
1.2	27	2.8	40
1.4	30	3.0	40

Source: Adapted from Zakieh (2007a), p. 18.

Table 11-4. Productivity Variations within a Crew and between Crews for Bricklayers

Productivity Index = 1.02 for Crew 1				Productivity Index = 0.8 for Crew 2			
Workday	PI	Workday	PI	Workday	PI	Workday	PI
1	1.0	23	1.0	1	1.2	23	0.8
3	1.5	25	1.1	3	0.7	25	0.7
5	0.7	27	0.7	5	0.7	27	1.0
7	1.4	29	1.2	7	1.0	29	0.5
9	1.0	31	0.8	9	1.6	31	0.5
11	0.9	33	0.6	11	1.0	33	1.1
13	0.9	35	0.7	13	1.1	35	1.1
15	1.3	37	0.7	15	0.6	37	0.6
17	0.8	39	0.9	17	1.6	39	0.7
21	1.5	41	0.6	21	1.3	41	0.4

Source: Adapted from Zakieh (2007a), Figure 4.

During construction at one jobsite, three work crews performing similar work started working overtime, and the work hours were increased from 60 h per week to 72 h per week for four weeks. This represented 20% in overtime work hours. During the weeks of overtime hours, production increased by 6% and productivity increased by 9%, but there was a 5% loss of productivity for every 5 h of overtime worked per week (Zakieh 2007a). While the crews were working overtime, over 20% of the crew time was expended on moving scaffolding. This time delay could have been avoided if the contractor had used portable scaffolding. This was a technology issue, and productivity could have been increased by 15% if different technology had been used, eliminating the requirement for overtime. According to Horner and Talhouni (1994), doubling a labor force causes productivity to be reduced by 37% due to more disruptions and interruptions, more absenteeism, and insufficient or inadequate supervision to deal with larger numbers of workers.

The following are major findings of the Saudi Arabian research investigation (Zakieh 2007a, p. 29):

- Over 60% of working days suffered from avoidable interruptions, causing a loss in work hours of 10% to 14%.
- Productivity varied by up to 40% within the same crews and over 25% among crews performing similar activities.
- Overtime work, and overmanning of work crews, caused a significant loss in productivity.
- Productivity on Tuesdays, Wednesdays, and Thursdays was significantly lower than the productivity rates achieved during the remaining part of the week [weekends are on Friday and Saturday in Saudi Arabia].

Interruptions occurred because the client was responsible for "design information, materials, inspections, and permits and the contractor was responsible for design information, sequencing, scheduling, materials inspections, instructions/supervision, equipment, tools, and permits" (Zakieh 2007a, p. 29). The barriers to improving productivity during this research investigation were "a lack of management focus at all times, a lack of alignment of goals, contractual conflicts, difficulties in measuring productivity, a lack of labor force focus, and weak commitment to continuous improvement" (Zakieh 2007a, p. 30). Suggestions for improving productivity that resulted from this research investigation were "focusing primarily on a commitment to productivity improvement, select and train the labor force, plan work in detail, avoid overtime work, and keep the size of the labor force as small as possible" (Zakieh 2007a, p. 31).

11.8 Summary

Studying global sources of productivity variability helps project management personnel develop more efficient productivity improvement programs. The International Labor Organization and the International Bank for Reconstruction and Development have conducted a number of studies related to the quality of labor in developing countries, and labor quality is improving in nations where there are health, medical-care, housing, and educational programs.

The effects of education and training on productivity are interlinked, as education increases the ability of laborers to learn from practical on-the-job training sessions. Health and nutrition combine in a complex way that has a synergistic effect on labor productivity; therefore, it is difficult to identify a direct relationship among nutrition, health, and productivity rates.

Conducting statistical studies on labor productivity yields data that are useful for managers to have access to when they are planning production and forecasting in global contracting. However, when developing cost estimates for global construction projects, labor productivity factors provide more accurate results. Examples of statistical applications presented in this chapter demonstrated how labor productivity factors might be used for planning and forecasting on construction projects.

Managers should use labor productivity factors when they are determining appropriate construction technology to be used in developing countries. Various methods have been developed for this purpose ranging from the use of available productivity data to elaborate mathematical models. When reliable productivity data are not available, simpler techniques suffice for developing cost estimates that compare different techniques for accomplishing a task. More elaborate methods should be used in global contracting because risks are greater and accurate data are required to develop bid estimates. Education, nutrition, management, work ethics, national transportation techniques, unemployment rates, and religion all affect labor production rates.

This chapter discussed many of the factors that affect productivity in the global area and also provided a case study that compared productivity issues in the United States and Nigeria. A mathematical model was provided that could be used to determine the relative importance of productivity variables for different countries. The last part of the chapter discussed a case study of a productivity improvement research investigation conducted in Saudi Arabia.

11.9 Key Terms

artisans
base labor productivity rates
business agents
coefficient of linear correlation
correlation coefficient
dependent variable
face
fixers
gratuities
hygiene factors
independent variable
International Labor Organization
job size/activity correction curves
joiners
mafia
paternalistic societies
pilferage
poojas
processing fees
productivity index
rank-agreement factor
rank-order correlation coefficient
socioeconomic factors
steel fixers
workweek correction curves

11.10 Exercises

11.1 Of the five primary sources of global labor productivity differences discussed in this chapter, which is the most important and why?

11.2 What items cause variations in global productivity rates?

11.3 How could management help increase global productivity rates?

11.4 What are the factors that influence productivity rates in the United States versus Nigeria related to lack of materials?

11.5 What are the labor productivity factors in Table 11-1 for Iran, Spain, Venezuela, the Philippines, and Ceylon? If the labor on a project in Washington,

D.C., took 970 h to complete a task, how many hours would the same task take in each of these countries?

11.6 What are the eight items that can be used to adjust base labor productivity rates?

11.7 How does nutrition affect construction workers and their productivity rates?

11.8 How could the difference between Protestant, Islamic, and paternalistic cultures be used to motivate workers on a project where the workers are from all three types of societies?

11.9 If construction workers are forced to share tools, how could productivity be increased at jobsites when no additional tools are available?

11.10 Explain the difference between business agents used in the United States and those used in newly emerging Eastern bloc countries.

References

Borcherding, J. D., Samelson, N. M., and Sebastian, S. M. (1980). "Improving motivation and productivity on large projects." *J. Constr. Div.*, 106(1), 73–89.

Coukis, B. (1983). *Labor-based construction programs*, World Bank, Washington, DC.

Herzberg, F., and Russell. D. (1953). "The effects of experience and change of job interest on the Kuder Preference Record." *Journal of Applied Psychology*, 37(6), 478.

Horner, R., and Talhouni, B. (1994). *Effects of accelerated working, delays, and disruptions on labour productivity*, Chartered Institute of Building, Bracknell, UK.

International Bank for Reconstruction and Development. (1974). "Study of the substitution of labor and capital in civil construction projects." *Staff Working Paper No. 172*, Washington, DC.

International Labor Organization (ILO). (1969). "Measuring productivity." *Studies and Report New Series No. 75*, Geneva.

Koehn, E., and Brown, G. (1986). "International labor productivity factors." *J. Constr. Eng. Manage.*, 112(2), 299–302.

Koehn, E., and Caplan, S. (1987). "Work improvement data for small and medium-size contractors." *J. Constr. Eng. Manage.*, 113(2), 327–339.

Moavenzadeh, F. (1978). "Construction industry in developing countries." *World Development*, 6(1), 99–116.

Olomolaiye, P., Wahab, K., and Price, A. (1987). "Problems influencing craftsmen's productivity in Nigeria." *Build. Environ.*, 22(4), 317–323.

U.S. Government. (1975). "The planning of control, production, productivity and costs in civil construction projects." *Technical Memorandum No. 15*, Library of Congress, World Bank Group, Washington, DC.

U.S. Government. (1977). "The relationship of nutrition and health to worker productivity in Kenya." *Technical Memorandum No. 26*, National Institutes of Health, Washington, DC.

U.S. Government. (1981). "Labor productivity: Un Tour d'Horizon." *Staff Working Paper No. 497*, World Bank, Washington, DC.

Yates, J., and Battersby, L. (2003). "Master builder project delivery system and designer construction knowledge." *J. Constr. Eng. Manage.*, 129(6), 1–10.

Yates, J., and Guhathakurta, S. (1993). "International labor productivity." *Journal of the American Association of Cost Engineers International*, 35(1), 15–25.

Zakieh, R. (2007a). *The essence of improving construction productivity in practice*, Saudi Aramco, Jedah, Saudi Arabia.

Zakieh, R. (2007b). *Improving the GCC construction productivity (Part I)*, Saudi Aramco, Jedah, Saudi Arabia.

CHAPTER 12

Sustainability in Engineering and Construction

One area that is increasing in importance in the engineering and construction (E&C) industry is the incorporation of sustainability practices into E&C projects. This chapter discusses sustainability practices that could potentially affect productivity and productivity improvement studies. If construction workers are required to incorporate sustainable practices or techniques, productivity rates may initially decline while workers adjust to the new methods and procedures or become familiar with sustainable materials. But as workers become comfortable with using and installing sustainable methods and materials, productivity rates should increase, as was shown in Figure 5-1, a learning curve for workers. Providing training sessions on sustainable methods, techniques, and materials influences how quickly workers will master the new procedures. If the workers are not provided with proper training on sustainable methods and materials, they may not understand the importance of adapting their abilities to effectively implement sustainable practices.

This chapter provides background information on **sustainable development** practices to assist productivity improvement team members, engineers, and project and construction management personnel in explaining the importance of incorporating sustainable practices into construction projects. The first part of the chapter covers **sustainability** in construction and explains how **corporate social responsibility** (CSR) is driving the implementation of sustainable practices. It also mentions life-cycle environmental and cost analysis techniques to provide a basis for discussing the benefits of sustainable practices and to show how these techniques are used to evaluate whether to use sustainable practices. Information on estimating the loss of worker productivity in the surrounding community during a construction project is also included in this chapter. In addition, this chapter explains the importance of preventing pollution, recycling, and conserving energy during construction.

The second part of the chapter introduces the U.S. Green Building Council (USGBC) Leadership in Energy and Environmental Design (LEED) certification process. The LEED certification process and the LEED categories used to evaluate structures are also included in this chapter. The LEED certification process is used to assess the sustainability of structures on the basis of the incorporation of sustainable practices into the design of the structures. In addition to creating sustainable structures, engineers and constructors should also incorporate sustainable practices into construction processes and use sustainable construction materials. Other sustainable organizations, publications, and certification systems are provided at the end of the chapter.

12.1 Sustainable Development

In the 21st century, there is increasing concern for the environment and interest in implementing sustainable development policies. Analyzing sustainability as it applies to construction projects requires examining it from both an environmental and a social impact perspective. There are three ways to avoid damaging or mitigating the negative effects of construction projects on the global environment:

1. Reduce energy use.
2. Minimize pollution to reduce environmental damage and health risks.
3. Minimize the amount of resources used in order to reduce **embodied energy** and resource depletion in
    - The direct and indirect methods used to extract materials used in manufacturing construction materials,
    - Transportation, and
    - The generation of material waste at jobsites.

The aspects of construction that are directly related to sustainability issues are listed in Table 12-1.

Corporate social responsibility is one of the major driving forces for the incorporation of sustainable practices into the corporate strategies of firms. **Socially responsible investment** (SRI) communities are another force driving the implementation of sustainable practices, as reflected by the increasing use of the **Dow Jones Global Sustainability Index** (DJGSI). Environmental and social credibility is also

Table 12-1. Construction Issues Related to Sustainability

Sustainability Issues in Construction
Compliance with government regulations
Environmental impacts of production operations
Resource efficiency
Responsible supply chains and procurement
Social and community impact of projects
Supplier and vendor environmental and social responsibility
Sustainable designs and materials
Environmental footprint of structures

influencing whether firms are able to secure investments or receive preferential treatment on bids.

One definition of sustainable development is "development that meets the needs of the present without compromising the ability of future generations to meet their own needs" (Samaras 2004, p. 1). **Corporate sustainability** has been defined as "a business approach that creates long-term shareholder value by embracing opportunities and managing risks deriving from economic, environmental, and social developments" (Samaras 2004, p. 1). The **triple bottom line** (economic, environmental, and social value) in design and construction includes approaches such as "design for the environment, context sensitive designs, value engineering, life cycle cost analysis, and Leadership in Energy and Environmental Design (LEED) certification for projects. Sustainable construction techniques include implementing a sustainable design, meeting or exceeding sustainable design specifications, developing strategies to minimize and reuse construction waste and spoils, optimizing asset efficiency, and pursuing the highest level of LEED certification possible" (Samaras 2004, p. 1).

Sustainable practices in construction are being used mostly in the area of materials. Selecting construction materials that are reusable, that use fewer resources during their manufacture or transporting, or that are recyclable is now a consideration during the design stage of projects. Sustainable practices also include reducing energy consumption during construction and operation and using renewable-energy alternative technologies during the construction of projects. In the area of waste management, project considerations include producing less waste and recycling more waste. In the area of pollution, the goals are now to use less toxic products and to reduce noise and spatial pollution. Planning for deconstruction requires considering whether materials that will be removed from structures are recyclable or reused when a structure is demolished at the end of its useful life.

Building construction and demolition generate about 25% of municipal solid waste and 50% of the hazardous wastes in the United States. Buildings use 40% of the total energy resources and 16% of available water (USGBC 2014). Indoor air pollution is one of the top five environmental risks to public health. Building-related activities are responsible for 35–45% of the total **carbon dioxide** (CO_2) generated in the United States. Construction uses large quantities of stone, aggregate, sand, and steel and approximately 25% of all of the virgin wood. Buildings use 75% of the **polyvinyl chloride** (PVC) manufactured worldwide. Manufacturing and fires are linked to emissions of a wide range of persistent **bioaccumulative** (builds up in the environment) and toxic materials, including the release of **dioxin**, which is a highly toxic **carcinogen** (cancer-causing agent) (USGBC 2014).

The construction industry should work closely with government agencies, owners, designers, and the manufacturing industry to help reduce the amount of pollution created in the industry. When engineers are designing projects, they should now consider incorporating sustainable practices into their designs in order to create **green structures**. Engineers also need to perform **life-cycle cost analyses** that not only include the cost impacts of their designs but also evaluate the amount of pollution created during the manufacture, transportation, and installation of different construction materials.

12.2 Life-Cycle Environmental and Cost Analysis

Life-cycle cost analysis includes how a structure is designed, how it is constructed, what is required to operate and maintain it, and how it affects human health and environmental quality. Before materials are ordered, they should be evaluated to determine whether they are available locally to reduce the CO_2 emitted by transporting materials long distances. Efficiently using materials, eliminating excessive waste, and reducing the amount of excess material buried in landfills are goals that should be incorporated into plans for site development and construction projects. Renewable materials should be incorporated into projects. Construction activities should include pollution control measures and safety procedures. Incorporating sustainability practices provides measurable benefits to construction workers and the surrounding community and its citizens (Yates 2008).

The EPA's (1996) *Office of Research and Development Strategic Plan* lists U.S. environmental priorities as combatting global climate change, loss of biodiversity, habitat destruction, and stratospheric ozone depletion. In addition, rising global temperatures, increasing exposure to humans to ultraviolet radiation, and the diminishing availability of natural resources are also major concerns to environmentalists. Members of the E&C industry should be aware of the life cycle of resources, establish life-cycle health and environmental considerations, and integrate these considerations into material and product specifications.

In the United States, 40% of the energy consumed is consumed by the construction industry during the construction, operation, and disposal of construction materials. This estimate was developed using life-cycle environmental and cost analysis techniques that account for the extraction, processing, manufacturing, demolition, and disposal of construction materials (Munier 2005).

Not only the durability of materials, but also the service life of materials should be evaluated because service life provides a better indicator for selecting the right alternative on the basis of life-cycle costs (Hodges 2005). According to Hodges (2005), green and sustainable do not automatically mean long lasting; sustainable design processes do not always consider material durability. When incorporating sustainable practices, organizations should include an evaluation of both the durability and service life of materials in their overall strategies. Incorporating sustainable materials or practices in a project might result in lower operating and maintenance costs, even if these materials and practices are more expensive to implement during construction (Hodges 2005).

Several major components should be used for assessing life-cycle costs in the construction industry:

- Operational energy,
- Embodied energy,
- Transport energy,
- Waste,
- Water, and
- Biodiversity.

During each phase of a project, the environmental performance of each component, along with the waste that it generates, is assessed, and the total of all of the components provides an indication of the environmental performance of a structure. Life-cycle costs include "initial cost, maintenance costs, operating costs, replacement or refurbishment cost, retirement and disposal (decommissioning) cost, and other costs such as taxes, depreciation, and additional management costs. However, for infrastructure facilities, life-cycle cost may also include in addition to the ownership costs, the costs to the users of the structure as well as costs to others who are not direct users of the structure but are impacted by the infrastructure facility" (Hastak et al. 2003, p. 1409).

Engineers and constructors should consider life-cycle cost assessments (LCCAs) because they are an important measure of the investment an individual, a corporation, or a government agency will make in a structure or infrastructure from project initiation to disposal. LCCAs include initial, maintenance, operating, replacement, renovation, retirement, disposal, and decommissioning costs. In addition, they include direct and indirect costs and depreciation.

Other methods for accounting for externalities are **emergy** accounting and **emdollars**. *Emergy* means embodied energy, and *emdollars* are the economic equivalent to emergy. These terms are used to measure the value of an activity, not according to its market value, but according to the amount of available energy required for its manufacture, production, marketing, and other activities (Munier 2005). Embodied energy is defined as "the energy consumed in all activities necessary to support a process, including upstream processes" (Treloar 1997, p. 375). Embodied energy is "divided into two components, the direct energy requirement and the indirect energy requirement. Direct energy includes the inputs of energy purchased from producers used directly in a process (including, in the case of a building, the energy to construct it). Indirect energy includes the energy embodied in inputs of goods and services to a process, as well as the energy embodied in upstream inputs to those processes" (Treloar 1997, p. 375).

In addition to the methods mentioned previously for life-cycle environmental and cost analysis, there are standard methods for preparing a **social cost-benefit analysis**, including the following:

1. Define the project. Explain the rationale and the objectives behind the proposal and identify the beneficiaries of the project.
2. Identify the constraints. These may pertain to financial, physical, legal, administrative, distributional, environmental, or other matters.
3. Identify the alternatives, including the "do-nothing" alternative.
4. Assess the project life and discount rate. Public sector projects typically use a discount rate of 5–10% per annum.
5. Identify the costs and benefits. The costs and benefits are incremental and accrue both to the providing authority and to all external parties. The use of a balance sheet is the preferred method as it eliminates the possibility of double counting.

6. Evaluate the costs and benefits. All costs and benefits should be converted into monetary terms wherever possible.
7. Calculate the net present value using the time value of money. The costs and benefits are formatted into yearly cash flows, and the net benefits (benefits minus cost) are calculated for each year and discounted. The sum of the discounted values is the net present worth.
8. Analyze the risk. Test the sensitivity of each alternative to changes in key variables or assumptions by using sensitivity analysis.
9. Identify distributional effects and other issues. Assess the impact of the alternatives on different groups in the community or in the region. Explore environmental issues.
10. Present the analysis, including reasons for the recommendation with adequate support.
11. Calculate the emissions produced while transporting materials. Transporting materials long distances by ship or rail may produce lower emissions than using regional materials transported by diesel-powered trucks.
12. Determine the life-cycle cost of the facility. Evaluate the cost of purchasing raw materials, transporting the raw materials to where they will be processed into building materials, transporting the building materials to the facility where they will be installed, and salvaging any products as waste.

Eqs. (12-1), (12-2), and (12-3) could be used to estimate total emissions generated while transporting materials (Gerilla et al. 2007, p. 2781):

$$EF_c = (E_s \times W_u \times V_a \times P_u)/Y \tag{12-1}$$

$$EF_m = \{(E_s \times W_u \times V_a \times P_u \times K) \times [(Y_m/Y)+1]\}/100 \tag{12-2}$$

$$EF_a = (E_s \times W_u \times V_a \times D_p) \times [(1/Y) - K/(Y_m \times 100)] \tag{12-3}$$

EF_c = **pollutant emission factor** for construction (kg-pollutant/year in m^2); EF_m = pollutant emission factor for maintenance (kg-pollutant/year in m^2); EF_a = pollutant emission factor for disposal (kg-pollutant/year in m^2); E_s = specific emission (kg-pollutant); W_u = unit weight (kg/m^3); V_a = material volume/unit area (m^3/m^2); P_u = unit price of material ($/kg); Y = design life; Y_m = refurbishing/rebuilding cycle (year); K = temporary repair rate for preventive maintenance (%/year); and D_p = unit price of discarded material ($/kg).

12.3 Implementing Sustainable Construction Practices

The construction sector impacts the environment during every stage of construction, from the mining of raw materials (quarry, operation, and production) to the construction of structures (noise, dust, and the generation of hazardous materials) to the operation of facilities (the disposal of wastewater, energy consumption, and toxic emissions) (Basu and Van Zyl 2006).

On construction projects, the following areas would benefit the most from implementing sustainability practices:

- Planning and financing modifications,
- Design modifications,
- Material production,
- Material transportation,
- Resource efficiency,
- Material selection,
- Energy sources,
- Construction equipment and fuel sources,
- Methods for complying with government regulations,
- Supplier and vendor social responsibility,
- Social and community impacts of projects, and
- Production operations.

Implementing sustainable development practices—especially in material selection in relation to the embodied energy involved in extracting, producing, and transporting materials during construction projects—benefits the environment. If designers incorporate materials into their designs that are reusable, are recyclable, or require fewer resources to produce and transport, this reduces the need for disposing of unused construction materials. During the deconstruction of projects at the end of their useful life, any salvageable building materials should be recycled or reused rather than disposed of in landfills. Another concern during the construction phase is energy and water conservation.

The term *sustainable development* emerged from the 1987 publication *Our Common Future*, written by the World Commission on Environment and Development (also known as the Brundtland Commission). This publication defined sustainable development as "meeting the needs of the present, without jeopardizing the ability of future generations to meet their needs" (World Commission on Environment and Development 1987, p. 1).

Other definitions for sustainable development include "a system of changes in public attitude and policy through which the population and vital activities of a community may be continued into the indefinite future without robbing the community of its usable resources" (Cywinski 2001, p. 13). Another definition is "a process of change in which the direction of investment, the orientation of technology, the allocation of resources, and the development and functioning of institutions meet present needs and aspirations without endangering the capacity of natural systems to absorb the effects of human activities, and without compromising the ability of future generations to meet their own needs and aspirations" (Cywinski 2001, p. 14).

Sustainable construction involves "creating construction items using best-practice clean and resource-efficient techniques from the extraction of the raw materials to the demolition and disposal of its components" (Ofori 2000, p. 196). According to the European Commission's (2001) *Proposals for a Response to the Challenges of Sustainable Construction*, sustainable construction

> is the set of processes by which a profitable and competitive industry delivers built assets (buildings, structures, supporting infrastructure and their immediate surroundings) which in turn:

- Enhance the quality of life and offer customer satisfaction.
- Offer flexibility and the potential to cater to user changes in the future.
- Provide and support desirable natural and social environments.
- Increase investment in people and equipment for a competitive economy.
- Achieve higher growth whilst reducing pollution and maximizing the efficient use of resources.
- Share the benefits of growth more widely and more fairly.
- Improve towns and protecting the quality of the countryside.
- Contribute to sustainable development internationally.

The U.S. Department of Energy's (2008) *Energy and the Environmental Guidelines for Construction* mentions that to promote the use of sustainable construction practices at construction jobsites, it is important to

- Explain methods for protecting vegetation, such as designating access routes and parking;
- Require methods for clearing and grading sites that lower the impact as much as possible;
- Examine how runoff during construction may affect sites, and consider creating storm-water management practices, such as piping systems or retention ponds or tanks that can be used after the building is complete;
- Be sure that the infrastructure for recycling construction and demolition materials is ready and operating from the beginning of projects, set up an on-site system to collect and sort waste for recycling or reuse, and monitor the system consistently throughout all phases of construction;
- Create plans for recycling that set goals to recycle or salvage a minimum of 50% (by weight) of construction, demolition, and land-clearing waste from construction sites, and aim for a minimum of 75%;
- If possible, choose products and materials with minimal or no packaging;
- Obtain materials in the sizes that are necessary, instead of cutting materials to size at jobsites; and
- Always monitor the amount of waste being produced during construction and compare it with preexisting goals and guidelines.

Sustainable practices, including selecting environmentally friendly materials and technologies and using construction processes that require less toxic materials, consume less energy, and produce less waste, should be incorporated into construction projects during the planning stage.

12.4 Loss of Productivity of Workers in Surrounding Areas

Construction activities may negatively affect members of the community surrounding construction projects, particularly in the following areas (Gilchrist and Allouche 2005, p. 93):

- Loss of income,
- Productivity reduction,
- Reduction in tax revenue, and
- Property damage.

The **loss of productivity** (LOP) experienced by workers in the surrounding community who are unable to perform their assigned work function due to an ongoing construction project is estimated using the following formula (Gilchrist and Allouche 2005, p. 97):

$$\begin{aligned}\text{LOP} = {} & \text{number of employees affected in the} \\ & \text{surrounding community} \\ & \times \text{(average hourly output of} \\ & \text{employees in dollars/hour)} \\ & \times \text{productivity reduction factor} \\ & \times \text{project duration in hours}\end{aligned} \tag{12-4}$$

For example, 10 employees are working in an area affected by the construction project, they have an average hourly output of \$50.00 per hour, the productivity reduction factor is 0.60 (the productivity of the 10 workers is reduced by 40% during construction), and the construction project lasts for one year. For these parameters, the LOP is calculated as follows:

$$\begin{aligned}\text{LOP} = {} & 10 \times 50 \times 0.60 \times 40 \text{ h per week} \\ & \times 52 \text{ weeks} = \$624{,}000.\end{aligned}$$

Therefore, the firm with the 10 employees affected by the construction project will sustain an LOP of approximately \$624,000 for the year that the project is under construction.

The impact on nearby property values during a construction project is estimated considering "neighborhood accessibility and environmental variables instead of merely considering property market value" (Gilchrist and Allouche 2005, p. 98). User delay costs and average traffic delay costs attributable to construction activities are quantified using a method introduced by Gilchrist and Allouche (2005, p. 99):

$$\begin{aligned}\text{average user delay costs} = {} & \\ & \text{average number of passengers per car} \\ & \times \text{average delay per car} \\ & \times \text{average hourly wage} \times \text{percentage of wage}\end{aligned} \tag{12-5}$$

The average traffic delay cost could be calculated on the basis of project duration, average annual daily traffic (AADT), the peak-hour factor (k), the number of passengers per car, and the average hourly wage:

$$\begin{aligned}\text{average traffic delay costs} &= \text{AADT}\\ &\times k \text{ factor} \times \text{lane closure duration}\\ &\times \text{average user delay costs}\end{aligned} \tag{12-6}$$

12.5 Energy Consumption during Construction Projects

Construction projects consume the most energy during the generation of construction materials. Transporting materials accounts for between 15% and 20% of the energy consumed at jobsites. Industrial construction projects use asphalt and concrete to build access roads to jobsites and for permanent roadways. Producing asphalt requires 57% of the total energy consumed when generating materials for construction projects, cement consumes 25%, and the remaining processes consume 18% (Moroueh et al. 2001).

Several strategies for reducing energy consumption during construction are recommended in the article "Sustainable Development and the Construction Industry" (Spence and Mulligan 1995, p. 281), including

- Improved energy efficiency in kiln processes,
- Using cheaper or non-premium fuels in kiln processes,
- Use of recycled materials in production processes,
- Use of low-energy additives or extenders,
- Using fewer materials,
- Selection of low-energy materials and structural systems,
- Designing low-rise buildings in place of high-rise buildings,
- Selection whenever possible of waste or recycled materials, and
- Design for recycling, long life, and adaptability to varying requirements.

It is estimated that 8–20% of the emissions generated throughout the world are attributable to construction activities and materials production, and approximately 3% of emissions are attributable to the production of cement and lime. Some suggestions provided by Spence and Mulligan (1995, p. 283) for helping to reduce atmospheric pollution during construction include

- Reducing avoidable transportation of materials,
- Improving site-management efficiency,
- Reducing of the quantity of site wastes produced, and
- Systematic separation of all unavoidable construction wastes, to facilitate recycling.

One impediment to the use of renewable energy sources on construction projects is the short duration of projects. The capital investment required to provide alternative renewable energy sources prohibits firms from being awarded competitive bids. Renewable energy sources have been used successfully in other industries—such as the oil industry, which uses wind turbines and solar panels on its unmanned oil production platforms in the North Sea—but these sources of energy are used on stationary projects of a longer duration than typical construction projects.

Construction firms are able to buy energy from firms using renewable energy sources, but currently, the cost for energy created by renewable methods is higher than the cost of energy from nonrenewable sources. Some construction firms are selectively choosing to use renewable energy, depending on the pricing structure where projects are being built.

12.6 Pollution Prevention and Recycling

To determine the techniques being implemented in the construction industry, a research investigation funded by the Construction Industry Institute surveyed executives from construction firms. The survey participants were asked what types of processes they used to help prevent pollution during construction. The pollution-prevention techniques mentioned by the respondents included reducing or eliminating noise, scheduling deliveries for early in the day to avoid truck use on high ozone days, minimizing the idling of equipment, installing scrubbers and mufflers on equipment, limiting activities that cause noise at night, preplanning traffic routes, protecting against water runoff and erosion, and treating effluent and nonpotable water and reusing it for dust suppression and landscape irrigation (Yates 2008).

Examples of the processes used by the survey respondents to reduce waste during construction include selling or reusing material by-products where scrap metal is separated and resold, sending materials back into corporate inventory to sell to recyclers, refurbishing transformers and meters, recycling by-products, sharing materials with other jobsites, selling usable materials to marketers who resell it for its original intended purpose, and aggregating disposable waste to minimize energy expended in its final disposition (Yates 2008).

The survey respondents indicated that they use a variety of processes to recycle wastes, including donating to local community organizations, using scrap-metal dumpsters, putting items back into corporate inventory, advertising surplus in the organization, returning materials to dealers and vendors, setting zero-waste-to-landfill targets, using materials as feedstock for reuse, and establishing recycling pathways for excess materials (Yates 2008).

To reduce the amount of waste generated at jobsites, the survey participants suggested correctly sizing equipment, materials, and components; precutting drywall, pipe, and conduit; controlling take-off and material ordering; choosing durable and reusable materials and products; using off-specification concrete or remnants of concrete to make curbstone barrier blocks; using reusable concrete forms; and incorporating modularization (Yates 2008).

All of the techniques mentioned in the previous sections could be implemented on construction projects to improve the sustainability of construction processes.

Although productivity improvement studies are conducted primarily to increase production rates, when clients stipulate that they want to include sustainability techniques and processes during construction, productivity improvement team members should consider the sustainable practices discussed in this chapter and Chapter 13 in their evaluations of construction processes.

12.7 Leadership in Energy and Environmental Design

The Leadership in Energy and Environmental Design (LEED) Green-Building Rating System is a voluntary, consensus-based national rating system and certification program for sustainable structures that was developed by members of the U.S. Green Building Council (USGBC). Many segments of the building industry are represented by members of the USGBC, including architects; building owners; contractors; engineers; federal, state, and local code and regulatory officials; financiers; product manufacturers; real estate developers; and utility providers.

There is a Construction Rating System (LEED NC), and LEED standards are also currently available for (USGBC 2012):

- Laboratories,
- Commercial interiors (LEED-CI),
- **Core and shell**,
- Existing building operations and maintenance (LEED-EB),
- Health care facilities,
- Homes,
- Neighborhood developments,
- New commercial construction and major renovations,
- Retail stores, and
- Schools.

Leadership in Energy and Environmental Design certification promotes integrated building design, construction, operation, and maintenance practices that minimize the environmental impact of structures. The LEED certification program

- Defines *green building* by establishing a common standard of measurement,
- Recognizes environmental leadership in the building industry and prevents false claims,
- Stimulates green competition,
- Raises consumer awareness of green-building benefits, and
- Transforms the building market. (USGBC 2008)

The LEED organization emphasizes using integrated technologies that promote

- Water savings,
- Energy efficiency,

- Sustainable materials selection, and
- Improved indoor environmental quality.

The LEED certification process recognizes structures that meet the green-building requirements of the USGBC. The USGBC LEED certification process promotes expertise in green building by offering project certification, professional accreditation, training programs, and other resources.

Members of the USGBC evaluate and update the LEED certification process. The LEED-NC 3.0 includes **regionally weighted credits**, seamless online registration and certification processes, and planned integration with three-dimensional computer-aided drafting software (building information modeling software) to help monitor the viability of various sustainability strategies and technologies.

Many government agencies are incorporating LEED initiatives or equivalent processes into their projects. By 2014, in the U.S. there were more than 6,412.6 million gross ft^2 (595.73 million gross m^2) of LEED-registered projects, and over 44,270 projects were registered to LEED (USGBC 2014). The top ten countries with LEED certified buildings are (USGBC 2014)

1. United States
2. Canada
3. China
4. United Arab Emirates
5. Brazil
6. India
7. Mexico
8. Germany
9. Turkey
10. Republic of Korea

The LEED initiative provides information that design team members are able to use to create sustainable projects and as an evaluation system to assess sustainability achievements according to industry standards. The LEED checklist for projects includes different sustainability categories and a scoring system. The total number of points achieved is used to determine an overall LEED rating of certified, silver, gold, or platinum. Examples of LEED categories with related sustainable strategies are as follows (USGBC 2012):

1. Sustainable Sites (SS)
    - Site selection
    - Urban or brownfield redevelopment
    - Alternative transportation
    - Storm-water management
2. Water Efficiency (WE)
    - Landscaping
    - Wastewater technologies

3. Energy and Atmosphere (EA)
   - Energy optimization
   - Chlorofluorocarbon (CFC) reduction
   - Green power
4. Materials and Resources (MR)
   - Building reuse
   - Construction waste management
   - Certified wood
5. Indoor Environmental Quality (IEQ)
   - Low-emitting materials
   - Ventilation effectiveness
6. Innovation and Design Processes (IDP)
   - LEED accredited professional on design team
   - Development of a sustainable education program
   - Achievement of sustainability goals far in excess of stated LEED requirements

The four rating systems and the point ranges required for each level are as follows (USGBC 2012):

- Certified: 26–32 points
- Silver: 33–38 points
- Gold: 39–51 points
- Platinum: 52–69 points

Additional costs are associated with incorporating sustainable elements into a structure to achieve a LEED certification. In one study, the additional cost was estimated as between \$2 and \$5 per square foot (\$2 and \$5 per 0.0929 m^2) for basic certification (Kibert 2008, p. 327). For the higher levels of LEED certification, the cost premiums were estimated by reviewing 33 buildings (Kibert 2008, p. 327):

- Platinum: 6.5%
- Gold: 1.82%
- Silver: 1.11%
- Certified: 0.66%
- Average: 2.52%

Since this study was conducted, the cost of integrating sustainable elements into structures has declined because of process improvements. The additional costs associated with each of the LEED certification levels vary with the type of structure and other variables, and it is hoped that the costs will continue to decline in the future.

Firms may either compile their own documentation for the assessment or hire a trained assessor. If a LEED-accredited professional (LEED AP) is part of the design team, then credits are awarded for his or her participation. The USGBC is the organization that performs assessments and determines the LEED score. Each credit is worth one point and is awarded on the basis of actions that help reduce environmental impacts.

LEED Credits, Prerequisites, Subcategories, and Possible Points

The LEED rating system provides a method for obtaining points based on the six LEED categories. In addition, there are prerequisites, subcategories, and credits for each category that total to the possible points. The categories are as follows (Haselbach 2008, p. 9):

1. Sustainable Sites
   - Prerequisites: 1
   - Subcategories: 8
   - Possible points: 14
2. Water Efficiency
   - Prerequisites: none
   - Subcategories: 3
   - Possible points: 5
3. Energy and Atmosphere
   - Prerequisites: 3
   - Subcategories: 6
   - Possible points: 17
4. Materials and Resources
   - Prerequisites: 1
   - Subcategories: 7
   - Possible points: 13
5. Indoor Environmental Quality
   - Prerequisites: 2
   - Subcategories: 8
   - Possible points: 15
6. Innovation and Design Processes
   - Prerequisites: none
   - Subcategories: 2
   - Possible points: 5

When members of a firm decide that they would like to seek LEED certification for a potential project, the project team first registers with the USGBC at the level of certification they hope to achieve during the project. Registering with the USGBC at the inception of a project allows design team members access to the USGBC website and the appropriate templates for tracking the project. The certification process does not proceed until the end of construction, although in the newer version of the LEED certification process credits are awarded for activities that occur during the design phase.

LEED Certification Checklist

The USGBC uses a checklist when it evaluates a project that is under review for LEED certification. The checklist allows the USGBC to determine the level of certification on the basis of the total number of points awarded to the project. Table 12-2 provides a list of the prerequisites, credits, and points possible for a LEED-NC 2.2 project.

Benefits of LEED Certification

According to the USGBC, some of the benefits of green structures are

- Recover higher first costs, if there are any;
- Designed for cost effectiveness;
- Boost employee productivity;
- Reduce liability;
- Create value for tenants;
- Increase property value;
- Take advantage of incentive programs;
- Benefit the community;
- Achieve more predictable results. (Kibert 2008, p. 330)

The USGBC has also suggested that green structures help address other issues, such as

- High electric power costs;
- Deteriorating power grid problems, such as power quality and availability;
- Possible water shortages and waste disposal issues;
- State and federal pressure to reduce criteria pollutants;
- Global warming;
- Rising incidence of allergies and asthma, especially in children;
- The health and productivity of workers;
- The effect of school environments on children's ability to learn;
- Increases in operating and maintenance costs for state facilities. (Kibert 2008, pp. 330–331)

The following list provides some of the benefits of having a LEED-certified structure (USGBC 2012):

- Third-party validation of green features and degree of sustainability;
- Enforcement of complete implementation of designed green features;
- LEED brand association; and
- Incentives or requirements from public agencies, including
  - San Jose, California, offering an array of resources to projects pursuing LEED certification such as financial incentives, awards, and streamlined permitting processes;

Table 12-2. Example of LEED-NC 2.2 Checklist

Yes	No	Category	Credit Descriptions	Points
			Sustainable Sites	
Y		Prerequisite 1	Construction activity pollution prevention	Required
		Credit 1	Site selection	1
		Credit 2	Development and community connectivity	1
		Credit 3	Brownfield redevelopment	1
		Credit 4.1	Alternative transportation—public transportation access	1
		Credit 4.2	Alternative transportation—bicycle storage and changing rooms	1
		Credit 4.3	Alternative transportation—low-emitting and fuel-efficient vehicles	1
		Credit 4.4	Alternative transportation—parking capacity	1
		Credit 5.1	Site development—protect or restore habitat	1
		Credit 5.2	Site development—maximize open space	1
		Credit 6.1	Storm-water design—quantity control	1
		Credit 6.2	Storm-water design—quality control	1
		Credit 7.1	Heat island effect—nonroof	1
		Credit 7.2	Heat island effect—roof	1
		Credit 8	Light pollution reduction	1
			Water Resources	
		Credit 1.1	Water-efficient landscaping—reduce by 50%	1
		Credit 1.2	Water-efficient landscaping—no potable use or no irrigation	1
		Credit 2	Innovative wastewater technologies	1
		Credit 3.1	Water use reduction—20% reduction	1
		Credit 3.2	Water use reduction—30% reduction	1
			Energy and Atmosphere	
Y		Prerequisite 1	Fundamental commissioning of the building energy systems	Required
Y		Prerequisite 2	Minimum energy performance	Required
Y		Prerequisite 3	Fundamental refrigerant management	Required
		Credit 1	Optimize energy performance	1 to 10
			10.5% new building or 3.5% existing building renovations	1
			14% new buildings or 7% existing building renovations	2
			17.5% new buildings or 10.5% existing building renovations	3
			21% new buildings or 14% existing building renovations	4
			24.5% new buildings or 17.5% existing building renovations	5
			28% new buildings or 21% existing building renovations	6

continued

Table 12-2. *(continued)*

Yes	No	Category	Credit Descriptions	Points
			31.5% new buildings or 24.5% existing building renovations	7
			35% new buildings or 28% existing building renovations	8
			38.5% new buildings or 31.5% existing building renovations	9
			42% new buildings or 35% existing building renovations	10
		Credit 2	On-site renewable energy	1 to 3
			2.5% renewable energy	1
			7.5% renewable energy	2
			12.5% renewable energy	3
		Credit 3	Enhanced commissioning	1
		Credit 4	Enhanced refrigerant management	1
		Credit 5	Measurement and verification	1
		Credit 6	Green power	1
			Materials and Resources	
		Prerequisite 1	Storage and collection of recyclables	Required
		Credit 1.1	Building reuse—maintain 75% of existing walls, floors, and roof	1
		Credit 1.2	Building reuse—maintain 100% of existing walls, floors, and roof	1
		Credit 1.3	Building reuse—maintain 50% of interior nonstructural elements	1
		Credit 2.1	Construction waste management—divert 50% from disposal	1
		Credit 2.2	Construction waste management—divert 75% from disposal	1
		Credit 3.1	Materials reuse—5%	1
		Credit 3.2	Materials reuse—10%	1
		Credit 4.1	Recycled content—10% (postconsumer + 1/2 preconsumer)	1
		Credit 4.2	Recycled content—20% (postconsumer + 1/2 preconsumer)	1
		Credit 5.1	Regional materials—10% extracted, processed, and manufactured regionally	1
		Credit 5.2	Regional materials—20% extracted, processed, and manufactured regionally	1
		Credit 6	Rapidly renewable materials	1
		Credit 7	Certified wood	1
			Indoor Environmental Quality	
Y		Prerequisite 1	Minimum indoor air quality (IAQ) performance	Required
Y		Prerequisite 2	Environmental tobacco smoke control	1
		Credit 1	Outdoor air delivery monitoring	1

Table 12-2. *(continued)*

Yes	No	Category	Credit Descriptions	Points
		Credit 2	Increased ventilation	1
		Credit 3.1	Construction IAQ management plan—during construction	1
		Credit 3.2	Construction IAQ management plan—before occupancy	1
		Credit 4.1	Low-emitting materials—adhesives and sealants	1
		Credit 4.2	Low-emitting materials—paints and coatings	1
		Credit 4.3	Low-emitting materials—carpet systems	1
		Credit 4.4	Low-emitting materials—composite wood and agrifiber products	1
		Credit 5	Indoor chemical and pollutant source control	1
		Credit 6.1	Controllability of systems—lighting	1
		Credit 6.2	Controllability of systems—ventilation	1
		Credit 7.1	Thermal comfort—design	1
		Credit 7.2	Thermal comfort—verification	1
		Credit 8.1	Daylight and views—daylight for 75% of spaces	1
		Credit 8.2	Daylight and views—views for 90% of spaces	1
			Innovation and Design Process	
		Credit 1.1	Innovation in design—provide specific title	1
		Credit 1.2	Innovation in design—provide specific title	1
		Credit 1.3	Innovation in design—provide specific title	1
		Credit 1.4	Innovation in design—provide specific title	1
		Credit 2	LEED accredited professional	1
			Project Total (precertification estimates)	69 points

Source: Adapted from USGBC (2012).
Note: Certified, 26–32 points; Silver, 33–38 points; Gold, 39–51 points; Platinum, 52–69 points.

- Oregon having a business energy tax credit program for projects that achieve a LEED silver rating or higher;
- Arlington, Virginia, waiving height or density limitations for LEED-certified projects; and
- Many cities, states, and federal agencies, including the Government Services Administration, having mandated LEED for public buildings.

Impact of LEED on Productivity

The initial impact of LEED certification procedures on the productivity of the workers on construction projects could be negative while the construction workers become familiar with new LEED-related methods, techniques, and materials. If more projects seek LEED certification, then workers will become familiar with the requirements for LEED-certified projects, and their learning curve will not be as steep as it was for the first few LEED-certified projects. The number of LEED-certified projects is increasing rapidly in the United States; therefore, the impact of requiring LEED certification on projects and the resulting reduction in productivity will decline in the future.

12.8 Sustainability Organizations, Publications, and Certification Programs

In addition to LEED, there are numerous other sustainability organizations, publications, and certification programs that provide information on sustainable development practices for buildings, including

- BEES Stars (National Institute of Standards and Technology [NIST]),
- BES 601 and 602—Responsible Sourcing of Construction Products (Building Research Establishment),
- Building Research Establishment Environmental Assessment Method (BREEAM) (Building Research Establishment Trust),
- Civil Engineering Environmental Quality Assessment Award Scheme (CEEQUAL) (Institute of Civil Engineers [ICE]),
- *Codes for Sustainable Homes* (CSH) (Department for Communities and Local Government U.K.),
- Comprehensive Assessment System for Building Environmental Efficiency (CASBEE) (Japan Sustainable Building Consortium),
- Council on Tall Buildings and Urban Habitat (CTBUH),
- Design Quality Indicators (DQI) (Construction Industry Council),
- *Energy and the Environmental Guidelines for Construction*—Department of Engineering Building Technology Program (U.S. Department of Energy),
- *Environmental Performance of Building Guidelines* (Environmental Protection Agency),
- *Envision*—Sustainable Infrastructure Rating System (Institute for Sustainable Infrastructure),
- Forest Stewardship Council,
- Green Globes Building (Owners and Managers Association [BOMA] in Canada and the Green Building Initiative [GBI] in the United States. Accredited by the American National Standards Institute [ANSI]),
- *Green Guide to Specifications* (Building Research Establishment),
- Green Star (Green Building Council of Australia [GBCA]),
- *Greenroads* (U.S. Federal Highway Administration),
- International Green Construction Code (International Code Council),
- *ISO 14,000 Series of Environmental Management Standards* (International Organization for Standardization),
- *Standard 189.1—Design of High Performance Green Buildings* (American Society of Heating, Refrigeration, and Air Conditioning Engineers [ASHRAE]),
- *Sustainability and the Construction Industry in the United Kingdom* (Chartered Institute of Building 2004),
- *Sustainability Design Guide* (Los Alamos National Laboratory),
- Sustainable Buildings and Construction Initiative of the United Nations (United Nations Environmental Program),
- *Sustainable Sites Initiative: Guidelines for Performance Benchmarks 2009* and *The Case for Sustainable Landscapes*,

- United Stated Department of Engineering Building Technology Program (U.S. Department of Energy), and
- World Green Building Council (WGBC).

All of the organizations, publications, and certification programs listed provide detailed information on a variety of different sustainability subjects, practices, and certification systems.

12.9 Summary

This chapter covered a variety of topics related to sustainability that impact the productivity of construction operations, including sustainable practices in construction, corporate social responsibility, life-cycle environmental and cost analysis, loss of productivity to members in the community surrounding construction jobsites, energy consumption, pollution prevention, and recycling.

This chapter also discussed the LEED certification process and the role the USGBC plays in this process. It provided an overview of the LEED rating system and certification process, which is designed to assess the sustainability of structures. The types of credits, prerequisites, subcategories, and possible points for the rating system were discussed, and a LEED-NC 2.2 sample checklist, detailing the credits for each LEED category, was provided in this chapter. Some of the benefits of having LEED certification for structures were listed that were provided by the U.S. Green Building Council. Other sustainability organizations, publications, and certification systems were also mentioned in this chapter.

All of the topics discussed in this chapter have a major impact on productivity at construction jobsites. When sustainability practices are being incorporated at construction jobsites, productivity rates may decrease while workers learn how to incorporate sustainable techniques, processes, or materials. But once workers become familiar with sustainable practices and their proper implementation, productivity rates should return to previous rates.

12.10 Key Terms

bioaccumulative
carbon dioxide
carcinogen
core and shell
corporate social responsibility
corporate sustainability
dioxin
Dow Jones Global Sustainability Index
embodied energy
emdollars
emergy
green structures
life-cycle cost analyses
loss of productivity
pollutant emission factor
polyvinyl chloride
regionally weighted credits
social cost-benefit analysis
socially responsible investment
sustainability
sustainable construction
sustainable development
triple bottom line

12.11 Exercises

12.1 Discuss what is required for a structure to achieve one of the four different LEED certification levels.

12.2 Discuss the three most important benefits of a structure having LEED certification.

12.3 Explain what the LEED Green Building Rating System is and how it is used to determine the sustainability of a structure.

12.4 What are the main LEED categories that each contain related to sustainable strategies included in the LEED certification system?

12.5 What is the objective of having LEED certification?

12.6 Which segments of the building industry are represented in the USGBC?

12.7 What are the benefits of registering with the USGBC at the inception of a project that is attempting to achieve LEED certification?

12.8 What aspects of the LEED rating system would directly affect the implementation of productivity improvement techniques?

12.9 Do any of the formulas provided in this chapter apply to the productivity of construction workers on projects seeking LEED certifications, and if so, which ones?

12.10 How would a manager motivate construction workers to adopt sustainable methods to reduce waste on construction projects?

12.11 Explain sustainable construction and its relationship to sustainable development.

References

Basu, A., and Van Zyl, D. (2006) "Industrial ecology framework for achieving cleaner production in the mining and minerals industry." *J. Cleaner Prod.*, 14(3), 299–304.

Cywinski, Z. (2001). "Current philosophy of sustainability in civil engineering." *J. Prof. Issues Eng. Educ. Pract.*, 127(1), 12–16.

EPA. (1996). *Office of Research and Development (ORD) strategic plan.* <http://epa.gov/ncer/publications/archive/stratplan.pdf> (March 2014).

European Commission. (2001). *Proposals for a response to the challenges of sustainable construction.* <http://ec.europa.eu/enterprise/construction/suscon/finrepsus/sucop2.htm> (March 2014).

Gerilla, G., Teknomo, K., and Hokao, K. (2007). "An environmental assessment of wood and steel reinforced concrete housing construction." *Build. Environ.*, 42(7), 2778–2784.

Gilchrist, A., and Allouche, N. (2005). "Quantification of social costs associated with construction projects: State-of-the-art review." *Tunneling Underground Space Technol.*, 20(1), 89–104.

Haselbach, L. (2008). *The engineering guide to LEED—New construction*, McGraw Hill, New York.

Hastak, M., Mirmiran, A., and Richard, D. (2003). "A framework for life-cycle cost assessment of composites in construction." *J. Reinf. Plast. Compos.*, 22(15), 1409–1430.

Hodges, P. (2005). "A facility manager's approach to sustainability." *Journal of Facilities Management*, no. 4, 312–324.

Kibert, C. (2008). *Sustainable construction: Green building design and delivery*, Wiley, New York.

Moroueh, U., Eskola, P., and Laine-Ylijoki, J. (2001). "Life-cycle impacts of the use of industrial by-products in road and earth construction." *Waste Manage. (Oxford)*, 21(3), 271–277.

Munier, N. (2005). *Introduction to sustainability: Road to a better future*, Springer, Dordrecht, Netherlands.

Ofori, G. (2000). "Greening the construction supply chain in Singapore." *European Journal of Purchasing and Supply Management*, 6(3–4), 195–206.

Samaras, C. (2004). *Sustainable development and the construction industry: Status and implementation*, Carnegie Mellon University, Pittsburgh, PA.

Spence, R., and Mulligan, H. (1995). "Sustainable development and the construction industry." *Habitat International*, 19(3), 279–292.

Sustainability and the Construction Industry in the United Kingdom (Chartered Institute of Building 2004), London.

Treloar, G. (1997). "An input-output model of the embodied energy pathways of Australian residential buildings." *Journal of Economic Systems Research*, 9(4), 375–391.

U.S. Department of Energy. (2008) *Energy and the environmental guidelines for construction*, Washington, DC.

U.S. Green Building Council (USGBC). (2008). *Leadership in energy and environmental design*, <http://www.usgbc.org/leed> (March 2014).

U.S. Green Building Council (USGBC). (2012). *Leadership in energy and environmental design*, <http://www.usgbc.org/leed> (March 2014).

U.S. Green Building Council (USGBC). (2014). *Infograhic: LEED in the world*, <http://www.usgbc.org/articles/infographic-leed-world> (April 2014).

World Commission on Environment and Development. (1987). *Our common future*, Oxford University Press, Oxford, U.K.

Yates, J. (2008). "Sustainability in industrial construction." *Research Report 250-11*, Construction Industry Institute, Austin, TX.

CHAPTER 13

Sustainable Construction Materials

This chapter discusses **sustainable construction materials** and introduces information from various sources, including the *LANL Sustainable Design Guide* (LANL 2002), about sustainable construction materials. Information is also included on the processes for manufacturing different construction materials to demonstrate the cradle-to-grave consequences of construction materials. Sustainable alternative construction materials are reviewed in this chapter—including paints, sealants, steel, cement, fly ash concrete, **concrete canvas**, **porous concrete**, Hardie board, asphalt, masonry products, fiber-reinforced polymeric composite materials, wood products, **polyvinyl chloride** (PVC) products, thermoplastic products, and metal products.

13.1 Sustainable Construction Materials

This section discusses sustainable construction materials being incorporated into construction projects. Incorporating many of the materials discussed in the following sections should not affect worker productivity rates because the materials are similar to traditional construction materials; they are only being manufactured using more sustainable processes or with more sustainable materials. If workers are not familiar with these particular materials, there may be a learning curve effect—that is, production rates may be lower in the beginning, but they will increase as the workers become more familiar with the materials.

Sustainable construction materials "minimize resource use, have low ecological impacts, pose no or low human and environmental health risks, and assist with sustainable site strategies" (Calkins 2009, p. 3). In addition to incorporating sustainable materials, structures would be more sustainable by using fewer materials or by reducing the size of the structures. The reuse of existing structures or structural elements from existing structures also leads to more sustainable structures. Sustainable structures include materials that will not only last for the life of a structure but be reclaimed and reused in future structures (Calkins 2009). Another method for reducing the environmental impact of construction materials is to use materials that have been **sustainably harvested** or mined in such as manner as to reduce or not create air, water, or soil pollution.

Suggestions for reducing resource use when selecting materials and products include the following (Calkins 2009, pp. 3–5):

- Reuse existing structures in place,
- Reduce material use,
- Use durable materials,
- Reclaim and reuse materials or products in whole forms,
- Use reclaimed materials from other sources,
- Reprocess existing structures and materials for use on site,
- Use reprocessed materials from other sites,
- Specify materials and products with reuse potential and design for disassembly,
- Specify recycled-content materials and products,
- Use materials and products with recycling potential,
- Specify materials and products made from renewable resources, and
- Specify materials or products from manufacturers with product take-back programs.

Suggestions for choosing materials or products that minimize environmental impacts include the following (Calkins 2009, pp. 6–7):

- Use minimally processed materials,
- Specify low embodied energy materials,
- Specify materials produced with energy from renewable sources,
- Use local materials,
- Specify low-polluting materials,
- Specify low water-use and low water-polluting materials,
- Specify low-emitting materials or products, and
- Specify materials or products that avoid toxic chemicals or by-products.

Materials or products that assist with sustainable site design strategies include the following (Calkins 2009, p. 8):

- Products that promote a site's hydrologic health,
- Materials or products that sequester carbon,
- Products that reduce the urban **heat island** effect,
- Products that reduce energy consumption of site operations, and
- Products that reduce water consumption of site operations.

According to the *LANL Sustainable Design Guide* (LANL 2002), certain characteristics are preferable in construction materials. Table 13-1 lists the preferred material characteristics. The *LANL Sustainable Design Guide* (LANL 2002) also includes a table of design evaluations for materials and resources, and it is provided in Table 13-2.

Table 13-1. Sample Characteristics of Environmentally Preferable Materials

Category	Characteristic
Life-cycle cost impact (LCCI)	Relative impact of life-cycle cost of building operations (not to be confused with environmental life-cycle assessment, which measures environmental burdens, not financial impact)
Energy efficiency (EE)	Construction materials that directly influence building energy use
Water efficiency	Construction materials that directly influence building water use
Locally manufactured (LM)	Construction materials manufactured within a defined radius (500 mi [804.74 km] for the LEED rating system)
Material reduction (MR)	Products or materials that serve a defined function using less material than is typically used
Locally derived raw materials (LRM)	Construction materials that are locally manufactured using raw materials obtained within the defined radius
Nontoxic (NT)	Construction materials that release relatively low levels of emissions of odorous, irritating, toxic, or hazardous substances (volatile organic compounds [VOCs], formaldehydes, particulates, and fibers are examples of substances emitted from construction materials that could adversely affect human health [allergens, carcinogens, and irritants])
Recycled content (RC)	Amount of reprocessed material contained within a construction product that originated from postconsumer use or postindustrial use (including the reuse of existing building structures, equipment, and furnishings)
Salvages (S)	Construction materials that are reused as is (or with minor refurbishing) without having undergone any reprocessing to change the intended use (including the reuse of existing building structures, equipment, and furnishings)
Rapidly renewable (RR)	Construction materials that replenish themselves faster (within 10 years) than traditional extraction demand and do not result in adverse environmental impacts
Certified wood (CW)	Construction materials manufactured all or in part from wood that has been certified by the Forest Stewardship Council as originating from a well-managed forest

Source: Adapted from LANL (2002), p. 237.

The rest of this chapter examines sustainable construction materials and processes, including paint products; steel production; cement and concrete production; masonry; asphalt production; fiber-reinforced polymeric composite materials; wood products; polyvinylchloride and thermoplastic products; mining, minerals, and metal products.

Table 13-2. Sustainable Design Evaluations for Materials and Resources

Material	Material Cost	Life-Cycle Cost Impact	Energy Efficiency	Water Efficiency	Material Reduction	Locally Manufactured	Locally Derived Raw Material	Nontoxic	Recycled Content	Rapidly Renewable	Certified Wood
Ceiling tiles	= +	–							X		
Carpet	=	=			X			X	X		
Fabrics	= +	= –						X	X		
Resilient flooring	= +	= –						X	X	X	
Interior/exterior paints	=	=						X	X		
Sealants and adhesives	=	=						X			
Steel	=	=			X				X		
Cement concrete	=	=	X		X	X	X		X		
Insulation	=	–	X			X	O	X	X		
Bathroom cubicles	=	=							X		
Wood products	= +	=			X			X		X	X
Gypsum wallboard	=	=				X	X		X		
Furniture	= +	=							X	X	X
Brick concrete masonry unit	=	=				X	X				
Roofing	=	=	X						X		
Windows	+	–	X						X		
Doors	= +	–	X						X		
Ceramic tile	=	=						X		X	X
Insulating concrete forms	+	–	X						X		
Structural insulated panels	+	–	X						X		
Aerated autoclaved concrete	+	–							X		
Exterior finishes						X	O				
Permeable paving	=	–		X					X		

Source: Adapted from LANL (2002), p. 237.

Note: O indicates potentially applicable material and resource issue, research ongoing; X indicates applicable material and resource issue; = indicates equivalent; – indicates generally less expensive; + indicates generally more expensive.

13.2 Painting Products

Latex paints with some, or almost all (99%), content from recycled materials are now available. One environmental concern regarding paints is the amount of **volatile organic compound** (VOC) emissions that result from their use. Volatile organic compounds are "carbon-containing compounds that readily evaporate at room temperature and are found in many housekeeping, maintenance, and building products made with organic (carbon-based) chemicals. Paint, glues, paint strippers, solvents, wood preservatives, aerosol sprays, cleansers, and disinfectants, air fresheners, stored fuels, automotive products, and even dry-cleaned clothing and perfume are all sources of VOCs. There are six major classes of VOCs: aldehydes (formaldehyde), alcohols (ethanol, methanol), aliphatic hydrocarbons (propane, butane, hexane), aromatic hydrocarbons (benzene, toluene, xylene), ketone (acetone), and halogenated hydrocarbons (methyl chloroform, methylene chloride)" (Kibert 2008, p. 284).

Formaldehyde is the most commonly found VOC in construction, as it is used in "paints, wood products, and floor finishes, glues, binders, particleboard, interior-grade plywood, wallboard, some paper products, fertilizers, chemicals, glass, and packaging materials" (Kibert 2008, p. 294). Formaldehyde has been known to irritate eyes, the upper respiratory tract, and other body surfaces.

The independent nonprofit organization **Green Seal** (GS) certifies paint products that meet ISO 14,024 environmental label standards and its GS-11 standard for paints and coatings. The GS-11 standard was developed to restrict the creation of VOC emissions and to ban the use of toxic chemicals in paints (LANL 2002). Table 13-3 provides the emissions limits for paints according to LANL (2002).

Green Seal also has a standard for the emissions of VOC from sealants and adhesives, GS-46. Tables 13-4 and 13-5 show the allowable emissions limits for these products.

Table 13-3. Volatile Organic Compound Emission Limits for Paints

Paint Applications	Type	VOC Content Limit[a] (grams of VOC per liter)
Interior coatings (GS-11)	Flat	<150
	Nonflat	<50
Exterior coatings (GS-11)	Flat	<200
	Nonflat	<100
Anticorrosive (GS-03)	Gloss	<250
	Semigloss	<250
	Flat	<250

Source: Adapted from LANL (2002), p. 238.
[a] Excluding water and tinting added at the point of sale.

Table 13-4. Volatile Organic Compound Emission Limits for Sealants

Type of Sealant Application	VOC Content Limit[a] (grams of VOC per liter)
Sealant Application	
Architectural	250
Roadways	250
Single-ply roof material installation/repair	450
Nonmembrane roof installation/repair	300
Other	420
Sealant Primer Applications	
Architectural—Nonporous	250
Architectural—Porous	775
Other	750

Source: Adapted from LANL (2002), p. 238.
[a]Water; acetone; parachlorobenzotrifluoride (PCBTF); cyclic, branched, or linear, fully methylated siloxanes (VMS); and difluoroethene (HCF-152a) are not considered part of this product.

Table 13-5. Allowable Volatile Organic Compound Emissions for Adhesives

Adhesive Applications—Architectural	VOC Content Limit[a] (grams of VOC per liter)	Adhesive Applications—Specialty	VOC Content Limit[a] (grams of VOC per liter)
Indoor carpet	50	PVC welding	285
Carpet pad	50	Chlorinated PVC welding	270
Outdoor carpet	150	ABS welding	400
Wood flooring	100	Plastic cement welding	250
Rubber flooring	60	Adhesive primer for plastic	250
Subfloor	50	Contact adhesive	80
Ceramic tile	65	Special purpose contact adhesive	250
Vitrified clay tile and asphalt tile	50	Adhesive for traffic marking tape	150
Dry wall and panel	50	Structural wood member adhesive	140
Cove base	50	Sheet-applied rubber lining	850
Multipurpose construction	70	*Substrate Specific*	
Structural glazing	100	Metal to metal	30
Single-ply roof membrane	250	Plastic foams	50
		Porous material (except wood)	50
		Wood	30
		Fiberglass	80

Source: Adapted from LANL (2002), p. 238.
[a]Water; acetone; PCBTF; cyclic, branched, or linear, fully methylated siloxanes (VMS); and difluoroethene (HCF-152a) are not considered part of this product.

13.3 Steel Production

In the United States, the steel industry produces approximately 7% of the **anthropogenic** (human-caused) emissions of carbon dioxide (CO_2). If the mining and transportation of iron ore are included in the calculations, the emissions increase to approximately 10%. Large portions of the emissions are from the coke and coal used in producing iron. Emissions are also generated by the electric power used for scrap melting and the natural gas used for producing iron. Energy costs constitute approximately 15–20% of the overall cost of steel production (International Iron and Steel Institute 2011).

In 2011, China, Japan, the United States, India, Russia, and South Korea produced 77.3% of the steel in the world. To produce 1 ton of steel, 19 **gigajoules** of energy (equivalent to three barrels of crude oil) is required during the production process. The U.S. steel industry has reached 95% material efficiency ratings, which indicates that only 5% of the by-products of steel-production processes are sent to landfills for incineration; therefore, the steel industry is reaching its maximum capacity for efficiency in reducing wastes (International Iron and Steel Institute 2011).

Table 13-6 provides a list of the major steel-producing countries in the world in 2011, along with their rank and levels of production.

Table 13-6. Major Steel-Producing Countries in 2011

Country	Rank	Production (millions of metric tons)	Country	Rank	Production (millions of metric tons)
China	1	683.9	Iran	17	13.0
Japan	2	107.6	United Kingdom	18	9.5
United States	3	86.2	Poland	19	8.8
India	4	71.3	Belgium	20	8.1
Russia	5	68.9	Austria	21	7.5
South Korea	6	68.5	Netherlands	22	6.9
Germany	7	44.3	South Africa	23	6.7
Ukraine	8	35.3	Egypt	24	6.5
Brazil	9	35.2	Australia	25	6.4
Turkey	10	34.1	Argentina	26	5.7
Italy	11	28.7	Czech Republic	27	5.6
Taiwan	12	22.7	Saudi Arabia	28	5.3
Mexico	13	18.1	Sweden	29	4.9
France	14	17.8	Kazakhstan	30	4.7
Spain	15	15.8	Slovakia	31	4.2
Canada	16	13.1	Finland	32	4.0
			Total	*1,424.1*	

Source: Adapted from International Iron and Steel Institute (2011), p. 37.

Producing 1 metric ton of metal products requires megajoules of embodied energy and kilograms of **embodied carbon**. The amounts of energy and carbon produced for various metals are as follows (Calkins 2009, p. 340):

- Aluminum, cast products—167,500 and 9,210
- Aluminum, extruded—153,500 and 8,490
- Aluminum, rolled—150,200 and 8,450
- Brass—44,000 and 3,710
- Copper—47,500 and 3,780
- Lead—25,000 and 1,290
- Stainless steel—51,500 and 6,150
- Steel, bar and rod—19,700 and 1,720
- Steel, galvanized sheet—35,800 and 2,820
- Steel, pipe—23,000 and 1,800
- Steel, section—22,700 and 1,790
- Steel, sheet—20,900 and 1,640
- Steel, wire—36,000 and 2,830
- Titanium—298,000 and unknown
- Zinc—61,900 and 3,200

Not only does steel production require large amounts of energy, but it also releases toxins into the environment. In 2003, the steel industry was faced with disposing of, treating, and releasing 636 million lb (288.48 million kg) of toxins. "Sixty-two percent of these were managed (usually recycled) and 38%, 242 million pounds (109.77 million kg), were disposed of or released into the environment. Of this approximately 4.8 million pounds (2.18 million kg) were released into the air, 4.8 million pounds (2.18 million kg) were released into water, and the remainder was released on land" (Calkins 2009, p. 335).

In 2005, "fossil fuel combustion accounted for 94% of CO_2 emissions, with the remainder from sources such as chemical conversions (e.g. cement, iron, and steel production), forestry, and land clearing for development" (Calkins 2009, p. 15). Table 13-7 lists the greenhouse gas (GHG) emissions by industrial sector in the

Table 13-7. U.S. GHG Emissions from Industral Processes in 2008

Industry	CO_2 Equivalent (millions of metric tons)
Steel manufacture	62.8
Cement manufacture	31.6
Lime manufacture	13.8
Aluminum manufacture	8.8
Petrochemical production	3.5
Aluminum production	3.3
Glass	1.3
Zinc	1.3
Lead	0.5

Source: Adapted from EPA (2008).

United States for 2008. Table 13-8 lists CO_2 emissions for the different steel-manufacturing processes, including basic oxygen furnaces (BOF), electric arc furnaces (EAF), and directly reduced iron basic electric arc furnaces for 2005.

A major energy consumer related to the use of steel products in the construction industry is transportation. Transporting materials by sea requires 0.2 MJ/km/ton, which means an emissions factor of 0.0269 million tons (0.0244 million metric tons) of CO_2 per billion ton-miles. Transporting ore, coal, and steel products requires 105 million tons (95.26 million metric tons) of CO_2 per year or 0.14 tons of CO_2 per ton of steel (Braathen 2003).

During the previous two decades, the steel-manufacturing industry has reduced its CO_2 emissions. Table 13-9 shows the CO_2 emissions for several industries from 1990 to 2010. As Table 13-9 indicates, the steel industry has reduced its CO_2 emissions to below 1990 levels.

Table 13-8. Emissions of CO_2 in the Steel Industry in 2005

Iron and Steel Production	Basic Oxygen Furnace (parts per million)	Standard Electric Arc Furnace (parts per million)	Directly Reduced Iron Basic Electric Arc Furnace (parts per million)	Total (parts per milion)
Coal	1,115	9	2	1,126
Power	18	59	16	93
Natural gas	12	0	21	33
Rolling and Finishing				
Fuel oil	16	0	0	16
Power	44	17	3	64
Fossil fuels	87	35	7	129
Total	1,292	120	50	1,462
CO_2 per ton of steel	2.5	0.6	1.2	1.9

Source: Adapted from International Iron and Steel Institute (2005), p. 35.

Table 13-9. CO_2 Emissions by Industry from 1990 to 2010

CO_2 Source	1990 (MMtpy)	2005 (MMtpy)	2010 (MMtpy)
Total CO_2	4,998.5	5,305.9	5,840
Cement manufacture	33.3	45.2	30.5
Lime production	11.5	14.4	13.2
Aluminum production	6.8	4.1	3.0
Iron and steel production	97.1	64.0	52.5
Ammonia	13.0	9.2	8.7
Ferroalloy production	1.2	1.4	1.7
Petrochemical production	3.3	4.2	3.7

Source: Adapted from EPA (2012).
Note: MMtpy = million metric tons per year of CO_2 equivalent.

13.4 Cement and Concrete Production

This section discusses cement production and how cement is used in the construction industry in concrete. Cement production is an energy-intensive manufacturing process that produces high levels of air pollution.

Amano and Ebihara (2005) evaluated 16 industrial categories using data from numerous sources—such as the national physical distribution census, national and regional input-output tables, and comprehensive energy statistics for Japan for the year 1995—to determine the environmental intensity in local regions and industrial sectors. The categories used for evaluation were

- Agriculture,
- Cement,
- Chemical,
- Coal and petrol,
- Commercial aspects,
- Construction,
- Energy supply,
- Fiber,
- Food,
- Metal,
- Mining,
- Nonferrous metals,
- Pulp,
- Service,
- Steel, and
- Transport.

The objective environmental load items included CO_2, nitric oxide, sulfuric oxide, and suspended particulate matter (PM) emissions for 47 Japanese regions. The study determined that the cement industry generates the most CO_2 per primary energy input of any of the other industry segments. One method for measuring industrial eco-intensity is the ratio of environmental load to energy flow. Figures 13-1 and 13-2 summarize the CO_2 and nitric oxide emissions for various industries examined in the Japanese study. The cement industry produced the highest level of emissions because of the energy required to process the large quantities of limestone necessary for cement production (Amano and Ebihara 2005).

13.5 Fly Ash Concrete and Other Cement Substitutes

The processes used to manufacture cement produce the highest level of CO_2 emissions of any construction material manufacturing process. One alternative that helps reduce the GHG emissions caused by cement production is replacing some of the

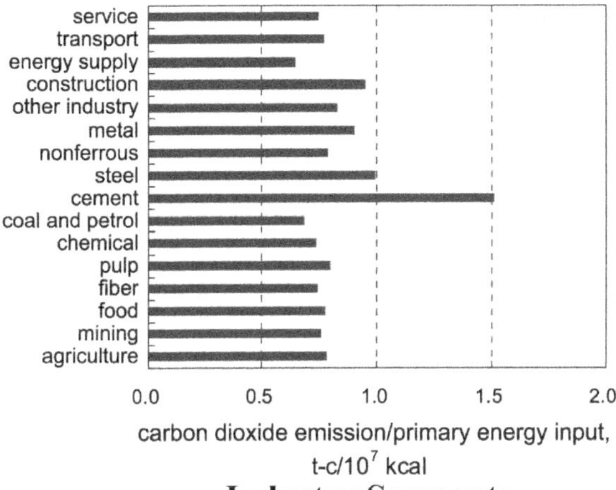

Figure 13-1. CO_2 Emissions versus Primary Energy Inputs for Different Industry Sectors (tons of carbon dioxide per 10^7 kcal)

Source: Adapted from Amano and Ebihara (2005), p. 163.

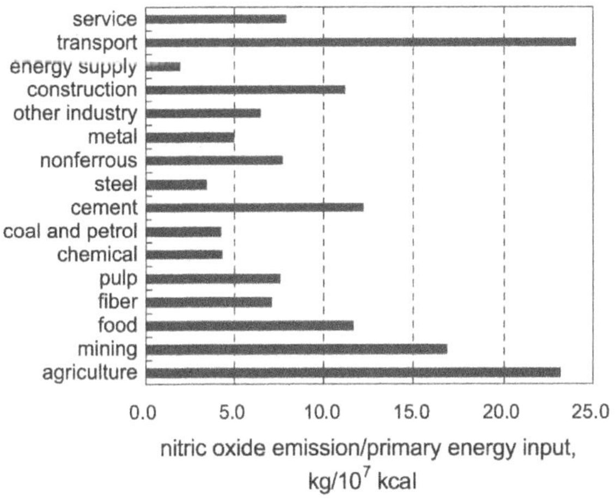

Figure 13-2. Nitric Oxide Emissions versus Primary Energy Inputs for Different Industry Sectors (kilograms of nitric oxide per 10^7 kcal)

Source: Adapted from Amano and Ebihara (2005), p. 163.

cement in concrete with **coal fly ash** (a residual produced during the burning of coal) or **granulated blast furnace slag**. Fly ash is sometimes used to replace 15–30% of the cement, and large structures, such as girders, road bases, major walls, and dams, are sometimes made with up to 70% fly ash. Many state departments of transportation have built concrete road systems using 30% fly ash. Another product being developed is **ashcrete**, which is created by using almost 100% fly ash. Although fly ash is a viable alternative to cement, it contains natural **radioisotopes**; therefore, its use should be monitored by **radio analytic** laboratories to determine the **residual radioactivity** (LANL 2002).

Other materials that may be substituted for Portland cement **clinker** (fused stony matter from a furnace) in concrete production are rice-husk ash, wood ash, natural **pozzolans**, and silica fume (Naik and Mariconi 2006). Manufacturing clinker is the most energy-intensive part of producing cement. The large kilns used to process the raw materials, to evaporate the water in the materials, and to **calcine** (heat to a high temperature to drive off wastes and produce a powder) the **carbonate constituents** (calcinations) consume 90% of the energy used to produce cement (Naik and Mariconi 2006).

Naik and Mariconi (2006, p. 7) have also indicated that crushed glass "is highly reactive with cement (alkali silica reaction). But Class F fly ash was used as a replacement for cement by mass of 45% or more, which helped in controlling alkali-silica reaction. However, ground waste glass was used as aggregate for mortars and no reaction was detected with particle size up to 100 meters."

According to Naik and Mariconi (2006, p. 13), "Wood fly ash has substantial potential for use as a **pozzolanic mineral admixture** and as an activator in cement-based materials. Wood ash has been used in the making of structural-grade concrete, bricks/blocks/paving stones, flowable slurry, and blended cements. Air entrained concrete can be achieved by using wood fly ash up to 35%. Structural-grade concrete can be made using wood fly ash and its blends with Class C fly ash to achieve a compressive strength of 50 MPa or higher."

An alternative aggregate to crushed rocks is **glass-reinforced plastic** scrap, which is ground into a fine powder and mixed with cement. Additional substitutes for natural aggregates in concrete include reclaimed concrete aggregate, air-cooled blast furnace slag, expanded blast furnace slag, **palletized blast furnace slag**, tires or crumb rubber pellets, plastics, and crushed bricks (Calkins 2009).

Worldwide, the concrete production industry consumes 1 trillion L (0.22702 trillion gal.) of water and 8 billion tons (7.2576 billion metric tons) of sand and gravel per year, but "recycled-aggregate fractions up to 15 mm (0.5905 in.), although containing masonry rubble up to 25–30 percent, proved to be suitable for manufacturing structural concrete even if employed as a total substitution of the fine and coarse natural aggregate fractions" (Naik and Mariconi 2006, p. 14).

In Sweden, there is concern that the by-products of slag produced by the blast-furnace process and **bottom ash** from municipal waste incineration plants could leach toxic substances; therefore, the government in Sweden restricts the use of these by-products in concrete production (Roth and Eklund 2003).

13.6 Porous Concrete

In some situations, it is beneficial to install porous concrete, which allows surface storm water to permeate the concrete and settle into the ground layer below. Porous concrete is "concrete with uniformly graded course aggregate, usually No. 89 or No. 8 with no fines. The uniformly sized aggregate creates pore spaces between 11% and 21% of the mix for water to flow through the pavement. The typical porous pavement is six inches (15.24 cm) thick with a minimum subbase of four inches (10.16 cm) of open graded aggregate. This can support a 2,000 pounds per square inch (psi) (140.65 kg/cm^2) load. Thickening the slab and subbase may support heavier loads. A thickened subbase will also accommodate soft subgrade and/or provide greater storm water storage for slower percolating soils" (Calkins 2009, p. 133).

13.7 Concrete Formwork

Concrete formwork is an expensive element of concrete production, but if the formwork is reusable this substantially reduces costs. The main types of formwork are wood, steel, aluminum, plastic, earth, and fabric. Wood forms may be used multiple times if adequate amounts of form release agents (form oil) are properly applied before the concrete is placed in the formwork. One form release agent that does not damage wood is plant oil. Steel and plastic forms are also reusable, and earth forms are sometimes used for footing forms.

13.8 Concrete Canvas

One innovative use for cement is concrete canvas, an application developed by two students at the London Royal College of Art. Concrete canvas is "made of cement-impregnated fabric folded into a plastic sack. After the fabric is saturated with water, the structure is inflated, and dries to form an impermeable shell. The shelters could be sterilized (for use as operating theaters), secured with a locking door, insulated with earth or sandbags, or ventilated with windows cut out of the skin. They come in sizes ranging from 53 to 177 ft^2 (4.924 to 16.44 m^2) of floor space, and could be joined to form larger structures" (Yabroff 2008).

13.9 Masonry Products

Traditional clay bricks are made of clay and shale. Clay contains feldspar, quartz, and other impurities, including iron oxide. Shale is a **sedimentary rock** that includes clay, mud, and silt. A large percentage of the waste produced during the manufacturing of bricks is ground into **grog** (also known as firesand or chamotte clay, which is mostly silica and alumina produced by firing clay and grinding it into specific particle sizes) and added to the mix. Waste products reused in this manner are sold for landscaping or aggregate base.

In addition to clay, some brick manufacturers now use bottom ash from coal-fired power plants and petroleum-contaminated soils because the firing process for bricks burns off excess **hydrocarbon** (HC). Other elements used for brick production include "fly ash, sewage sludge, waste treatment incinerator ash, recycled iron oxides, metallurgical wastes, papermaking sludge, rice husks, slag, and recycled glass" (Calkins 2009, p. 181).

The most expensive elements of brick production are the mining of the materials used in the bricks and the cost of energy for firing the bricks at temperatures of 100–400°F (37.78°C–204.4°C) for 15–50 h. The most common fuels used for firing bricks are natural gas, coal, and sawdust. Manufacturing bricks creates several types of pollution, including "sulfur dioxide, sulfur trioxide, nitrogen oxide, carbon monoxide, carbon dioxide, metals, methane, ethane, hydrochloric acid, and fluoride compounds" (Calkins 2009, p. 183).

If manufactured properly, bricks are durable materials for residential and commercial construction. The most important aspect of masonry construction is the mortar and joints. If water is not able to penetrate into the joints, then the bricks will retain their structural integrity. Because bricks expand and contract, expansion joints need to be added every 20–35 ft (6.1–10.67 m) and at "points of stress or weakness such as level changes, openings, and between panels and columns" (Calkins 2009, p. 191). If brick walls are constructed correctly, they do not need to be sealed or coated in any manner because they are already water resistant.

Brick pavements are not as durable as other types of pavement, especially in conditions subject to freezing and thawing, because the bricks will shift. The shrinking and swelling of the soil causes the bricks to become offset from their original locations and frequent repairs may be required to realign the bricks.

Rock is another construction material that is durable if it is constructed using the proper type of mortar mix and if the gaps between the rocks are properly filled with mortar. The benefit associated with the durability of rock as a construction material is offset by the initial cost of the material and construction. Marble and granite are also initially expensive to purchase and install, but their longevity is demonstrated by all of the marble and granite structures from prior centuries that are still standing.

The embodied energy in megajoules per metric ton required for producing masonry products are as follows (Calkins 2009, p. 239):

- Aggregate: 150
- Granular base (50/50 fine and course aggregate): 90
- Stone/gravel chippings: 300
- Local granite: 5,900
- Imported granite: 13,900
- Limestone: 240
- Sand: 100

13.10 Asphalt Pavement

Asphalt production creates pollution and consumes large amounts of energy during the processing of the feedstock and mixing of the asphalt. Extracting the raw materials used in asphalt also pollutes the environment. In addition, the darkness of asphalt pavements creates heat islands by absorbing solar radiation and releasing it back into the environment as heat.

Asphalt production requires crude oil, which has to be mined by drilling into the surface of the Earth. The oil-extraction process requires large amounts of water, which is contaminated during extraction with oil, sulfides, ammonia, phenols, heavy metals, and suspended and dissolved solids (Calkins 2009). Heating asphalt binders releases emissions that may cause health problems in workers. Asphalt is composed of approximately 85% course and fine aggregates by volume and 94% by weight. The aggregate production process requires mining and crushing, both of which consume large amounts of fuel and also cause pollution.

Methods for reducing the environmental impact of traditional hot-mix asphalt include lowering the production and placement temperatures of asphalt mixes. Lowering these temperatures results in the following benefits (Calkins 2009, p. 205):

- Energy savings,
- Reduced emissions,
- Decreased fumes,
- Reduced aging of the asphalt binder,
- Decreased wear of equipment, and
- Reduced drain down of asphalt.

To lower the temperature for warm-mix asphalt by 50–100°F, (10–37.78°C) emulsions, foam processes, or additives are added to the mixes to improve the workability of the asphalt. Hot-mix asphalt requires paving temperatures of between 275 and 325°F (135 and 162.8°C), warm-mix requires temperatures between 275 and 300°F (135 and 148.89°C), and cold-mix requires temperatures of approximately 60°F (15.56°C) (Calkins 2009).

One method for improving the sustainability of asphalt is to use recycled aggregate. The recycled materials that can be used for aggregate include recycled asphalt, tires, roofing shingles, glass, slag, and concrete. According to the Asphalt Recycling and Reclaiming Association, in 2001 approximately 80% of asphalt was recycled into new asphalt (Calkins 2009).

In the Leadership in Energy and Environmental Design (LEED) rating system, credits are provided for asphalt that does not absorb ultraviolet rays and return them to the environment as heat. The ability of a material to reflect, rather than absorb heat, is called solar reflectance, or **albedo**:

> An albedo of 0.0 indicates total absorption of solar radiation and a 1.0 value represents total reflectivity. Generally, albedo is associated with color, and lighter colors being more reflective.
>
> The **solar reflective index** (SRI) combines albedo and emittance into a single value expressed as a fraction (0.0 to 1.0) or percentage. A source for SRI data on basic paving is SS credit 7.1 of the U.S. Green Building Council's LEED for New Construction Version 2.2 (2008). The reference guide states that new asphalt has an SRI of 0, meaning that all solar radiation is absorbed, while new white Portland cement concrete has an SRI of 86. Other pavement types generally range between these values with a 35 SRI for new gray concrete. The LEED credit requires an SRI of at least 29 for 50% of the paving. (Calkins 2009, p. 213)

13.11 Fiber-Reinforced Polymeric Composite Materials

Fiber-reinforced polymeric (FRP) **composite materials** are being used in the following industries: heavy construction, highway construction, oil and gas, chemical, petrochemical, power, and mining and processing. Carbon-fiber composites may be used for repairing and rehabilitating systems because they are "ten times as strong as steel, at less than a quarter the density, and they are corrosion resistant. Application of composite repairs involves layers of carbon fiber impregnated with epoxy resin being built up to the specified requirements in terms of thickness, overlap onto good metal, fiber orientation, gradient at the ends or edges of the repair and so on, in line with the repair design specifications. The repairs could be designed for the lifetime required—from just a couple of years, to permanent (25 years plus)" (Engineer Live 2007, p. 1).

According to Hastak et al. (2003, p. 1409), "Composites offer several advantages over conventional materials such as superior strength/weight and stiffness/weight ratios, a higher degree of chemical inertness, and design flexibility. Some of the potential down-stream benefits include lower life-cycle costs, lighter members, high corrosion and fatigue resistance, and higher live load capacity."

Architectural and structural elements manufactured from highly durable FRP composite materials are increasingly being specified in highway bridge decks, bridge superstructures, commercial building architectural façades, beams, columns, and marine structures (Federal Highway Administration 2011; MDA 2004). When evaluating the sustainability of FRP composite materials used on construction projects, the extraction and processing of raw materials and the manufacturing processes should be considered, along with the service life, which often far exceeds a comparable component, element, or structure constructed using mild steel or traditional reinforced concrete. When analyzing the energy requirements for the manufacturing and disposal of FRP materials, each of the individual constituent materials should be evaluated separately.

Fiber-reinforced polymeric composite materials contain two main constituent materials, fibers and matrix material, both of which require different manufacturing processes. Many types of fibers, fiber architecture, and matrix materials are available; therefore, their life-cycle performance and mechanical properties may be designed for specific applications. The two fibers most frequently incorporated into construction material composites are glass and carbon. Glass fibers are manufactured from melted silica (sand) and carbon typically derived from precursor materials such as polyacrylonitrile (PAN) fibers, rayon fibers, or pitch. Other fibers are produced from renewable or recyclable materials, such as hemp, flax, and mild and stainless steel (Burgueno et al. 2004; Fu et al. 2008). A process called sizing is used to coat the fibers with a chemical compound to allow greater adhesion to the matrix material.

Matrix materials should be analyzed as part of the life cycle of composites because they are a major element of FRP composites. Fiber-reinforced polymeric matrix materials typically used in construction include thermoplastic resins, which may be reshaped upon heating, and the more common thermoset resins, whose cross-linking (curing) process does not allow reshaping. Typical thermoset resins include polyester resins, which are produced by condensation polymerization of dicarboxylic acids and difunctional alcohols (glycols). Unsaturated polyester resins use an unsaturated material, such as maleic anyhride or fumaric acid, and styrene to produce a low-viscosity liquid. Other thermoset resins are corrosion-resistant vinylesters; epoxies; high-temperature-resistant, low-smoke phenolic resins; and polyurethanes (MDA 2005).

Fibers and resins are combined in a wide range of manufacturing processes, including hand lay-up, bag molding, autoclave curing, compression molding, resin-transfer molding, pultrusion, filament winding, and vacuum infusion (Yuhazri et al. 2010). Volatiles such as styrene are released in differing amounts during FRP component manufacturing and processing and vary depending on resin selection. Emissions should be monitored and controlled during the individual processes (vacuum infusion, bag molding, compression molding) or by compliant heating, ventilating, and air-conditioning systems used to collect and process emissions from processes such as hand lay-up and filament winding.

Fiber-reinforced polymeric composite materials often possess a superior service life and require less maintenance as compared with many building materials, such as steel or conventionally reinforced concrete or masonry. However, specifications for FRP composite materials used in construction should consider various fibers, resins, and manufacturing processes to minimize environmental impacts. Designers and constructors should consult with local authorities about the proper disposal of FRP materials because they may not be biodegradable.

13.12 Wood Products

Forest products are becoming scarce as forests throughout the world are being harvested to produce construction materials, such as dimension lumber, plywood, and beams. To continually regenerate forests, some forest product companies are planting their own trees. Only 5% of the total forest cover in the world is planted forests, but these forests contribute 35% of the commercial wood in the world (International Paper 2006).

According to Nogueron and Laestadius (2007, p. 1), "Forests and paper products are used and reused by society over long periods of time, which represents an expanding reservoir of carbon removed from the atmosphere. On average, one ton of paper contains about 1.33 metric tons of carbon equivalent CO_2. Forests contribute to net carbon emissions when they are logged, converted, or burned at a faster rate than they grow back. An estimated 24% of global carbon dioxide emissions are attributable to land-use change and forestry."

Wood procurement is monitored and certified by third parties, such as Bureau Veritas Quality International (BVQI), the Sustainable Forestry Initiative (SFI), American Tree Farm System, Canadian Standards Association, and the Forest Stewardship Council. Wood procured through the SFI program has to come from legal sources and have been harvested using techniques that protect water quality in the surrounding area and adhere to the principles of responsible forest management, as stated by the SFI (International Paper 2006). The Forest Stewardship Council also monitors wood products by issuing chain of custody certificate numbers (LANL 2002).

The standards followed for sustainable development in the forest products industry were developed as a result of the 1992 United Nations Conference on the Environment and Development (UNCED) Rio Convention. These standards were further refined at the Helsinki and Montreal meetings. The forestry criteria and indicators provide guidelines on the following (United Nations Framework Convention on Climate Change 2005):

- Water quality,
- Biodiversity,
- Wildlife protection,
- Habitat provision,
- Sustained yield harvests,
- Economic viability,
- Legality, and
- Social and economic issues.

According to Calkins (2009, p. 6), "Environmentally responsible forest management includes practices that protect the functional integrity and diversity of tree stands, minimize clear cutting, protect old-growth forests, and minimize wasteful harvesting and milling techniques" as prescribed by the Forest Stewardship Council. When forests are eliminated, "they no longer provide ecological services such as carbon sequestration, habitat, erosion control, and regulation of the **hydrologic cycle**. Forests play a vital role in stabilizing the climate by sequestering atmospheric carbon. The Food and Agriculture Organization of the United Nations (FAO) estimates that between 1990 and 2005, the carbon storage capacity of forests declined by more than 5%" (Calkins 2009, p. 19).

Sustainably developing forest products requires using every part of a tree. After raw wood has been processed into dimension lumber, plywood, and beams, the leftover bark, sawdust, shavings, and resin might be used to create additional products, such as fiberboard, **plystrand** (created by fusing wood chips together and encasing them in veneer to resemble plywood), and fireplace logs. Wood shavings are not considered to be recycled materials because they are disposed of in landfills. Shavings and bark are also used for **biorenewable fuel** in industrial boilers and heavy machinery (International Paper 2006).

The bonding resins that are used in making wood products might emit toxic substances such as formaldehyde (LANL 2002). The resins that come from trees include sterols used to make perfume, fabric, toothpaste, tires, and pharmaceuticals. Another by-product is a cholesterol-lowering ingredient used in drinks, milk, yogurt, and other foods.

Unfortunately, the trees that provide the most sustainable decay-resistant wood are becoming scarce. Decay resistant trees include redwood (California), western red cedar (Washington, Oregon, Idaho, and Montana), white cedar (eastern United States), incense cedar (California, Nevada, and Oregon), bald cypress (southern states), black locust (Tennessee, Kentucky, West Virginia, and Virginia), ipe (Latin America), jarrah (Australia), teak (Southeast Asia), American mahogany (southern Mexico to Bolivia), and African mahogany (West Central Africa).

To compensate for the declining availability of decay-resistant wood, wood products are being treated with wood preservatives to increase their durability and resistance to decay, insects, and weathering. Since the 1930s, a variety of chemicals have been used to preserve wood products, many of which are hazardous to both humans and the environment. The utility, railroad, and agricultural industries were the first industries to widely use chemically treated wood.

Petroleum-based creosote was replaced by a water-based wood treatment called **chromated copper arsenate** (CCA), commonly known as **pressure-treated wood**, during the twentieth century. Chromated copper arsenate treatment is applied to wood building materials, and it provides wood with a combination of fungicide, pesticide, herbicide, and insecticide protection. The use of CCA prolongs the service life of wood exposed to water, soil, fungi, mold, or insects. In some cases, treating wood with CCA extends its useful life from five years to 30 or 40 years. However, there are carcinogens in CCA.

Hundreds of thousands of tons of **arsenic** and **chromium** are used each year to preserve wood. At the end of its useful life, preserved wood is buried in unlined and unmonitored landfills throughout the United States. Even though the government requires that customer information sheets (CISs) and warning labels be attached to CCA construction materials, the labels may no longer be attached by the disposal stage (EPA 2002). The toxic chemicals in CCA leach out of wood when they are exposed to water or soil and migrate from landfills into water supplies (Khan et al. 2006). Waste from construction sites is typically buried directly under and on top of unprotected soil in landfills (EPA 2003).

During the past 40 years, the Occupational Safety and Health Administration (OSHA) has implemented new, more protective standards for occupational exposure to arsenic and chromium. To comply with the OSHA policy of limiting employee exposure to carcinogens to the lowest feasible level, many new laws are in effect to protect employees in the wood-treatment industries. Whereas arsenic and chromium are carefully regulated as individual chemicals because of their toxicity, the regulation of CCA has historically been much less restrictive for all occupations using CCA once the treatment has been applied to wood. Currently, there are precautions and suggested **personal protective equipment** (PPE) for CCA use supplied by registrants of CCA, but neither the EPA nor OSHA requires occupational protection beyond the treatment phase.

Arsenic and chromium are human carcinogens used in the manufacturing of industrial products. Arsenic occurs naturally in soil or rocks and traces of it are found in water, food, trees, and plants. Arsenic and chromium are elemental heavy metals, and they exist in varying valence states and containment matrices. Natural events, such as volcanic eruptions, erosion of rocks, and even forest fires, release arsenic into the environment. Arsenic discharge is also a direct result of activities such as smelting, mining, using combustion engines, burning fossil fuels, incinerating waste, producing pulp and paper, treating wood, and manufacturing cement.

When arsenic is manufactured for industrial use, it is primarily produced as a by-product in the smelting of nonferrous metal ores, such as gold, silver, lead, nickel, and cobalt (Bleiwas 2000). Due to the toxic nature of arsenic and the expenses associated with containing the production emissions, arsenic production in the United States was essentially eliminated with the implementation of the Clean Air Act Extension of 1970.

The recovery of arsenic from the smelting of nonferrous metals takes place in 17 countries throughout the world, with the bulk of the imported arsenic used in the United States coming from China (Bleiwas 2000). Prevalent in past agriculture use, arsenic was once employed to kill weeds, fungi, and other pests. It is still found in common products, including wood preservatives, rat poison, paints, dyes, pharmaceuticals, fungicides, pesticides, semiconductors, and some medicinal tonics. The inorganic forms of arsenic are much more toxic to humans than the organic types found in food (World Health Organization 2004).

According to the California Environmental Protection Agency (2006), arsenic is one of the highest ranked chemicals of the 164 developmental and reproductive toxicants. It is also seen as a Group 1 carcinogen by the International Agency for Research on Cancer (IARC) and the EPA. The Committee on Medical and Biological Effects of Environmental Pollutants (1977, p. 176) wrote, "Evidence of significant systematic concentrations of arsenic has been found in several studies of the incidence of lung cancer in populations exposed to arsenic dust." In addition, direct contact between arsenic-laden dust and the mucus membranes of the nose can cause a perforation of the nasal septum after only a few weeks of exposure.

The chemicals contained in CCA migrate to water supplies and drinking water. According to the California Environmental Protection Agency (2004, p. 55): "At relatively low acute intake levels, arsenic provokes mild gastrointestinal effects. The Feinglass 1973 Report showed the acute gastrointestinal effects ... (nausea or vomiting, dryness or burning of the mouth and throat, abdominal pain, and diarrhea). One of the most common long-term indicators of acute arsenic exposure is Mees' lines, which are ridges that appear on the fingernails six to eight weeks after the exposure."

Exposure to continuous doses of chromium in drinking water or through accidents or occupations also occurs. According to OSHA (U.S. Department of Labor 2006, p. 9), "Human data would place hexavalent chromium compounds into Group 1, meaning there is decisive evidence of the carcinogen properties of those compounds in humans." The EPA (2007, p. 2) wrote, "Skin exposures to hexavalent chromium for children contacting treated wood surfaces exceed the OSHA level of concern for skin sensitization."

The EPA and OSHA regulate arsenic and occupational exposure hazards. In 1978, the U.S. Department of Labor produced new rules for permanent exposure to inorganic arsenic and reduced the **permissible exposure limits** (PELs). A directive published by OSHA in 1978 states that PEL "include arsenic, all arsenic-containing, inorganic compounds and arsine among the substances in the 'High Hazard Health' category.... Respiratory protection is required against any of the substances included or specified in the list that follows: (i) arsenic trichloride, (ii) arsenic trifluoride, (iii) arsenic pentafluoride, (iv) arsenic tribromide, (v) arsenic triiodide, (vi) arsenic monophosphide" (p. 2).

Material safety data sheets (MSDSs) inform interested parties about products and possible hazards associated with the handling, use, and storage of products, and they provide safety and emergency information. Material safety data sheets became federally mandated in the mid-1980s in their present form. They have to accompany all products with hazardous constituents, and employers should have them available for workers at jobsites and manufacturing facilities. The MSDS for CCA has changed content many times since the 1970s as new discoveries were made about the hazards of arsenic and chromium.

Because exposure to CCA wood products is toxic to humans, research has been conducted to determine whether there are any viable alternatives to using CCA. Lebow (2004) presents some alternative wood treatments, including

- Acid copper chromate (ACC),
- Alkaline copper quaternary (ACQ),
- Copper azole (CBA-A and CA-B),
- Copper citrate (CC),
- Copper dimethyldithiocarbamate (CDDC), and
- Copper HDO (CX-A).

Lebow (2004) states that the retention rate for the chemicals in the alternatives is equivalent to that of CCA products. Because they are typically copper-based and the other components have not been identified as mammalian carcinogens, these alternatives may be used as replacements for CCA in residential applications.

Wood-treatment industry workers are exposed to concentrated levels of arsenic and chromium in CCA when they apply the treatment to lumber products. When construction personnel work with these products, they are being exposed to known carcinogens, and this exposure may cause illness and other negative health effects such as Bell's palsy (D. Johnloz, personal correspondence to David McCrea, June 10, 1995).

The cutting, nailing, and placement of treated wood by construction workers releases heavy metals that are absorbed through the skin, eyes, mouth, nose, and lungs. Demolition workers face many of the same hazards as construction workers, but they may be less aware of the proper handling procedures due to their inability to identify CCA-treated products.

Carpenters, electricians, plumbers, masons, and landscape professionals are exposed to carcinogens from CCA during the installation and maintenance of wood structures, and utility workers are exposed to CCA-treated poles and pilings. Road and bridge construction crews work with CCA, as do agricultural and railroad workers. Electricians and plumbers are exposed to CCA during material installations when holes and channels are cut in the wood for wire and pipe installations. Rain runoff after CCA roof installation dislodges small particles that contain chemicals from the shingles. After construction, smaller debris is washed into surrounding landscape areas and into storm drains.

Figure 13-3. Home with Hardie Board Siding

13.13 Hardie Board

One alternative to using wood planking is **Hardie board**. Hardie board is being substituted for wood siding because it offers sustainable benefits. It is constructed of concrete and stamped with an artificial wood grain to give it the appearance of wood siding. Hardie board is available in various thicknesses and lengths and may be cut to desired lengths. The benefits of using Hardie board instead of wood or aluminum siding are that it is durable and lasts for decades, it has to be painted only about every 20 years, it is an excellent insulator, it is termite resistant, and it has an appearance that resembles real wood. Figure 13-3 shows a home that has Hardie board siding.

13.14 Polyvinyl Chloride and Thermoplastic Products

This section discusses polyvinyl chloride (PVC) construction materials.

Polyvinyl Chloride Products

New techniques are being developed to recycle polyvinyl chloride plastic waste in order to reduce the consumption of biomass used to produce PVC products. The recycling process breaks PVC down into synthetic gas and hydrogen chloride (HCl), which are then available for use in the production of new PVC products.

Denmark has enacted a tax on some PVC products to pay for their incineration so that they are not disposed of in landfills. Barriers to recycling PVC include the high cost of recycling relative to producing new PVC products from raw materials. The European Plastic Pipes and Fittings Association (TEPPFA) and the European PVC Window Profile and Related Building Products Association (EPPA) have set up collection and recycling task forces around Europe to ensure that over 50% of recovered pipes and windows are recycled (Leadbitter 2002). One of the problems associated with the use of PVC is the toxic chemicals, such as organochlorine, furans, and dioxins, that are by-products of PVC processing.

Thermoplastic Products

Innovative materials are becoming more prevalent in the oil and gas industry due to the rising cost of traditional piping materials, such as wood, clay, concrete, and metal. Even though **thermoplastics** have been in widespread use in "residential drain/waste/vent, gas transmission, acid waste drainage, water lines, underground irrigation, swimming pools, and water theme parks," they are just beginning to gain acceptance for industrial uses (Thermoplastic Industrial Piping Systems 2007, p. 1). In the oil and gas industry,

> plastic pipe, itself a derivative of oil and natural gas, has successfully been applied in handling most crudes, salt water, and natural gases. Most natural gas distribution today uses millions of feet of plastic pipe. Polyethylene piping, colored beige or orange, is the preferred material for this application. In the mining industry, the most popular use of thermoplastics is in ore leaching, in which the ore is treated with dilute sulfuric acid or sulfides and then with ferric sulfate solutions. Polyvinylchloride, CPVC, ABS, and polyethylene piping are used in many of the leaching process stages. Plastics also are used for the movement of ore slurries and other piping applications in under and above ground mining. (Thermoplastic Industrial Piping Systems 2007, p. 1)

Many products in addition to piping are produced from plastics such as polyethylene terephthalate, high-density polyethylene, PVC, polypropylene, polystyrene, and other resins (Munier 2005, p. 184):

- Credit cards, clear plastic containers, and pharmaceutical bottles;
- Hard plastic for CD and DVD cases, television and computer frames and casing, carry-out containers, and packing foam;
- Milk cartons, snack bags, and microwavable containers;
- Nontransparent bottles;
- Plastic fibers for upholstery and luggage;
- Transparent bottles.

In the United States, over 113 billion lb (51.26 billion kg) of plastic resin were produced in 2006. Of this amount, 14.9 billion lb (6.76 billion kg) were PVC, of which approximately 75% was used in construction, and 38.6 billion lb (17.51 billion kg) were used in polyethylene production. Twenty-nine percent of polyethylene was used for packaging products, and 19% was used in the construction industry. Approximately 11.5 billion lb (5.22 billion kg) of PVC were used in the construction industry in "piping, siding, flooring, windows, electrical wire, cable and other products" (Calkins 2009, p. 374).

Plastics are derived from petroleum or natural gas, and approximately 10% of the products produced by the petroleum and gas industry are used for plastic products. The same toxins are released during the extraction of petroleum products as during oil and gas production. Chlorine is used to manufacture PVC, and it requires less embodied energy to produce than other plastic products.

Plastics are also being used to make single-resin plastic lumber, commingled plastic lumber, composite lumber, **biocomposite lumber**, and fiberglass-reinforced lumber that incorporate at least 50% plastic content measured by weight and other materials such as fiberglass. They are also used to make recycled rubber for sidewalk paving units. **Bioplastics** are being introduced to replace petroleum-based plastics and incorporate plant materials, such as "cornstarch, soy, polylactides, or cellulosic [made from cellulose] materials" (Calkins 2009, p. 404).

13.15 Mining, Mineral, and Metal Products

The mining and mineral (MM) industry produces over 80 types of materials. The countries that supply a large proportion of the mining materials and minerals worldwide are the United States, Canada, Australia, Russia, Brazil, South Africa, China, and countries in the European Union (Azapagic 2004). The MM industry employs over 30 million workers in large operations and 13 million in small-scale operations, which is approximately 1% of the workforce worldwide. Table 13-10 summarizes the sustainability issues that affect the MM sector.

Table 13-10. Sustainability Issues in the Mining and Minerals Sector

Economic Issues	Environmental Issues	Social Issues
Contribution to the gross domestic product and wealth creation	Biodiversity	Bribery and corruption
Costs, sales, and profits	Emissions	Creation of employment
Distribution of revenues and wealth	Energy use	Employee education and skills development
Investments (capital, employees, communities, pollution prevention, and time closure)	Global warming and other environmental impacts	Equal opportunities and nondiscrimination
Shareholder value	Land use, management, and rehabilitation	Health and safety
Value added	Nuisance	Human rights and business ethics
	Product toxicity	Labor-management relationships
	Resource use and availability	Stakeholder involvement
	Solid wastes	Wealth distribution
	Water use, effluents, and leachates, including acid mine drainage	

Source: Adapted from Azapagic (2004), p. 643.

A major concern of the mining industry is acid drainage, which could lead to the long-term contamination of waterways. Some discharge also contains large quantities of cyanides and heavy metals. The mining process itself may also dangerously affect the workers who mine materials, especially if they are being exposed to materials such as asbestos, lead, and **uranium**. Mining companies are now including decommissioning and rehabilitation plans in their proposals for new mining operations. Eighty-eight percent of the firms surveyed by PricewaterhouseCoopers (2004) indicated that they have environmental postclosure plans, but only 45% have socioeconomic plans. Table 13-11 lists categories of environmental indicators used by the MM industry.

Metal ore is extracted from the Earth through a variety of techniques, including strip mining, open-pit mining, mountaintop removal, and dredging. Processing mined materials requires "milling, crushing, consolidation, washing, leaching, flotation, separation, and thermal processes" (Calkins 2009, p. 329). The raw materials required to manufacture iron or steel are iron ore, coal, and limestone, but additional additives, such as chromium, nickel, zinc, manganese, and cadmium, are used for alloys and coatings. The main elements of steel production are extracted using strip mining.

Table 13-11. Categories of Environmental Indicators Used in Mining and Minerals Industry

Category	Measures
Mineral resources	Availability, resource efficiency, and rate of depletion of mineral resources
Land use	Land requirements for mineral-related activities
Materials	Use of chemicals, packaging, and other materials; recycling rate
Water	Water use and efficiency
Energy	Energy use and efficiency; use of fossil fuels and renewable energy
Closure and rehabilitation	Pace of restoration and the level of commitment to rehabilitation
Biodiversity	Extent to which the extractive activities affect habitats and species
Air emissions/liquid effluents	Contribution to air, water, and land pollution and related impacts
Nuisance	Level of nuisance for neighboring communities
Compliance and voluntary activities	Environmental responsibility demonstrated through compliance and voluntary activities
Transport and logistics	Transport distances for products and employees
Suppliers and contractors	Environmental performance of suppliers and contractors
Products	Life-cycle environmental impacts of products

Source: Adapted from Azapagic (2004), p. 645.

Table 13-12. Metal Recycling in 2005 in United States

Metal	Amount (metric tons)	Percentage of Metal Recycled
Aluminum	2,990,000	36.0
Chromium	124,000	24.0
Copper	951,000	30.0
Iron and steel	65,400,000	54.0
Lead	1,140,000	74.5
Magnesium	72,800	44.0
Tin	14,000	30.0
Titanium	25,700	50.0
Zinc	345,000	29.5

Source: Adapted from Calkins (2009), p. 363.

The process for mining copper is one of the least efficient, requiring 400 tons (362.88 metric tons) of waste and by-products to create 1 ton (0.9072 metric tons) of copper. In addition to creating overburden waste, copper mining also results in contaminated water runoff that is toxic to fish (Calkins 2009).

The U.S. metal recycling effort has resulted in various percentages of metals being recycled by the industry. The amount and the percentage of various metals recycled in 2005 are listed in Table 13-12.

Those who select metals for use in construction projects should ask several questions to determine which metals are sustainable (Calkins 2009, p. 368):

- What are the potential air, water, and soil pollution impacts of the metal in extraction, production, manufacture, and fabrication?
- Will the metal structure last for the expected duration of the landscape?
- Are the metal structures reusable or recyclable?
- Are corrosion-protective coatings required?
- Do they off-gas VOCs, pose health risks to workers or users, or contribute to air, water, or soil pollution?
- Is there a risk of coating loss to the environment attributable to wear or **spalling** (cracking, flaking, chipping, or edge breakage)?
- Does the coating limit the recyclability of the metal member?
- How much metal may enter the environment from corrosion carried by runoff? Is the corrosion hazardous?
- What are the maintenance requirements of the metal structure?
- Will hazardous cleaners or new protective coating applications be required to maintain the structure?

13.16 Unconventional Building Products

In addition to the conventional building products mentioned in the previous sections, many types of unconventional building products are being designed and manufactured each year. *Environmental Building News* cited the following in its list of the top-10 innovative products in 2007 (Kibert 2008, p. 252):

- Polished concrete—Polishes old and new concrete slabs into attractive, durable, finished floors;
- Timber-salvaging system—Harvests trees submerged in reservoirs created by hydroelectric dams;
- Electronically tintable glazing—The tinting of the glass is changed using an electrochromic control (changes with the amount of sunlight);
- Water-resistant composite—Solid composite material made from postconsumer paper;
- Interior panels—Panels for workstations, trim, or toilet partitions made with 40% preconsumer recycled copolymers;
- Interior molding—Molding profiles made with at least 90% recycled polystyrene;
- Water-efficient showerhead—Showerheads used just 1.6 gal. (6.1 L or 1.33 imperial gal.) of water per minute;
- Irrigation system controls—Irrigation control based on local weather data;
- Evaporative cooler—Indirect evaporative cooler.

13.17 Summary

This chapter introduced the concept of sustainable construction materials and presented information from the Los Alamos National Laboratory on the types of sustainable materials that should be considered for buildings and structures. Individual construction materials were reviewed, and information was provided on the processes required for manufacturing different construction materials. In addition to traditional construction materials, sustainable and nontraditional construction materials were discussed to provide alternatives that should be reviewed during material selection for construction projects. The materials reviewed in this chapter included paints, sealants, steel, cement, concrete, fly ash concrete, concrete canvas, porous concrete, asphalt, masonry, carbon-fiber composites, wood products, PVC products, thermoplastic products, and petrochemical products.

13.18 Key Terms

albedo
anthropogenic
arsenic
ashcrete
biocomposite lumber
bioplastics
biorenewable fuel
bottom ash
calcine
carbonate constituents
chromated copper arsenate
chromium
clinker
coal fly ash

concrete canvas
embodied carbon
fiber-reinforced polymeric composite material
gigajoules
glass-reinforced plastic
granulated blast furnace slag
Green Seal
grog
Hardie board
heat island
hydrocarbon
hydrologic cycle
material safety data sheets
palletized blast furnace slag
permissible exposure limits
personal protective equipment
plystrand
polyvinyl chloride
porous concrete
pozzolans
pozzolanic mineral admixture
pressure-treated wood
radio analytic
radioisotopes
residual radioactivity
sedimentary rock
solar reflective index
spalling
sustainable construction materials
sustainably harvested
thermoplastic
uranium
volatile organic compounds

13.19 Exercises

13.1 How would the introduction of sustainable materials affect the productivity of construction workers who have not worked with sustainable materials on a construction project before?

13.2 What are sustainable construction materials?

13.3 According to LANL (2002), what are locally manufactured materials?

13.4 According to LANL (2002), what are locally derived raw materials?

13.5 Explain what VOCs are and why they should be avoided in paint products.

13.6 Discuss the potential harm caused by the use of CCA as a preservative in pressure-treated wood.

13.7 Discuss the major concern regarding fly ash as a cement substitute in concrete.

13.8 Explain why the cement industry produces the highest level of CO_2 emissions per primary energy input.

13.9 Describe the major uses for plastic resin.

13.10 Which of the metal production processes requires the most megajoules of embodied energy and kilograms of embodied carbon, and which process requires the least?

13.11 Discuss the different types of pollution generated during the manufacturing of bricks.

13.12 Discuss what process could be used to make concrete formwork more sustainable.

13.13 Explain why it is beneficial to substitute a percentage of fly ash for cement in concrete production.

13.14 How does ashcrete differ from fly ash concrete?

13.15 Discuss why porous concrete might be used rather than standard concrete in construction.

13.16 Discuss techniques for reducing the environmental impact of traditional hot-mix asphalt production.

13.17 What are the requirements for wood procured through the SFI program?

13.18 In addition to paint products, where else is formaldehyde found in construction products?

13.19 Which material listed in the *LANL Sustainable Design Guide* (LANL 2002) receives the highest rating, and why does it receive this rating?

13.20 Explain how concrete canvas works and suggest three potential uses for concrete canvas.

13.21 Which masonry product requires the highest level of embodied energy, and which requires the lowest level?

13.22 Describe what arsenic is and where it comes from.

13.23 Would productivity increase if construction personnel were involved at the design stage in selecting sustainable methods and materials for a construction project?

13.24 Would it take more or less time to install porous concrete rather than standard concrete? Explain why.

13.25 What negative side effect might influence the productivity of workers who are installing concrete made with a percentage of fly ash?

References

Amano, K., and Ebihara, M. (2005). "Eco-intensity analysis as sustainability indicators related to energy and material flow." *Management of Environmental Quality: An International Journal*, 16(2), 160–166.

Azapagic, A. (2004). "Developing a framework for sustainable development indicators for the mining and minerals industry." *J. Cleaner Prod.*, 12(6), 639–662.

Bleiwas, L. (2000). *Arsenic and old waste*, USGS, Washington, DC. <http://minerals.usgs.gov/minerals/mflow/d00-0195> (March 2014).

Braathen, A. (2003). *Environmental policy in the steel industry: Using economic instruments*, Environment Directorate, Directorate for Financial Fiscal and Enterprise Affairs, Bergen, Norway.

Burgueno, R., Quarliata, M., Mohanty, A., Mehta, G., Lawrence, T., and Misra, M. (2004). "Load-bearing natural fiber composite cellular beams and panels." *Composites Part A*, 35(6), 645–656.

California Environmental Protection Agency. (2004). *Public health goals for chemicals in drinking water: Arsenic*, Office of Environmental Health Hazard, Sacramento, CA. <http://oehha.ca.gov/water/phg/pdf/asfinal.pdf> (Jan. 2005).

California Environmental Protection Agency. (2006). *Safe Drinking Water and Toxic Enforcement Act of 1986: Chemicals known to the state to cause cancer or reproductive toxicity*, Office of

Environmental Health Hazard, Sacramento, CA. <http://www.oehha.ca.gov/prop65/prop65_list/files/P65single120806.pdf> (Jan. 2005).

Calkins, M. (2009). *Materials for sustainable sites,* Wiley, Hoboken, NJ.

Committee on Medical and Biological Effects of Environmental Pollutants. (1977). "Biologic effects of arsenic on man." *Arsenic,* National Academies Press, Washington, DC. <http://books.nap.edu/openbook.php?record_id=9003&page=173> (March 2014).

Engineer Live. (July 2007). *Composites aid permanent asset rehabilitation without shutdown,* International Oil and Gas Engineer. <http://www.engineerlive.com/international-oil-and-gas-engineer/explorationdrilling/18334/composites-aid-permanent-asset-rehabilitation-without-shutdown.thtml>. (November 2007).

EPA. (2002). *Arsenic treatment technologies for soil, waste, and water.* U. S. Environmental Protection Agency, Washington, D.C. <http://www.epa.gov/tio/download/remed/542r02004/arsenic_report.pdf> (Jan. 2005).

EPA. (2003). "Chromated copper arsenate (CCA): EPA testimony on chromated copper arsenate (CCA) treated wood." U.S. Environmental Protection Agency, Washington, DC. <http://www.epa.gov/oppad001/reregistration/cca/ccatestimony1.htm> (Jan. 2005).

EPA. (2007). "Acid copper chromate (ACC) residential uses won't be registered." U.S. Environmental Protection Agency, Washington, DC. <http://www.epa.gov/pesticides/factsheets/chemicals/acid_copper_chromate.htm> (Jan. 2007).

EPA. (2008). *Inventory of the United States greenhouse gas emissions and sinks 1990-2005,* U.S. Environmental Protection Agency, Washington, DC.

EPA. (2012). *United States green house gas inventory chapter four industrial processes.* U.S. Environmental Production Agency, Washington, D.C. <http://www.epa.gov/climatechange/Downloads/ghgemissions/US-GHG-Inventory-2014-Chapter-4-Industrial-Processes.pdf> (April 2014).

Federal Highway Administration. (2011). *List of fiber reinforced polymeric materials,* Innovative Bridge Research Council, Bridge Technology Division, Washington D.C.

Fu, H., Liao, B., Qi, F., Sun, B., Liu, A., and Ren, D. (2008). "The application of PEEK in stainless steel fiber and carbon fiber reinforced composites." *Composites Part B: Engineering,* 39(4), 585–591.

Hastak, M., Mirmiran, A., and Richard, D. (2003). "A framework for life-cycle cost assessment of composites in construction." *Journal of Reinforced Plastics and Composites,* 22(15), 1409–1430.

International Iron and Steel Institute. (2005). *Sustainability Report of the World Steel Industry 2005,* Brussels, 3–50.

International Iron and Steel Institute. (2011). *Sustainability report of the world steel industry 2011,* Brussels.

International Paper. (2006). *An international paper journal 2004–2006 sustainability update,* Super Tree, South Carolina.

Khan, B., Solo-Gabriele, H., Townsend, G., and Yong, C. (2006). "Release of arsenic to the environment from CCA-treated wood. 1. Leaching and speciation during service." *Environmental Science and Technology,* No. 40, 988–993.

Kibert, C. (2008). *Sustainable construction: Green building design and delivery,* Wiley, New York.

Leadbitter, J. (2002). "PVC and sustainability." *Progress in Polymer Science,* 27(10), 2197–2226.

Lebow, S. (2004). "Alternatives to chromated copper arsenate (CCA) for residential construction." *Proc., Environmental Impacts of Preservative-Treated Wood Conf.*, Florida Center for Environmental Studies, Orlando. <http://www.ccaresearch.org/Pre-Conference/pdf/Lebow.pdf> (March 2014).

Los Alamos National Laboratory (LANL). (2002). *LANL sustainable design guide*, Los Alamos, NM.

Market Development Alliance (MDA). (2004). *Composite basic materials (Part 2)*, American Composites Manufacturers Association, Arlington, VA. <http://www.acmanet.org/resource-center> (April 2014).

Market Development Alliance (MDA). (2005). *Overview of the FRP composites industry: A historical perspective*, American Composites Manufacturers Association, Arlington, VA. <http://www.mdacomposites.org/mda/overview.html> (Aug. 2011).

Munier, N. (2005). *Introduction to sustainability: Road to a better future*, Springer, Dordrecht, Netherlands.

Naik, T., and Mariconi, G. (2006). *Environmental-friendly durable concrete made with recycled materials for sustainable concrete construction*, Center for By-Products Utilization, University of Wisconsin–Milwaukee.

Nogueron, R., and Lacstadius, L. (2007). *Sustainable procurement of wood and paper-based products*, World Resources Institute, Washington, DC.

PricewaterhouseCoopers. (2004). *Nothing but the truth: Best practices guide for sustainability reporting*, Berne, Switzerland.

Roth, L., and Eklund, M. (2003). "Environmental evaluation of reuse of by-products as road construction materials in Sweden." *Waste Manage. (Oxford)*, 23(3), 107–116.

Thermoplastic Industrial Piping Systems. (2007). "Join the revolution ... think plastics." Glenn Ellyn, IL. <http://www.ppfahome.org/tips/articles.aspx> (March 2014).

United Nations Framework Convention on Climate Change (UNFCCC). (2005). "Background on the UNFCCC: The international response to climate change." Bonn, Germany. <http://unfccc.int/essential_background/items/2877> (Nov. 2007).

U.S. Department of Labor. (1978). *Applicability of the inorganic arsenic standard to operations involving chromated copper arsenate (CCA) wood preservative*, Occupational Safety and Health Administration, Washington, DC. <https://www.osha.gov/pls/oshaweb/owadisp.show_document?p_table=INTERPRETATIONS&p_id=18754> (March 2014).

U.S. Department of Labor. (2006). *Occupational exposure to hexavalent chromium—71:10099–10385*, Occupational Safety and Health Administration, Washington, DC. <http://www.osha.gov/pls/oshaweb/owadisp.show_document?p_table=FEDERAL_REGISTER&p_id=18599> (Jan. 2007).

World Health Organization. (2004). "Some drinking-water disinfectants and contaminants, including arsenic." *Monographs on the Evaluation of Carcinogenic Risks to Humans No. 84.*, Geneva, Switzerland. <http://monographs.iarc.fr/ENG/Monographs/vol84/volume84.pdf> (Jan. 2005).

Yabroff, J. (2008). "These four walls won't fall down." *Newsweek*, 151(15), 72.

Yuhazri, M., Phongsakorn, Y., and Sihombing, H. (2010). "A comparison process between vacuum infusion and hand lay-up method toward KENAF/polyester composites." *International Journal of Basic and Applied Sciences*, 10(3), 54–57.

APPENDIX A

(This case study was provided by Guang Jiang, Amrita Mukherjee, Chakrit Raksamata, Iqbal Jeet Singh, Sorawut Srisakorn, and Milkias Lemma Tefera)

Stone-Panel Curtain-Wall Installation Case Study

This case study summarizes the strategies used and the results obtained during a productivity improvement research investigation that evaluated the installation processes for stone-panel **curtain walls** on a tall office building (Conrad 2009). The researchers analyzed existing work processes and developed recommendations for improving work processes for the installation of stone panels on the curtain wall. At the beginning of the case study, a crew of four workers could install seven stone panels on the curtain walls per day. The goal of the productivity improvement study was to determine alternative work processes that would increase productivity by at least 10%.

A.1 Scope of the Work

The stone-panel curtain-wall construction process included several components: metal, stone-panel curtain walls, windowpanes, and weatherproof sealing. The installation of each of these components could have been analyzed for potential increases in productivity, but due to time limitations, this case study focused on evaluating only the work processes required to install the stone panels on the curtain wall. The researchers investigated the potential for increasing efficiency by improving the production rate for the stone-panel curtain-wall installation process, reducing delays or minimizing the number of workers in the workforce, and reducing and eliminating procedures that create unsafe working conditions.

A.2 Curtain-Wall Systems

External curtain-wall systems are used for building façades and do not provide support for dead loads other than the weight of the curtain-wall panels. Curtain walls are constructed of glass panels, metal panels, thin stone, or other artificial materials. External curtain walls protect structures from the outside environment, provide architectural features, and contribute to the uniqueness of structures. Curtain walls also provide thermal protection and a lighting source for the interior of structures (Hong Kong Polytechnic University 2008).

Four types of curtain-wall systems are installed in structures, and they are classified according to the components used during installation:

1. *Stick systems*: In **stick systems**, the walls are installed piece by piece. The **mullion** members are installed first, followed by the **transoms**, the panels, and then the glazing or window units. Stick systems were used extensively during the early years of metal curtain-wall development, and they are still widely used in construction.
2. *Unit systems*: **Unit systems** are composed entirely of large framed units preassembled at factories. The vertical edges of the units join to form mullion members, and the top and bottom members join to form horizontal rails.
3. *Unit and mullion systems*: **Unit and mullion systems** are a compromise between stick and unit systems. First, the mullion members are installed separately, and then preassembled framed units are placed between them.
4. *Column cover and spandrel systems*: In **column cover and spandrel systems**, the **column cover** is installed first, and it may be one or two stories high. Then, the long **spandrel** panels that span between the column covers are installed. Finally, the **glazing infills** are fixed to the frame.

Stick systems and unit systems are the most frequently used systems in modern construction. In this case study, the stick system shown in Figure A-1 was used, and

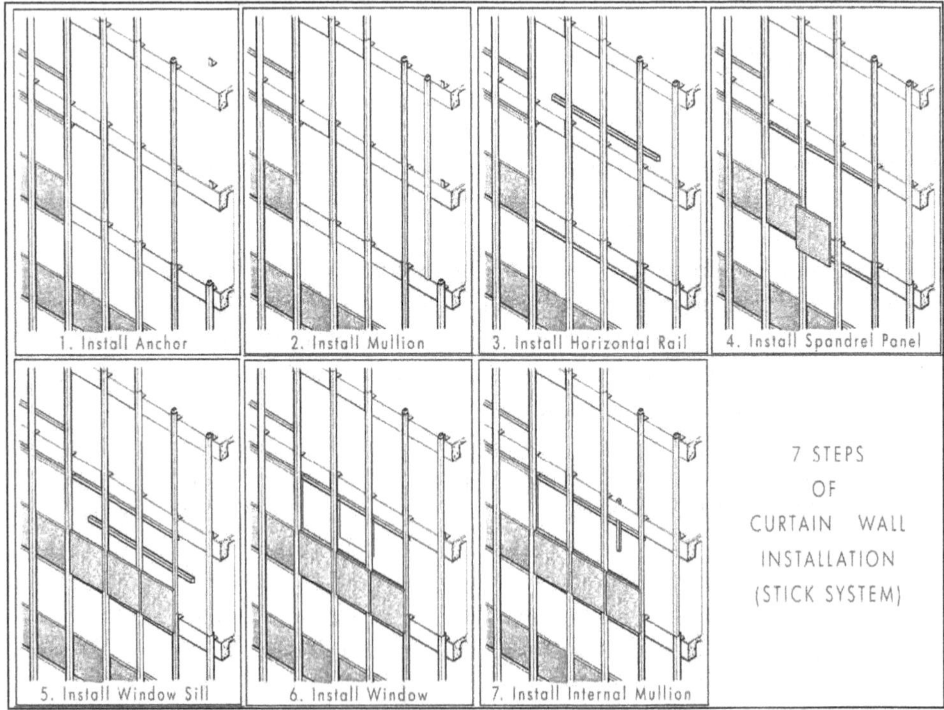

Figure A-1. Stick System Curtain-Wall Installation Steps

Source: Jiang et al. (n.d.), Figure 1; reproduced with permission.

it required three primary steps: (1) mullion installation, (2) stone-panel installation, and (3) glass-panel installation.

A.3 Case Study Schedule

The stone-panel curtain-wall installation case study was conducted based on the schedule shown in Figure A-2. The schedule included a work breakdown structure of the activities required to conduct the investigation, including planning the project, developing the work plan, investigating the jobsite, collecting data, analyzing the data, developing recommendations, preparing the productivity improvement report, and presenting the results of the investigation to the client.

A.4 Methodology

The methods used for this case study included collecting data by observing the stone-panel installation process, recording observations, interviewing jobsite personnel, photographing work processes, videotaping work processes, and conducting field-rating, 5-min-rating, and work-sampling studies.

Multiple iterations of the work processes were recorded to increase the accuracy of the analysis procedures. Digital photographs and video recordings of several cycles were reviewed to evaluate whether alternative work processes could improve production rates. The project manager provided current productivity data, detailed stone-work drawings, and the project schedule.

The researchers evaluated every detail of each operation, including the suitability of tools, equipment, and materials and the appropriateness of the size of the crews and work locations. First, existing installation methods were analyzed to identify whether all of the operations performed during the installations were necessary. Next, alternative work processes were developed and reviewed to determine which were applicable to the operation. The analysis methods used for this phase of the study included (1) work-process flow diagrams, (2) crew balance charts, and (3) work-process charts.

Work-process flow diagrams are sketches of how equipment, materials, and workers move around jobsites. Crew balance charts are diagrams or histograms used to record how each member of a work crew is expending his or her time on each activity. Crew balance charts are used to evaluate the effectiveness of workers, machines, and worker-machine combinations and to demonstrate the interrelationships between activities being performed by crew members. Work-process charts are used to monitor how materials are being moved and used by workers. After analyzing all of the work processes, the researchers provided recommendations for increasing productivity rates. The steps used to develop recommendations were as follows:

Observation/data collection → Analysis → Review analysis → Develop recommendations

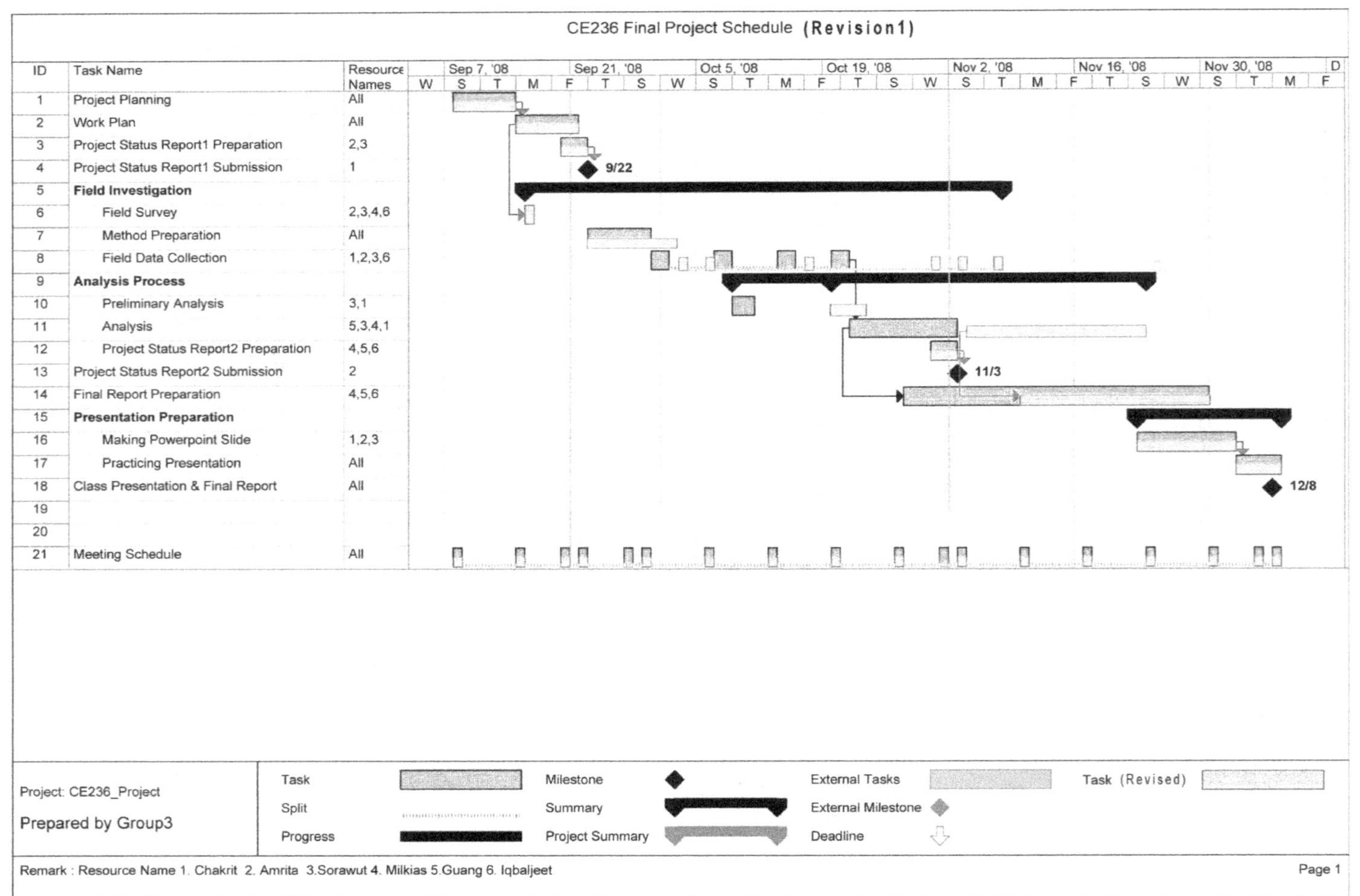

Figure A-2. Productivity Improvement Study Project Schedule

Source: Jiang et al. (n.d.), Figure 2; reproduced with permission.

A.5 Components of the Curtain-Wall Installation

If construction projects employ union workers, then the labor union trade classifications dictate which workers perform each function during construction operations. The three main trade unions involved in the installation of the curtain walls on this project were the **welders** who erected the metal sticks, the **stonemasons** who installed the stone panels, and the **glazers** who installed the glass panels.

The welders installed the aluminum sticks before the stone panels and glass were installed, and this activity was completed before the start of the productivity improvement study. The glazers installed the windowpanes after the masons had installed the stone panels. The aluminum-stick and glass-panel installation processes were not evaluated as part of this case study. The stone panels on the second to seventh floors were installed before the study commenced; therefore, the study focused on analyzing the productivity of the stone-panel installation process on floors higher than the seventh floor. After the stone panels and glass panes were installed, weather-resistant caulking sealants were injected into the gaps between the stone and glass panels to weatherproof the curtain wall. Figure A-3 shows the structure at the start of the productivity improvement study, and Figure A-4 provides the schedule for the installation of the mullions, stone panels, and glass panels.

Figure A-3. Curtain-Wall System in Midcompletion

Source: Jiang et al. (n.d.), Figure 3; reproduced with permission.

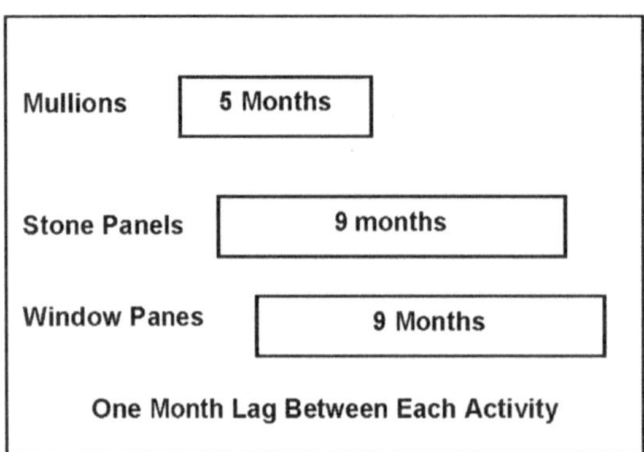

Figure A-4. Partial Schedule of Key Elements of Curtain-Wall-System Installation

Source: Jiang et al. (n.d.), Figure 4; reproduced with permission.

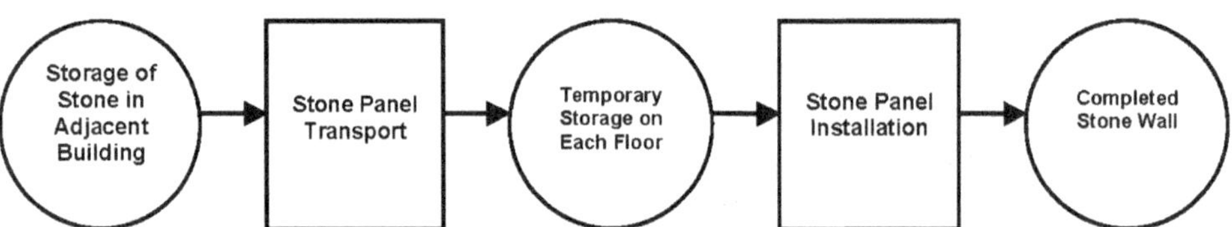

Figure A-5. Stone-Panel Installation Work-Process Flow Model

Source: Jiang et al. (n.d.), Figure 5; reproduced with permission.

Stone-Panel Installation Process

The stone-panel installation process required transporting and installing the stone panels on the curtain wall. The stone panels were transported to the work area from a storage area located on the roof of a parking garage adjacent to the construction jobsite. Figures A-5 and A-6 contain the stone-panel work-process flowcharts for this activity, and Figure A-7 shows where the stone panels were installed on the structure.

The stone panels were stored temporarily on the adjacent parking garage roof so that they would not interfere with construction. Crates containing six stone panels were moved using a forklift from the temporary storage area on the adjacent garage roof to the building under construction, and then they were distributed to the various floors in the building using the temporary lift, as shown in Figure A-7. Figure A-7 also shows the movement pattern of the stone transportation process from the temporary storage area on the adjacent parking garage roof. The stone panels were moved to one or two floors at a time. One crew, which consisted of one laborer and one driver, was required to transport the stone panels down from the garage roof to a temporary lift in the building under construction. Because the working space was limited at the jobsite, safety and security were primary concerns during the transporting of the stone panels.

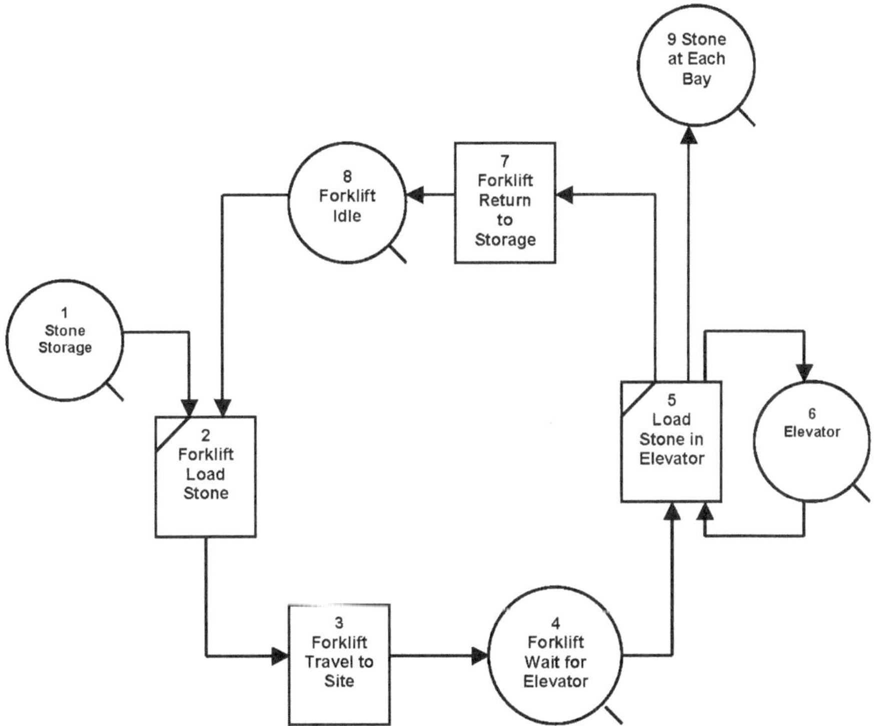

Figure A-6. Stone-Panel Transportation Flow Model

Source: Jiang et al. (n.d.), Figure 6; reproduced with permission.

Floor Plan and Bay Types

Each floor from the third floor to the top floor (the 16th floor) had repetitive floor plans, and the walls of the bays were repetitive sections. Three types of bays were on each floor, as shown in Figure A-8, and there were 12 stone panels for each bay. There were 12 middle bays per floor, and 12 Type A stone panels per bay. Drawings of these are shown in Figure A-9. There were 8 edge bays per floor with 12 Type B stone panels per bay, as shown in Figure A-10, and 4 corner bays per floor with 12 Type C stone panels per bay, as shown in Figure A-11.

Materials, Labor, and Equipment

The components required to install the stone panels on the curtain-wall system were marble granite stone, workers, movable platforms, and tools. The following types and numbers of stone panels were used:

Type A 12 × 12 = 144 stone panels per floor
Type B 15 × 8 = 120 stone panels per floor

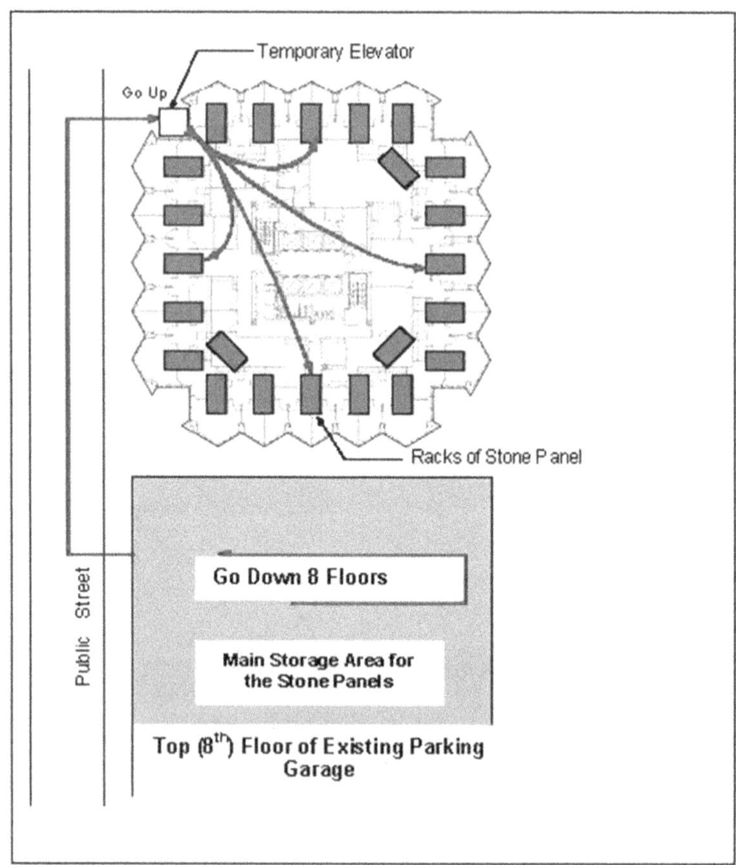

Figure A-7. Flow Diagram of Stone-Panel Transportation from Temporary Storage

Source: Jiang et al. (n.d.), Figure 7; reproduced with permission.

Type C 12 × 4 = 48 stone panels per floor
Total = 312 stone panels per floor (78 panels per side per floor)

From the 3rd to the 16th floor, 4,368 stone panels were required, and the assigned stone-panel identification numbers for the various sizes are shown in Figure A-12.

Two crews installed the stone panels: Crew A worked on the north and the east side of the structure and Crew B worked on the south and west side. Each crew had four workers, and for identification purposes, they were referred to as (1) upper lead, (2) upper runner, (3) stage lead, and (4) stage helper. The workers positioned themselves at their work locations during the stone-panel installation process. Workers 1 and 2 were on the building edge on the floor where the stone panels were being installed, and Workers 3 and 4 were outside on the same building floor, positioned on a movable platform. A photograph of this operation is provided in Figure A-13.

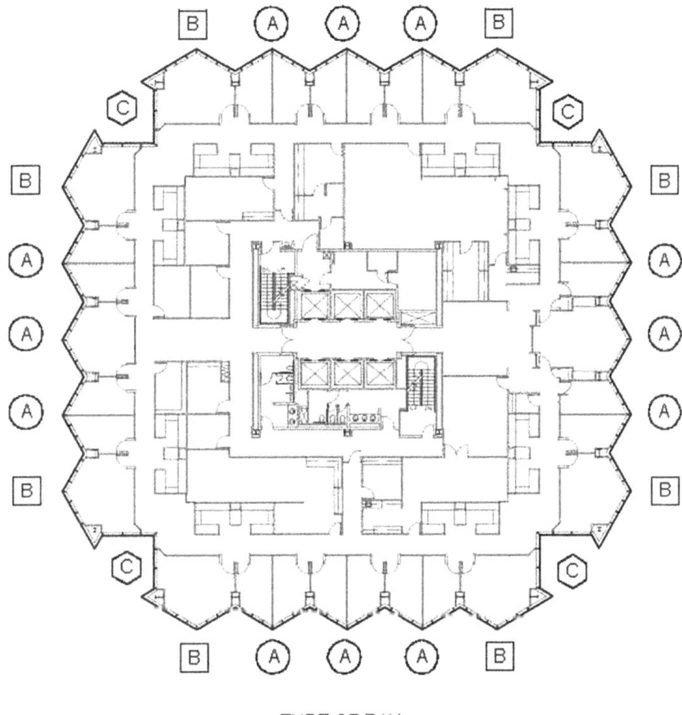

Figure A-8. Bay Types from 3rd Floor to 16th Floor

Source: Jiang et al. (n.d.), Figure 8; reproduced with permission.

Types of Platforms

Two types of platforms were used by the crew on the outside of the building to install the stone panels. The first type of platform was used for installing the middle bay Type A and edge bay Type B stone panels. The edge bays required a platform equipped with an extension platform, and the extension is shown in Figure A-14. A total of five Type 1 platforms were available for use by the various crews. The second type of platform was used for installing the corner bay Type C, as shown in Figure A-15. Three corners of the four-corner building used the second type of platform, and the fourth corner was by the temporary lift; therefore, a platform was not required at the fourth corner to install the stone panels because the crew could use the temporary lift.

Installation Times and Data Collection

Data were recorded on several occasions during the installation of the stone panels. The data gathered during the stone-panel installation process are summarized in

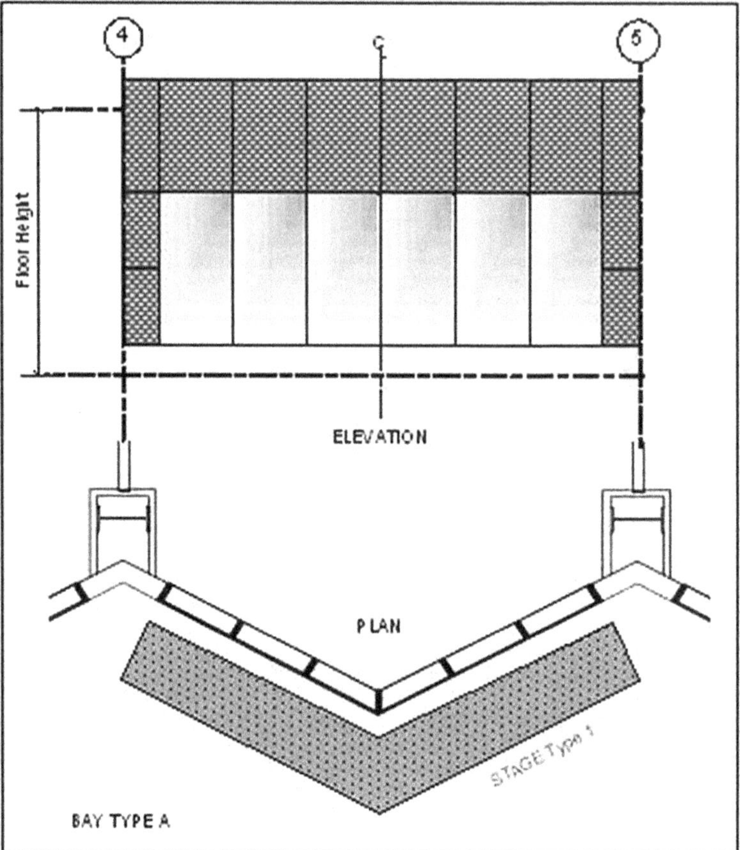

Figure A-9. Bay Type A—Middle Bay

Source: Jiang et al. (n.d.), Figure 9; reproduced with permission.

Table A-1. The field observations indicated that the size of the stone panels only slightly affected the installation times. The different positions or levels of stone panels did not affect the installation times, except for stone panels 11 and 12, which required significantly longer installation times.

Installation Process

Digital video recordings were made of the installation process for stone panels Types 1 and 6, and two photographs of this operation from the video recordings are shown in Figures A-16 and A-17.

Installation Sequence and Cycle

For bays Type A and B, the crew moved the platform in the same sequence every two floors. The crew first installed the stone panels from the left side to the right side of

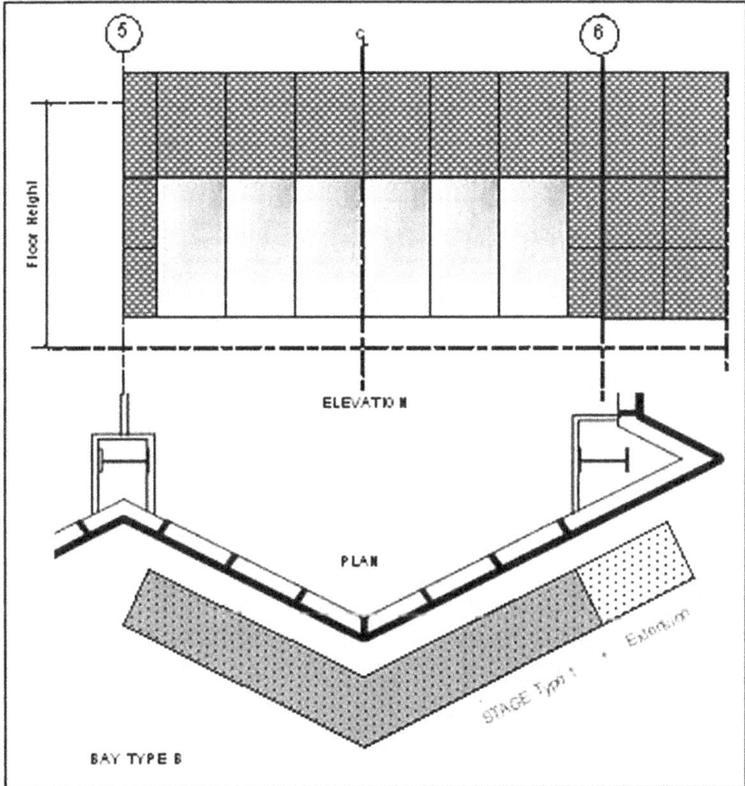

Figure A-10. Bay Type B—Edge Bay

Source: Jiang et al. (n.d.), Figure 10; reproduced with permission.

the bay, and then they returned to the left side again to install the stone panels on the next highest floor. This sequence is shown in Figures A-18 to A-20.

The corner panel bays Type C required a different independent crew, and Table A-2 shows the process this crew used and the time required for moving the platform.

A.6 Data Analysis

The installation time for one stone panel could be determined from the data provided in Table A-1. There were different installation times for each type of panel, and they were as follows:

Process Time

- General panel total installation time = 107.5 min ($T1$)
- Edge panel total installation time = 177.25 min ($T2$)
- Corner panel total installation time = 143 min ($T3$)

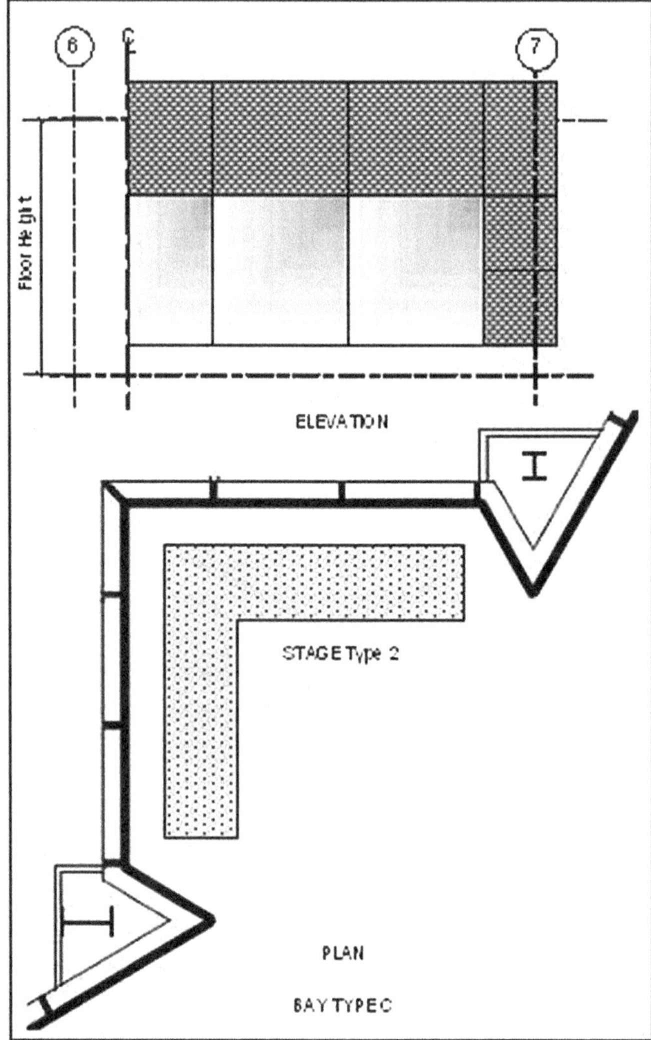

Figure A-11. Bay Type C—Corner Bay

Source: Jiang et al. (n.d.), Figure 11; reproduced with permission.

Transportation Time

- Total time to move the platform two floors = 256 min ($T4$)
- Total time to assemble the platform extension = 120 min ($T5$)
- Total time to disassemble the platform extension = 120 min ($T6$)

Average Installation Time for One Floor (on One Side of the Building)

$$T_{\text{total}} = \{[(T1 \times 3 \text{ bays}) + (T2 \times 2 \text{ bays}) + (T3 \times 1 \text{ bay}) + (T4 \times 2) + (T5 + T6) \times 2]/F\}/N \quad \text{(A-1)}$$

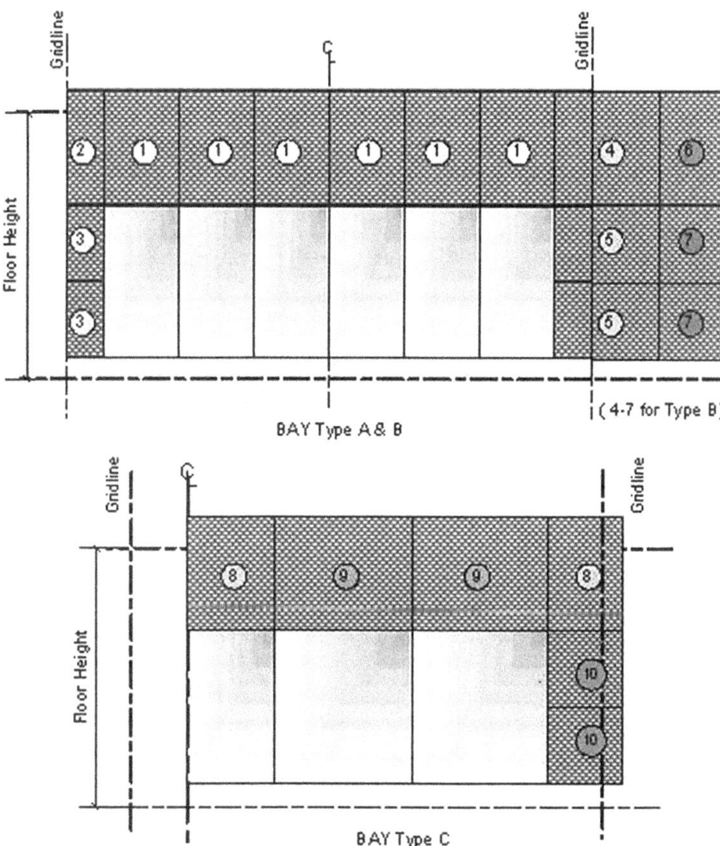

Figure A-12. Elevation of Curtain Wall for Bay Types A, B, and C—Stone-Panel Size Identification

Source: Jiang et al. (n.d.), Figure 12; reproduced with permission.

Figure A-13. Positions of Stone Crew Members and Identification Number

Source: Jiang et al. (n.d.), Figure 13; reproduced with permission.

Figure A-14. Type 1 Platform with Platform Extension for Type B Edge Bays

Source: Jiang et al. (n.d.), Figure 14; reproduced with permission.

Figure A-15. Type 2 Platform at Type C Corner Bays

Source: Jiang et al. (n.d.), Figure 15; reproduced with permission.

where F = number of floors or bays completed before the platform was moved; and N = number of panels for each floor on one side of the building.

There were 78 stone panels per floor on each side of the building. Because the platform was moved after the panels had been installed on two floors, the default number of the floor work is 2; thus, when 78 panel installations are used in Eq. (A-1), the result is

$$\begin{aligned} T_{\text{total per floor}} &= \{[(107.5\times 3)+(177.25\times 2)+(143\times 1) \\ &\quad +(256\times 2)+(120+120)\times 2]/2\}/78 \\ &= 20.15 \text{ min per panel, } 1{,}572 \text{ min per floor,} \\ &\quad 26.2 \text{ h per side of the building} \end{aligned}$$

Table A-1. Stone-Panel Installation Times

Stone-Panel Installation Time

Crew:	Stone-Panel Installation: Crew B					10/06/08 and
Job:	River Park II Project: Curtain-Wall Installation				Date:	10/17/08
Location		Bay Type A (Span 2 South) and Type B (Edge Span 1 South)				Remark

Stone Number	Position	Time to Complete (minutes)	Size			Remark
			Width	Height	Square feet (Square Meters)	
3	Left-bottom	8.0	1′ 6 5/8″ (0.473 m)	3′ 10 1/8″ (1.17 m)	5.97 (0.55)	
3	Left-top	8.0	1′ 6 5/8″ (0.473 m)	3′ 10 1/8″ (1.02 m)	5.97 (0.55)	
2	Left	8.5	1′ 6 5/8″ (0.473 m)	5′ 2 5/8″ (1.59 m)	8.10 (0.75)	
3	Right-bottom	7.5	1′ 6 5/8″ (0.473 m)	3′ 10 1/8″ (1.02 m)	5.97 (0.55)	
3	Right-top	7.5	1′ 6 5/8″ (0.473 m)	3′ 10 1/8″ (1.02 m)	5.97 (0.55)	
2	Right	8.0	1′ 6 5/8″ (0.473 m)	5′ 2 5/8″ (1.59 m)	8.10 (0.75)	
1	Left-left	9.5	3′ 4″ (1.02 m)	5′ 2 5/8″ (1.59 m)	17.40 (1.62)	
1	Left-middle	10.5	3′ 4″ (1.02 m)	5′ 2 5/8″ (1.59 m)	17.40 (1.62)	
1	Left-right	10.0	3′ 4″ (1.02 m)	5′ 2 5/8″ (1.59 m)	17.40 (1.62)	
1	Right-left	10.0	3′ 4″ (1.02 m)	5′ 2 5/8″ (1.59 m)	17.40 (1.62)	
1	Right-middle	9.5	3′ 4″ (1.02 m)	5′ 2 5/8″ (1.59 m)	17.40 (1.62)	
1	Right-right	10.5	3′ 4″ (1.02 m)	5′ 2 5/8″ (1.59 m)	17.40 (1.62)	
4		13.5	4′ 5 2/8″ (1.35 m)	5′ 2 5/8″ (1.59 m)	23.16 (2.15)	Edge Type 2
5	Top	12.5	4′ 5 2/8″ (1.35 m)	3′ 10 1/8″ (1.17 m)	17.06 (1.58)	Edge Type 2
5	Bottom	13.0	4′ 5 2/8″ (1.35 m)	3′ 10 1/8″ (1.02 m)	17.06 (1.58)	Edge Type 2
6		18.0	2′ 10 5/8″ (0.88 m)	5′ 2 5/8″ (1.59 m)	15.06 (1.34)	Edge Type 2
7	Top	18.5	2′ 10 5/8″ (0.88 m)	3′ 10 1/8″ (1.02 m)	11.09 (1.03)	Edge Type 2
7	Bottom	18.0	2′ 10 5/8″ (0.88 m)	3′ 10 1/8″ (1.02 m)	11.09 (1.03)	Edge Type 2

continued

Table A-1. *(continued)*

Location		Bay Type C (Corner South East)			Date:	10/31/08
Stone Number	Position	Time to Complete (minutes)	Size			Remark
			Width	Height	Square feet (Square Meters)	
8		10.0	3′ 6″ (1.02 m)	5′ 2 5/8″ (1.59 m)	18.27 (1.7)	
9		13.5	5′ 0″ (1.52 m)	5′ 2 5/8″ (1.59 m)	26.09 (2.42)	
10		14.0	5′ 5 4/8″ (1.66 m)	5′ 2 5/8″ (1.59 m)	28.49 (2.47)	
11		13.5	4′ 0 4/8″ (1.26 m)	5′ 2 5/8″ (1.59 m)	21.09 (1.60)	
12	Top	10.0	4′ 0 4/8″ (1.26 m)	3′ 10 1/8″ (1.02 m)	15.54 (1.44)	
12	Bottom	10.5	4′ 0 4/8″ (1.26 m)	3′ 10 1/8″ (1.02 m)	15.54 (1.54)	

Source: Jiang et al. (n.d.), Table 1; reproduced with permission.

Figure A-16. Installation of Stone Panel Type 1

Source: Jiang et al. (n.d.), Figure 16; reproduced with permission.

The total time per floor is based on the assumption that the platform was not moved up or down during the installation of the stone panels on the two floors. The average installation time was 20.15 min per panel. The time required to install each bay type, determined using Eq. (A-1), is shown in Figure A-21.

Figure A-17. Installation of Stone Panel Type 6

Source: Jiang et al. (n.d.), Figure 17; reproduced with permission.

Figure A-18. Two-Floor Installation Increments

Source: Jiang et al. (n.d.), Figure 18; reproduced with permission.

Accuracy of the Data

According to the data for the stone-panel installation times in Figure A-21, there is a direct relationship between the installation times and the size of the stone panels. It was observed that the installation times for stone panels 6 and 7 were significantly different from the other stone-panel installation times. A regression analysis was used

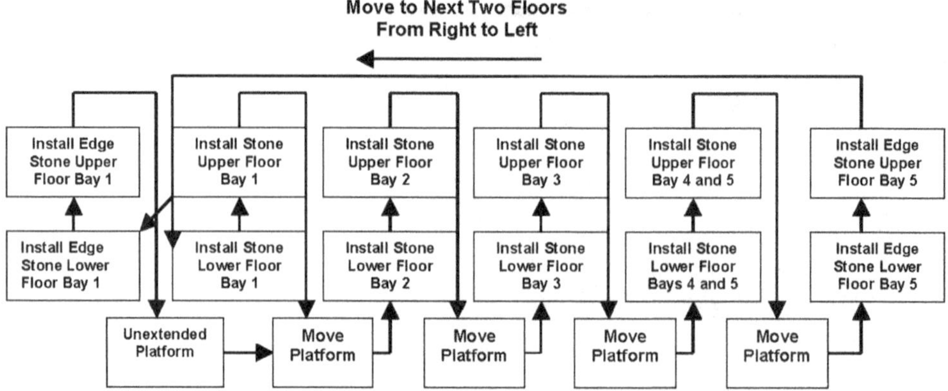

Cycle of Stone Panel Installation of Bays Type A and B

Figure A-19. Flowchart of Stone-Panel Installation Sequence

Source: Jiang et al. (n.d.), Figure 19; reproduced with permission.

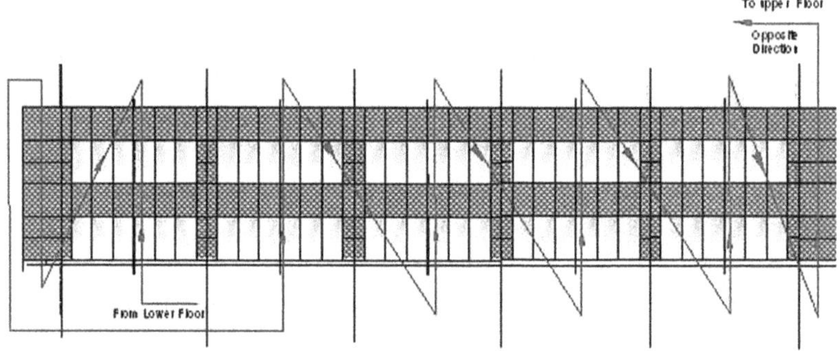

Figure A-20. Stone-Panel Installation Sequence over Two Floors

Source: Jiang et al. (n.d.), Figure 20; reproduced with permission.

to evaluate the correlation, and the results are shown in the Figure A-22. The *R*-square value for the relationship is 0.16, and the data for Panels 6 and 7 deviate from the trend line. The installation of stone panels 6 and 7 was affected by other factors; therefore, they were removed from the data set. As shown in Figure A-23, after eliminating the data for stone panels 6 and 7, the new *R*-square value was 0.79; therefore, there was a strong relationship between the size of the stone panel and installation time.

General Types of Stone Panels

Table A-3 shows the process chart for the installation of stone panels Type 1, and the process chart provides the sequencing, workers, and time required to complete the

Table A-2. Time Log for Moving Platform

Number	Flow Sequence	Time Recorded	Worker Number	Time Duration	Notes
1.0	**Lowering the platform**			**6 min**	**Stage at 12th floor level**
1.1	Lowering the platform to ground level at Bay 1	8:20	3 and 4	4 min	2 people (20 s per floor)
1.2	Loosen steel wire and prepare to move platform	8:24	3 and 4	2 min	
2.0	**Arranging the platform to work in Bay 2**			**4 h 16 min**	**4 people**
2.1	Prepare tools and area	8:30		25 min	At the same time as lowering the platform
2.2	Take off Wire 1 from the support to hook to the temporary beam	8:45	2	25 min	Worker 1 idle after 15 min
2.3	Move Support 1 to left from Point 1 on left	9:10	1 and 2	3 min	
2.4	Prepare Point 1 for fixing to Support 1 on left	9:13	1 and 2	7 min	
2.5	Install Support 1 at Point 2 on left	9:20	1, 2, and 3	11 min	
2.6	Take off Wire 2 from Support 2 to hook to the temporary support	9:31	1 and 2	25 min	Move the stone racks in position at the same time. Worker 4 and Worker 1 idle after 15 min
2.7	Move temporary Support 1 to next bay	9:56	1, 2, and 3	30 min	Move the platform at the same time Worker 3
2.8	Hook Wire 1 from the temporary support to Support 1 on the left	10:26	1 and 2	35 min	Move the stone racks in position at the same time. Workers 4 and 1 idle after 15 min
2.9	Move Support 2 on the right from Point 1 on the right	11:01	1, 2, and 4	3 min	

continued

Table A-2. *(continued)*

Number	Flow Sequence	Time Recorded	Worker Number	Time Duration	Notes
2.10	Prepare Point 2 on the right for fixing Support 1 on the right	11:04	1 and 2	6 min	
2.11	Install Support 2 at Point 2 on the right	11:10	1, 2, and 4	11 min	
2.12	Move temporary Support 2 to next bay	11:21	1, 2, and 3	30 min	Move the platform at the same time Worker 3, Worker 4 move the stone rack
2.13	Hook in Wire 2 from the temporary support to Support 2 at Point 2 on the right	13:15	1 and 2	25 min	Lunch break at 11:51 and return at 13:15
2.14	Clear the area	13:40	2	20 min	At the same time as pulling up the platform
3.0	**Pulling up the platform**			**7 min**	**Platform at 12th floor**
3.1	Tighten steel wire place platform properly with Bay 2	13:30	3 and 4	2 min	
3.2	Pulling up and installing the platform at the desired floor level	13:32	3 and 4	5 min	2 people (22 s per floor)
	Total	13:37	3 and 4	4 h 16 min	

Source: Jiang et al. (n.d.), Table 2; reproduced with permission.

installation process. Crew balance charts were used to analyze the crew sizes and associated tasks. Figure A-24 is the crew balance chart for installation of stone panels Type 1, Figure A-25 provides a revision to the crew balance chart, and Figure A-26 is a crew balance chart with additional revisions.

The crew balance chart for the installation of stone panels Type 2 is shown in Figure A-27. The chart demonstrates that the subcontractor could eliminate Worker 4, Worker 3 could perform the sealant operation, and Worker 2 could hold the

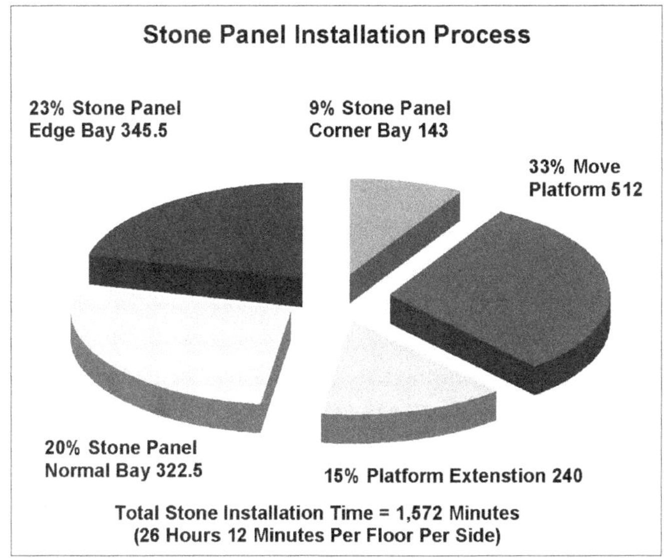

Figure A-21. Average Times to Install Stone Panels per Floor per Side

Source: Jiang et al. (n.d.), Figure 21; reproduced with permission.

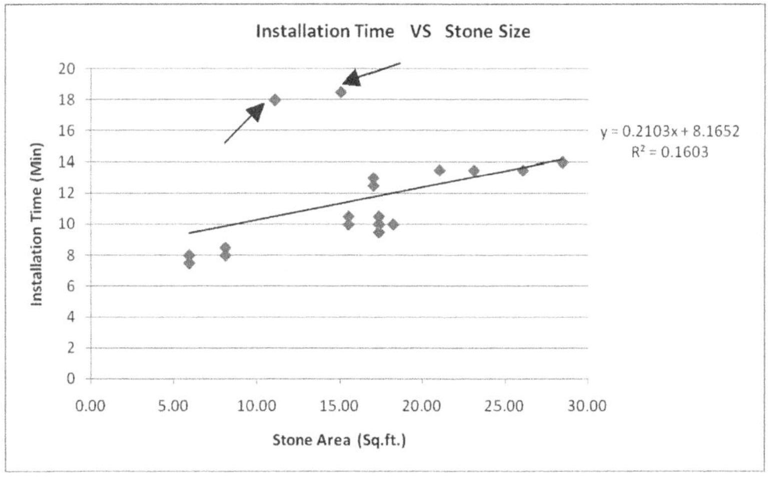

Figure A-22. Relationship between Stone-Panel Installation Times and Stone Size

Source: Jiang et al. (n.d.), Figure 22; reproduced with permission.

panels. Figure A-28 shows possible revisions to the crew balance chart, and Figure A-29 is the revised crew balance chart, including the following alterations:

- Worker 3 never adjusted the ropes; therefore, this procedure can be eliminated from the work processes.

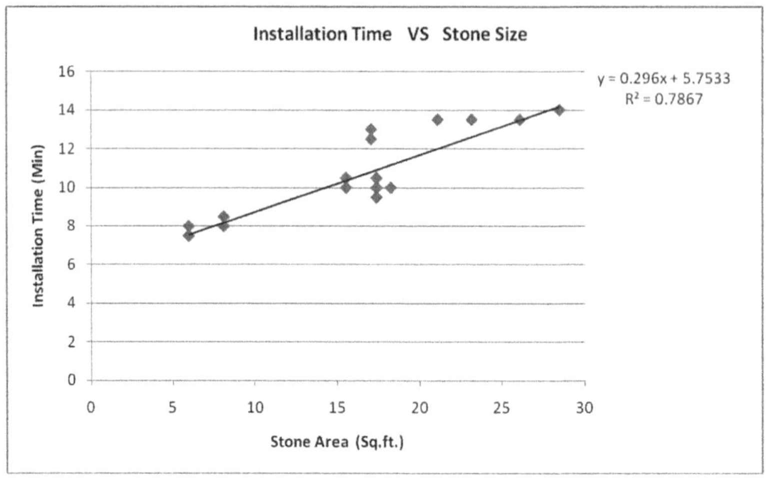

Figure A-23. Correlation of Times and Sizes for Stone Panel Types 1–5

Source: Jiang et al. (n.d.), Figure 23; reproduced with permission.

Analysis Crew Stone Panel Installation: Crew B — Date:___10/16/08______
Job: River ParkII Project : Currtain Wall Installation
Time Interval: — Page: ___1___of__1____
Remarks: Stone Type 1 (General) : Bay Type A (General Bay)

Time Interval (30 Seconds)	Start time	Man 1	Man 2	Man 3	Man 4
Video2					
1	0:00	Pull Pulley	Move to working floor	Idle	Idle
2	0:30		Bring a new panel		
3	1:00	Hold panel		Install panel	Hold Panel
4	1:30				
5	2:00		Idle		
6	2:30			Set Panel	Prepare tools
7	3:00				
8	3:30				
9	4:00				
10	4:30	Put sealant	Move to top floor		
11	5:00	Idle	Move chain & hook	Idle	Idle
12	5:30				
13	6:00				
14	6:30				
15	7:00		Install chain & hook		
16	7:30	Prepare tools	Move to working floor		Prepare tools
17	8:00	Idle			
18	8:30			Set panel	
19	9:00	Prepare tools			
20	9:30			Idle	Idle
21	10:00				
22	10:30				
23	11:00	Bring a new panel			

Figure A-24. Current Crew Balance Chart for Stone Panel Type 1 Installation

Source: Jiang et al. (n.d.), Figure 24; reproduced with permission.

Table A-3. Procedure for the Installation of Stone Panels Type One

Process Chart Worksheet

Project: River Park Tower II *Study Location:* Bay 3 West Side

Work Item: Installation of Stone-Panel No. 1 *Prepared By:* Chakrit Raksamata *Date:* 10/16/08

Evaluation No. 1

No.	Flow Sequence	Operation	Transportation	Inspection	Delay	Storage	Quantity	Distance in Feet	Time	Notes
1.	Fasten screw and level an existing bottom rail	●	⇨	☐	D	▽			1:00	Worker No. 3 and 4
2.	Seal silicone on the right bottom rail	●	⇨	☐	D	▽			0:30	Worker No. 3 and 4 Parallel with Step 3
3.	Transporting the stone panel	○	➡	☐	D	▽		25' max	2:30	Worker No. 2
4.	Get stone panel onto pulley and pull the chain to lower it	●	⇨	☐	D	▽			1:30	Worker No. 1 and 2
5.	Seal silicone onto the bottom of the stone panel	●	⇨	☐	D	▽			0:30	Worker No. 3 and 4
6.	Set stone panel to sit on bottom rail	●	⇨	☐	D	▽			1:30	Worker No. 1, 3, and 4
7.	Seal with silicone	●	⇨	☐	D	▽			0:30	Worker No. 1
8.	Install the bottom rail to secure the stone panel	●	⇨	☐	D	▽			1:30	Worker No. 2
9.	Move the stage up or down	●	⇨	☐	D	▽			0:30	Worker No. 3 and 4
10.	Remove clamp from the stone panel and raise the pulley up	●	⇨	☐	D	▽			1:00	Worker No. 1
	Total								11:00	

Source: Jiang et al. (n.d.), Table 3; reproduced with permission.

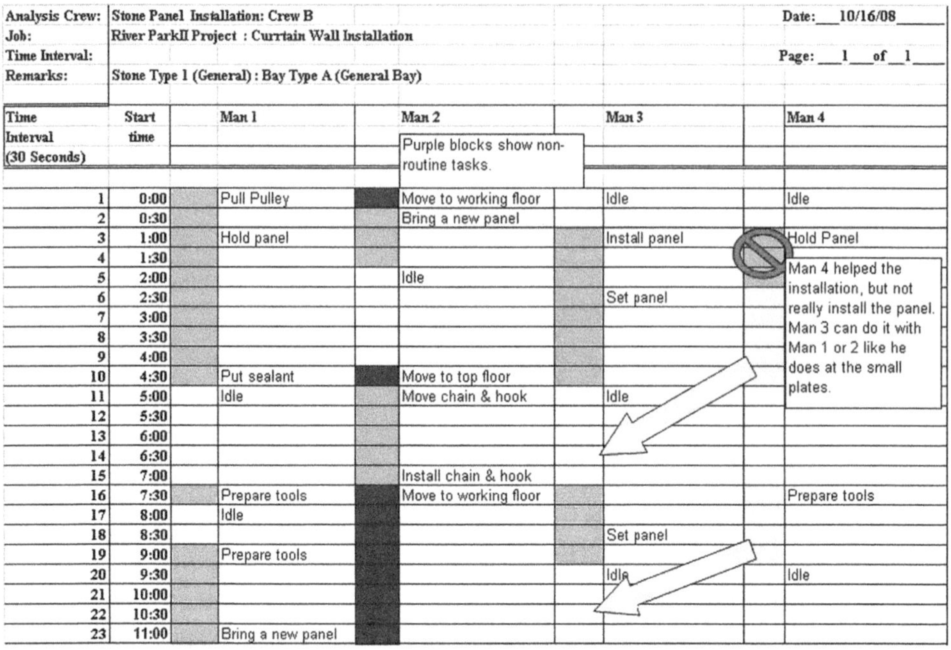

Analysis Crew:	Stone Panel Installation: Crew B	Date:	10/16/08
Job:	River ParkII Project : Currtain Wall Installation		
Time Interval:		Page:	1 of 1
Remarks:	Stone Type 1 (General) : Bay Type A (General Bay)		

Time Interval (30 Seconds)	Start time	Man 1	Man 2	Man 3	Man 4
1	0:00	Pull Pulley	Move to working floor	Idle	Idle
2	0:30		Bring a new panel		
3	1:00	Hold panel		Install panel	Hold Panel
4	1:30				
5	2:00		Idle		
6	2:30			Set panel	
7	3:00				
8	3:30				
9	4:00				
10	4:30	Put sealant	Move to top floor		
11	5:00	Idle	Move chain & hook	Idle	
12	5:30				
13	6:00				
14	6:30				
15	7:00		Install chain & hook		
16	7:30	Prepare tools	Move to working floor		Prepare tools
17	8:00	Idle			
18	8:30			Set panel	
19	9:00	Prepare tools			
20	9:30			Idle	Idle
21	10:00				
22	10:30				
23	11:00	Bring a new panel			

Purple blocks show non-routine tasks.

Man 4 helped the installation, but not really install the panel. Man 3 can do it with Man 1 or 2 like he does at the small plates.

Figure A-25. Proposed Crew Balance Chart for Stone Panel Type 1

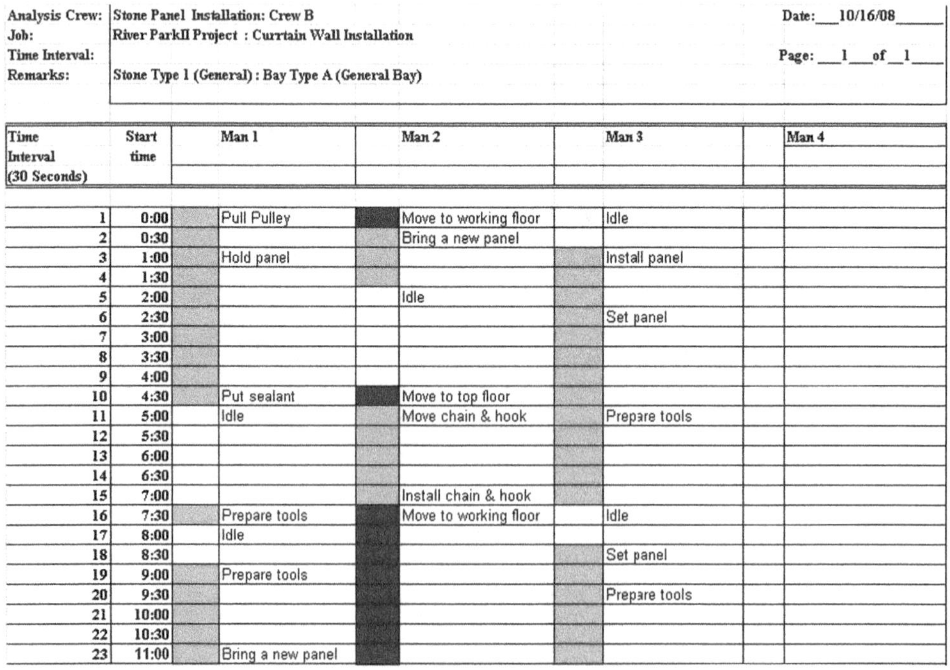

Source: Jiang et al. (n.d.), Figure 25; reproduced with permission.

Analysis Crew:	Stone Panel Installation: Crew B	Date:	10/16/08
Job:	River ParkII Project : Currtain Wall Installation		
Time Interval:		Page:	1 of 1
Remarks:	Stone Type 1 (General) : Bay Type A (General Bay)		

Time Interval (30 Seconds)	Start time	Man 1	Man 2	Man 3	Man 4
1	0:00	Pull Pulley	Move to working floor	Idle	
2	0:30		Bring a new panel		
3	1:00	Hold panel		Install panel	
4	1:30				
5	2:00		Idle		
6	2:30			Set panel	
7	3:00				
8	3:30				
9	4:00				
10	4:30	Put sealant	Move to top floor		
11	5:00	Idle	Move chain & hook	Prepare tools	
12	5:30				
13	6:00				
14	6:30				
15	7:00		Install chain & hook		
16	7:30	Prepare tools	Move to working floor	Idle	
17	8:00	Idle			
18	8:30			Set panel	
19	9:00	Prepare tools			
20	9:30			Prepare tools	
21	10:00				
22	10:30				
23	11:00	Bring a new panel			

Figure A-26. Improved Crew Balance Chart for Stone Panel Type 1

Source: Jiang et al. (n.d.), Figure 26; reproduced with permission.

Analysis Crew: Stone Panel Installation: Crew B Date: 10/16/08
Job: River ParkII Project : Currtain Wall Installation
Time Interval: Page: 1 of 1
Remarks: Stone Type 2 (Small) : Bay Type A (General Bay)

Time Interval (30 Seconds) Video2	Start time	Man 1	Man 2	Man 3	Man 4
1	0:00	Prepare for new panel	Idle	Adjust rope	Put sealant
2	0:30	Bring down a panel		Idle	Idle
3	1:00		Hand over tools		Prepare sealant
4	1:30		Idle		Put sealant
5	2:00			Tie the rope	Move inside
6	2:30	Adjust panel position		Install panel	Hold panel from inside
7	3:00			Bring tools	
8	3:30			Hand over tools	
9	4:00			Adjust panel	Adjust panel
10	4:30		Hold panel from inside		
11	5:00				Move to stage
12	5:30			Put sealant	Idle
13	6:00				Idle
14	6:30			Set panel	Hand over tools
15	7:00	Pull pulley		Put sealant	Prepare tools for Man 3
16	7:30	Prepare for new panel			

Figure A-27. Crew Balance Chart for Current Stone Panel Type 2 Bay Type A Installation

Source: Jiang et al. (n.d.), Figure 27; reproduced with permission.

Time Interval (30 Seconds) Video2	Start time	Man 1	Man 2	Man 3	Man 4
1	0:00	Prepare for new panel	Idle		Put sealant
2	0:30	Bring down a panel		Idle	Idle
3	1:00		Hand over tools		Prepare sealant
4	1:30		Idle		Put sealant
5	2:00			Tie the rope	Move inside
6	2:30	Adjust panel position		Install panel	Ho[illegible]de
7	3:00				
8	3:30			Hand over tools	
9	4:00			Adjust panel	Ad
10	4:30		Hold panel from inside		
11	5:00				Move to stage
12	5:30			Pu[illegible]nt	Idle
13	6:00				Idle
14	6:30			Set panel	
15	7:00	Pull pulley		Put sealant	
16	7:30	Prepare for new panel			

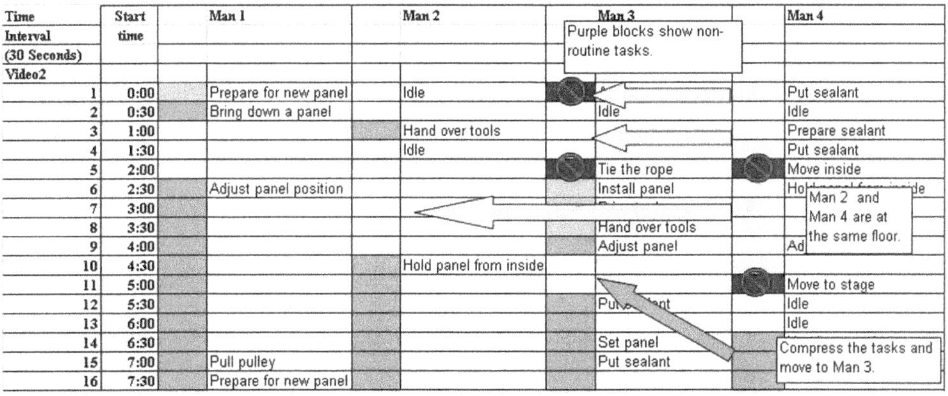

Figure A-28. Proposed Transfer of Work for Stone Panel Type 2 Installation

Source: Jiang et al. (n.d.), Figure 28; reproduced with permission.

- Between 0:00 and 1:30, Worker 4 can perform the work of Worker 3, who was idle.
- For the holding of the panel task from 2:30 to 4:30, Worker 4 held the panel in position, and Worker 2 was idle; however, after that period, Worker 2 held the same panel and let Worker 4 walk back to the platform. There was no need for the workers to share this task because Worker 2 could perform it alone.
- The tool preparation task occurred between 6:30 and 7:30. It could have been performed faster, and all of the work could have been performed by Worker 4 because Worker 3 was idle between 4:30 and 5:00.

Time Interval (30 Seconds) Video2	Start time	Man 1	Man 2	Man 3		Man 4
1	0:00	Prepare for new panel	Idle	Put sealant		
2	0:30	Bring down a panel		Idle		
3	1:00		Hand over tools	Prepare sealant		
4	1:30		Idle	Put sealant		
5	2:00		Hold panel from inside			
6	2:30	Adjust panel position		Install panel		
7	3:00			Bring tools		
8	3:30			Hand over tools		
9	4:00			Adjust panel		
10	4:30			Hand over tools		
11	5:00			Prepare tools		
12	5:30			Put sealant		
13	6:00					
14	6:30			Set panel		
15	7:00	Pull pulley		Put sealant		
16	7:30	Prepare for new panel				

Figure A-29. Improved Crew Balance Chart for Stone Panel Type 2 Installation

Source: Jiang et al. (n.d.), Figure 29; reproduced with permission.

Edge Panels

The work characteristics of the edge panels differed from those of the general panels. At the edge of the building, two workers did not have enough space to perform their tasks at the end of the platform. The installation times were longer because the stick system blocked the movement of the pulley and the stone panels. Thus, the workers had to be more cautious to prevent damaging the stone panels. After reviewing the crew balance chart shown in Figure A-30, it was discovered that Worker 4, who was a helper, could be eliminated because most of the work he performed was tool preparation for Worker 3, who was the main installer. The revised crew balance chart for this operation is shown in Figure A-31, and it includes the following changes:

- Between 0:00 and 10:00, Worker 1 performed the work of Worker 4 while Worker 4 was idle. Worker 1 was on the same floor as Worker 3, who was the installer; thus, it was still convenient for that worker to hand over the tools.
- After 10:00, both Worker 3 and Worker 4 were idle—especially Worker 3, who was free from 5:00 to 10:30, or approximately 6 min—and Worker 4 did not have a lot of work to prepare. Thus, Worker 3 could easily perform the work of Worker 4.
- Worker 3 had a 1-min break when Worker 4 was preparing the edge panel, as shown in Figure A-32.

These changes reduce the labor from four workers to three, or by 25%. The subcontractor could transfer one worker to work on other sections or eliminate that worker and save money on the cost of installation. Because moving the platform requires two workers to activate switches on each side of the platform at the same time, it was recommended that the subcontractor use a single-operator switch platform to reduce the number of workers to three.

Analysis Cre: Stone Panel Installation: Crew B — Date:__10/06/08_____
Job: River ParkII Project : Curttain Wall Installation
Time Interval: — Page:__1__of__2___
Remarks: Stone Type 4 (Edge) : Bay Type B (Edge Bay)

Time Interval (30 Seconds)	Start time	Man 1	Man 2	Man 3	Man 4
Video1					
1	0:00	Idle	Prepare tools	Prepare tools	Prepare tools
2	0:30				
3	1:00				
4	1:30	Pull pulley		Idle	
5	2:00		Bring stone panel		Idle
6	2:30	Idle	Move inside		Prepare tools
7	3:00			Prepare tools	
8	3:30				
9	4:00	Pull Pulley			
10	4:30				Idle
11	5:00	Idle		Put sealant	Prepare tools
12	5:30		Prepare stone panel	Idle	Idle
13	6:00				
14	6:30				
15	7:00	Pull pulley			
16	7:30		Set stone panel		
17	8:00				Prepare tools
18	8:30				
19	9:00				
20	9:30				Idle
21	10:00				
22	10:30				
23	11:00		Idle	Put sealant	Prepare tools
24	11:30			Idle	Idle
25	12:00	Pull Pulley			
26	12:30				
27	13:00		Prepare tools		
28	13:30			Set panel	
29	14:00				
30	14:30				Prepare tools
31	15:00		Idle		
32	15:30		Put sealant	Move stage	
35	16:00				Move stage
37	16:30	Remove clamp	Prepare tools	Prepare tools	Prepare tools
39	17:00	Hold stone		idle	
40	18:00	Bolt panel		Move stage	Move stage
41	18:30	Hold stone			Idle

Figure A-30. Proposed Crew Balance Chart for Stone Panel Type 4 Bay Type B Installation

Source: Jiang et al. (n.d.), Figure 30; reproduced with permission.

New Productivity Index for Output per Labor Hour

$$
\begin{aligned}
P_{2-1} &= (1 \text{ panel}/3 \text{ labor hours})/(1 \text{ panel}/4 \text{ labor hours}) \\
&= 4/3 \qquad\qquad\qquad\qquad\qquad\qquad\qquad\qquad\qquad\qquad \text{(A-2)} \\
&= 1.33
\end{aligned}
$$

Thus, by removing one worker from the process, productivity would increase by 33%.

Cause of the Time Differences for Stone Panels 6 and 7

The procedures for each stone-panel installation were similar, but the characteristics were different because the crews worked in confined spaces in different areas.

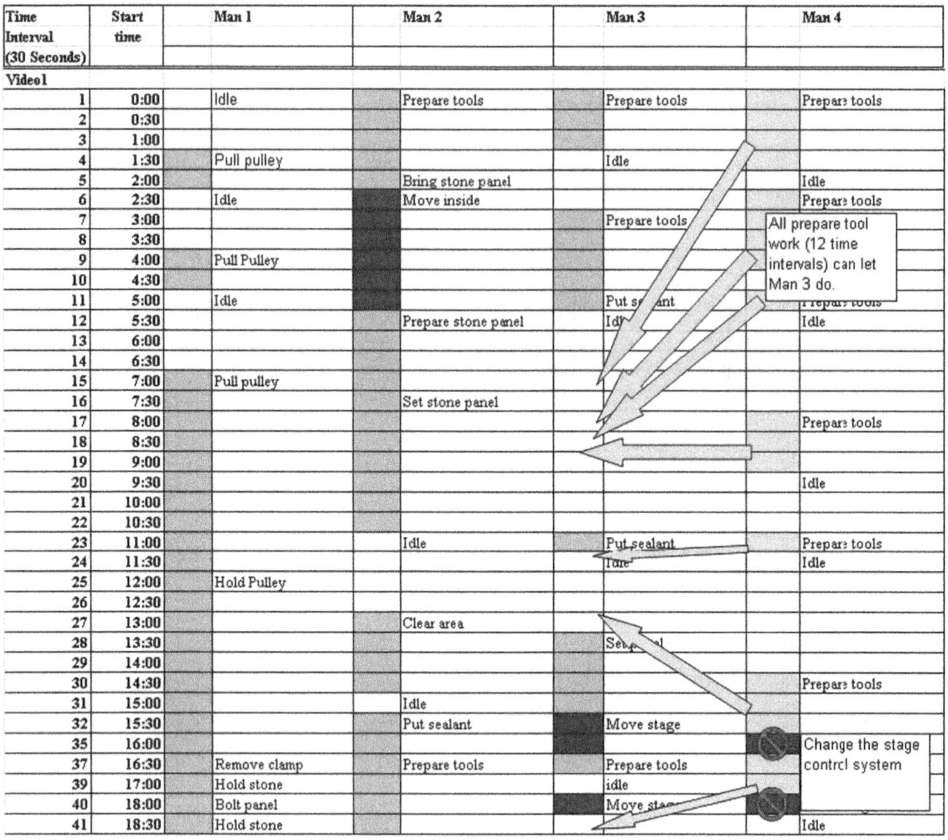

Time Interval (30 Seconds)	Start time	Man 1	Man 2	Man 3	Man 4
Video1					
1	0:00	Idle	Prepare tools	Prepare tools	Prepare tools
2	0:30				
3	1:00				
4	1:30	Pull pulley		Idle	
5	2:00		Bring stone panel		Idle
6	2:30	Idle	Move inside		Prepare tools
7	3:00			Prepare tools	
8	3:30				
9	4:00	Pull Pulley			
10	4:30				
11	5:00	Idle		Put s[illegible]ant	Prepare tools
12	5:30		Prepare stone panel	Id[illegible]	Idle
13	6:00				
14	6:30				
15	7:00	Pull pulley			
16	7:30		Set stone panel		
17	8:00				Prepare tools
18	8:30				
19	9:00				
20	9:30				Idle
21	10:00				
22	10:30				
23	11:00		Idle	Put sealant	Prepare tools
24	11:30			Idle	Idle
25	12:00	Hold Pulley			
26	12:30				
27	13:00		Clear area		
28	13:30			Se[illegible]l	
29	14:00				
30	14:30				Prepare tools
31	15:00		Idle		
32	15:30		Put sealant	Move stage	
35	16:00				Change the stage contrcl system
37	16:30	Remove clamp	Prepare tools	Prepare tools	
39	17:00	Hold stone		idle	
40	18:00	Bolt panel		Move sta[illegible]	
41	18:30	Hold stone			Idle

All prepare tool work (12 time intervals) can let Man 3 do.

Figure A-31. Crew Balance Chart of Stone Panel Type 4 Installation Eliminating One Worker

Source: Jiang et al. (n.d.), Figure 31; reproduced with permission.

When the researchers reviewed the photographs in Figures A-33 and A-34, they noticed that the platform being used was not appropriate for the edge-panel installation. The platform was not long enough for the workers to reach the installation position, and the workers had to carefully move the stone panels. This made the edge-panel installation process much longer than the general installation times. The subcontractor was able to extend the length of the stage approximately 2 ft 10 in. (0.86 m) so that Worker 3 could perform the work as he normally did and Worker 1 could control the pulley and the cramp with help from Worker 3. This change affected the performance time for Work Processes 4 and 6 listed in Table A-3. When the revised process was compared with the work-process chart for stone panels Type 1 and for stone panels Types 2, 4, and 6, most of the data were similar. Figures A-35 and A-36 show the process chart worksheets for this operation.

The duration of Work Processes 4 and 6 for Stone Panel 6 could be reduced to 1.30 min if the extension was added to the platform. With the platform extension, it would be possible to reduce the total installation time of stone panels 6 and 7:

Analysis Cre' Stone Panel Installation: Crew B Date:___10/06/08______
Job: River ParkII Project : Currtain Wall Installation
Time Interval: Page: ___1___of__2____
Remarks: Stone Type 4 (Edge) : Bay Type B (Edge Bay)

Time Interval (30 Seconds)	Start time	Man 1	Man 2	Man 3	Man 4
Video1					
1	0:00	Idle	Prepare tools	Prepare tools	
2	0:30				
3	1:00				
4	1:30	Pull pulley		Idle	
5	2:00		Bring stone panel		
6	2:30	Idle	Move inside		
7	3:00			Prepare tools	
8	3:30				
9	4:00	Pull Pulley			
10	4:30				
11	5:00	Idle		Put sealant	
12	5:30		Prepare stone panel	Idle	
13	6:00				
14	6:30				
15	7:00	Pull pulley			
16	7:30		Set stone panel		
17	8:00				
18	8:30				
19	9:00				
20	9:30				
21	10:00				
22	10:30				
23	11:00		Idle	Put sealant	
24	11:30			Prepare tools	
25	12:00	Hold Pulley			
26	12:30				
27	13:00		Clear area		
28	13:30			Set panel	
29	14:00				
30	14:30				
31	15:00		Idle		
32	15:30		Put sealant	Move stage	
35	16:00				
37	16:30	Remove clamp	Prepare tools	Prepare tools	
39	17:00	Hold stone		idle	
40	18:00	Bolt panel		Move stage	
41	18:30	Hold stone		Prepare tools	

Figure A-32. Improved Crew Balance Chart for Stone Panel Type 4 Installation

Source: Jiang et al. (n.d.), Figure 32; reproduced with permission.

$$\begin{aligned}&\text{installation time for stone panels 6 and 7}\\&=[18.5\text{ min}-(5.5\text{ min}+4.0\text{ min}-1.5\text{ min}-1.5\text{ min})]\\&=12\text{ min}\end{aligned}$$

The time required for Work Process 2 remained significantly higher because of the restricted area available for moving the stone panels into position.

A.7 Analysis of the Sequence for the Stone-Panel Installation

The workers planned to move the platform horizontally when they finished installing two vertical bays. It took approximately 251 min, or 4.2 h, to move the platform, and

Figure A-33. Worker 3 Idle

Source: Jiang et al. (n.d.), Figure 33; reproduced with permission.

Figure A-34. Worker 3 Reaching Over Edge of Platform to Perform Work

Source: Jiang et al. (n.d.), Figure 34; reproduced with permission.

the platform was moved six times. There were 15 floors and 4 sides to the building; therefore,

$$\begin{aligned}
&\text{total number of times platform was moved}\\
&\quad = (4 \text{ sides} \times 6 \text{ times moved})\\
&\quad = 48 \text{ times per floor}\\
&\quad = 48 \text{ times per floor} \times 15 \text{ floors}\\
&\quad = 720 \text{ times}
\end{aligned}$$

$$\begin{aligned}
&\text{total time required to move platform}\\
&\quad = (\text{number of times platform was moved}\\
&\quad \times \text{time required to move platform})\\
&\quad = (48 \text{ times per floor} \times 4 \text{ h per move})\\
&\quad = 192 \text{ h, or } 48 \text{ days per floor}
\end{aligned} \tag{A-3}$$

Process Chart Worksheet

Project	River park Tower II	**Study Location**	Bay1 West side
Work Item	Installation of Stone Panel No. 6	**Prepared By:**	Sorawut Srisakorn
Evaluation No.	1	**Date:**	6th October, 2008

No.	Flow Sequence	Operation / Transportation / Inspection / Delay / Storage	Quality	Distance in Ft.	Time	Notes
1	Fasten screw & Level a existing bottomRail	● ⇨ ▢ D ▽			1 : 30	Labor No. 3,4
2	Seal silicone on the bottom Rail	● ⇨ ▢ D ▽			0 : 30	Labor No. 3,4 Parallel with step3
3	Bring the stone panel	○ ➡ ▢ D ▽		25' Max.	3 : 30	Labor No. 2
4	Get stone onto pulley and Pull chain to lower	● ⇨ ▢ D ▽			5 : 30	Labor No. 1,2
5	Seal silicone onto bottom of stone	● ⇨ ▢ D ▽			0 : 30	Labor No. 3,4
6	Set stone to sit on bottom Rail	● ⇨ ▢ D ▽			4 : 00	Labor No. 1,3,4
7	Seal silicone	● ⇨ ▢ D ▽			0 : 30	Labor No. 2
8	Install the bottom Rail to secure panel	● ⇨ ▢ D ▽			1 : 30	Labor No. 2
9	Move stage Up/Down	● ⇨ ▢ D ▽		12' Max.	: 30	Labor No. 3,4
10	Remove clamp from stone and pull pulley up	● ⇨ ▢ D ▽			1 : 00	Labor No. 1
	Total				18 : 30	

Figure A-35. Process Chart Worksheet for Stone Panel Type 6

Source: Jiang et al. (n.d.), Figure 35; reproduced with permission.

To decrease the installation time, the subcontractor could install more than two vertical bays and then move horizontally, but the subcontractor still would have to install three bays at a time. The senior engineer on this project indicated that it was also possible to move the platform horizontally after the workers finished two vertical, consecutive bays. Further, the project engineer confirmed that four bays per move were possible. If the subcontractor had chosen to install three bays per move, the transportation time would have been reduced for this operation.

Figure A-37 shows how it is possible to work on three vertical bays per horizontal move. In reviewing this figure, the subcontractor determined that the parapet could be installed immediately after the 15th and 16th floors were installed, and the crews could work on the three vertical bays. This would shorten the scheduled time required to complete this operation.

The process for moving the platform was videotaped, and it required approximately 4 h, or half a day, to complete each move of the platform. To reduce the time needed to move the platform, it needed to be verified whether having only three workers move the platform with the two-operator platform switch was feasible. A crew balance chart was developed for moving the platform, and it is shown in Figure A-38. Figures A-39 and A-40 demonstrate that the platform could be moved with only three workers.

Process Chart Worksheet

Project	River park Tower II	**Study Location**	Bay3 West side
Work Item	Installation of Stone Panel No. 1	**Prepared By:**	Charkrit Raksamata
Evaluation No.	1	**Date:**	16th October, 2008

No.	Flow Sequence	Operation / Transportation / Inspection / Delay / Storage	Quality	Distance in Ft.	Time	Notes
1	Fasten screw & Level a existing bottomRail	● ⇨ □ D ▽			1 : 00	Labor No. 3,4
2	Seal silicone on the bottom Rail	● ⇨ □ D ▽			0 : 30	Labor No. 3,4 Parallel with step3
3	Bring the stone panel	○ ➡ □ D ▽		25' Max.	2 : 30	Labor No. 2
4	Get stone onto pulley and Pull chain to lower	● ⇨ □ D ▽			1 : 30	Labor No. 1,2
5	Seal silicone onto bottom of stone	● ⇨ □ D ▽			0 : 30	Labor No. 3,4
6	Set stone to sit on bottom Rail	● ⇨ □ D ▽			1 : 30	Labor No. 1,3,4
7	Seal silicone	● ⇨ □ D ▽			0 : 30	Labor No. 1
8	Install the bottom Rail to secure panel	● ⇨ □ D ▽			1 : 30	Labor No. 2
9	Move stage Up/Down	● ⇨ □ D ▽		5' Max.	0 : 30	Labor No. 3,4
10	Remove clamp from stone and pull pulley up	● ⇨ □ D ▽			1 : 00	Labor No. 1
	Total				11 : 00	

Figure A-36. Process Chart Worksheet for Stone Panel Type 1

Source: Jiang et al. (n.d.), Figure 36; reproduced with permission.

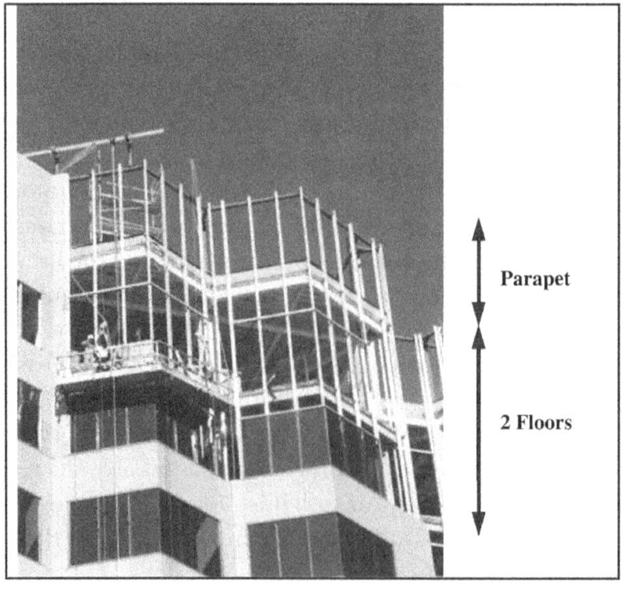

Figure A-37. Three-Floor Installation

Source: Jiang et al. (n.d.), Figure 37; reproduced with permission.

Time Interval (10 mins)	Start time	Man 1 Upper Lead	Man 2 Upper Runner	Man 3 Stage Leader	Man 4 Support Helpei (Stage Helper)
1	8.20	Idle	Idle	Move stage down	Move stage dow
2	8.30		Prepare Tool	idle	idle
3	8.40	Take off wire1 to temp	Take off wire1		
4	8.50	& Loose support1	& Loose support1	Hold wire	Move Stone in p
5	9.00	idle			idle
6	9.10	Move Support1	Move support	idle	Move support
7	9.20	Install Support1	Install Support1		Install Support1
8	9.30	Take off wire2 to temp			
9	9.40	& Loose support2	Take off wire2	Hold wire	Move Stone in p
10	9.50	idle			
11	10.00				
12	10.10	Move Temp Support	Move Temp Support	Move Stage	
13	10.20				idle
14	10.30	Hook in wire1	Hook in wire1		
15	10.40				
16	10.50	Idle	Prepare tools	idle	
17	11.00	Move support2	Move support2		Move support
18	11.10	Install Support1	Install Support1		Install Support1
19	11.20				
20	11.30	Move Temp Support	Move Temp Support	Move Stage	idle
21	11.40				
22	*13.15				
23	13.25	Hook in wire2	Hook in wire1		Move Stone in p
24	13.35			idle	
25	13.45				
26	13.55	Idle	Clear area		
27	14.05		Collect tool	Move stage down	Move stage dow

Break at 11.50 and then Start at 12:15

Figure A-38. Regular Crew Balance Chart for Platform Relocation Process

Source: Jiang et al. (n.d.), Figure 38; reproduced with permission.

Time Interval (30 Seconds) Video2	Start time	Man 1 Upper Lead	Man 2 Upper Runner	Man 3 Stage Leader	Man 4 Support Helper (Stage Helper)
1	8.20	Idle	Idle	Move stage down	Move stage down
2	8.30		Prepare Tool	idle	idle
3	8.40	Take off wire1 to temp	Take off wire1		
4	8.50	& Loose support1	& Loose support1	Hold wire	Move Stone in position
5	9.00	idle			idle
6	9.10	Move Support1	Move support	idle	Move support
7	9.20	Install Support1	Install Support1		Install Support1
8	9.30	Take off wire2 to temp			
9	9.40	& Loose support2	Take off wire2	Hold wire	Move Stone in position
10	9.50	idle			
11	10.00				
12	10.10	Move Temp Support	Move Temp Support	Move Stage	
13	10.20				idle
14	10.30	Hook in wire1	Hook in wire1		
15	10.40				
16	10.50	Idle	Prepare tools	idle	
17	11.00	Move support2	Move support2		Move support
18	11.10	Install Support1	Install Support1		Install Support1
19	11.20				
20	11.30	Move Temp Support	Move Temp Support	Move Stage	idle
21	11.40				
22	*13.15				
23	13.25	Hook in wire2	Hook in wire1		Move Stone in position
24	13.35			idle	
25	13.45				
26	13.55	Idle	Clear area		
27	14.05		Collect tool	Move stage down	Move stage down

Figure A-39. Crew Balance Chart with Elimination of One Worker

Source: Jiang et al. (n.d.), Figure 39; reproduced with permission.

Time Interval (10 mins)	Start time	Man 1 Upper Lead	Man 2 Upper Runner	Man 3 Stage Leader		Man 4 Support Helper (Stage Helper)
1	8.20	Move stage down	Idle	Move stage down		
2	8.30		Prepare Tool	idle		
3	8.40	Take off wire1 to temp	Take off wire1			
4	8.50	& Loose support1	& Loose support1	Hold wire		
5	9.00	idle				
6	9.10	Move Support1	Move support	Move support		
7	9.20	Install Support1	Install Support1	Install Support1		
8	9.30	Take off wire2 to temp				
9	9.40	& Loose support2	Take off wire2	Hold wire		
10	9.50	idle				
11	10.00					
12	10.10	Move Temp Support	Move Temp Support	Move Stage		
13	10.20					
14	10.30	Hook in wire1	Hook in wire1	Move Stone in position		
15	10.40					
16	10.50	Idle	Prepare tools	idle		
17	11.00	Move support2	Move support2	Move support		
18	11.10	Install Support1	Install Support1	Install Support1		
19	11.20					
20	11.30	Move Temp Support	Move Temp Support	Move Stage		
21	11.40					
22	*13.15					
23	13.25	Hook in wire2	Hook in wire1			
24	13.35			Move Stone in position		
25	13.45					
26	13.55	Idle	Clear area			
27	14.05	Move stage down	Collect tool	Move stage down		

Figure A-40. New Crew Balance Chart for Platform Relocation Process

Source: Jiang et al. (n.d.), Figure 40; reproduced with permission.

A.8 Productivity Index Calculations

The average time to complete the stone-panel installation was calculated as 20.15 min per panel, and the workers could install 0.049 panels per minute, 2.98 panels per crew hour, and 23.8 panels per crew day. The project schedule allowed 7 days for one floor installation with two crews, or 312 panels per 2 crews per 7 days = 22.3 panels per crew day. Three methods were considered for increasing the productivity rate for this installation process:

Reduce Input (I_2)—Recommendation 1 was to remove one laborer from the panel-installation team and change the platform control system to a one-worker operation.

$$\begin{aligned}\text{original productivity rate} &= \text{output per crew day} \\ &\quad /\text{number of laborers per crew for regular} \\ &\quad \text{installation per crew day} \\ &= O_1/I_1 \\ &= 23.8\ \text{panels}/4\ \text{laborers per crew day} \\ &= 5.95\ \text{panels/labor day}\end{aligned} \tag{A-4}$$

$$\begin{aligned}\text{improved installation rate per crew day} \\ &= O_1/I_2 \\ &= 23.8\ \text{panels}/3\ \text{laborers per crew day} \\ &= 7.93\ \text{panels per labor day}\end{aligned} \tag{A-5}$$

Therefore,

$$\begin{aligned}&\text{productivity index}\\&=(O_1/I_2)_t/(O_1/I_1)_t\\&=(7.93\text{ panels per labor-day}/5.95\text{ panels per labor day})\\&=133\%\end{aligned} \tag{A-6}$$

Increase Output (O_2)—Recommendation 2 was to improve the equipment by modifying the existing platform and adding a 2 ft 10 in. (0.86 m) extension to the edge of the platform. This extension would increase the work area for Worker 3 so that he or she could work more conveniently and faster when installing the edge panels, as shown in Figure A-41.

$$\begin{aligned}&\text{estimated time reduction} = \text{regular time}\\&\quad-\text{new estimated time}\\&\quad=18.5\text{ min}-12\text{ min}\\&\quad=6.5\text{ min per edge stone panel with 6 panels per floor}\end{aligned} \tag{A-7}$$

Thus, the time reduction = 6.5 min × 6 min – 39 min per floor.

$$\begin{aligned}&\text{total installation savings (new time per 1 floor installation)}\\&\quad=T_{\text{total per floor}}-\text{estimated time reduction}\\&\quad=1{,}572\text{ min}-39\text{ min}\\&\quad=1{,}533\text{ min, or }25.55\text{ h}\end{aligned} \tag{A-8}$$

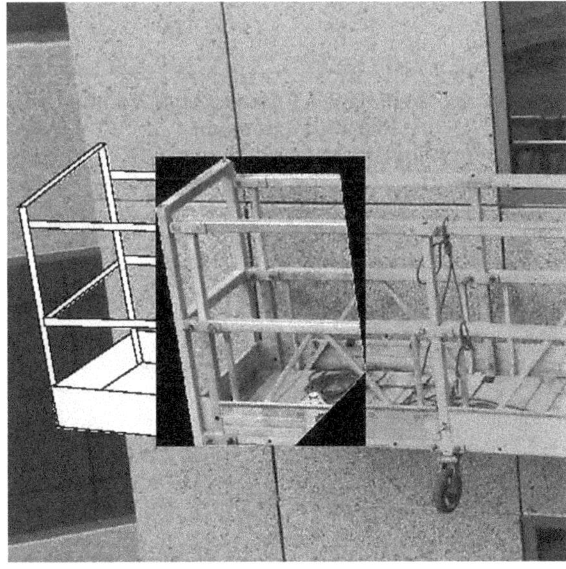

Figure A-41. Modified Platform Extension

Source: Jiang et al. (n.d.), Figure 41; reproduced with permission.

average time per panel for 78 panels per floor
= number of minutes per floor per side of
building/panels per floor (A-9)
= 1,533 min/78
= 19.65 min per panel

The **improvement ratio** is shown in Figures A-42 to A-43.

new productivity rate after installing the platform extension
= number of minutes per panel/60
= 19.65/60 min per hour (A-10)
= 3.05 panels per crew hour,
or 24.4 panels per crew day

Because the crew had four members per crew,

productivity rate for one worker
= 24.4 panels per crew day/4 laborers per crew
= 6.1 panels per labor day

The revised productivity index shows that a minor improvement to the total process resulted from the significantly different number of edge panels and the total panel count.

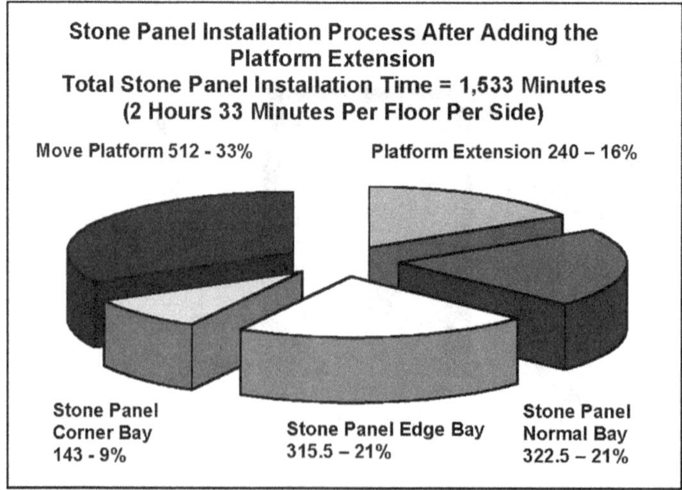

Figure A-42. Stone-Panel Installation Process with Platform Extension Improvement

Source: Jiang et al. (n.d.), Figure 42; reproduced with permission.

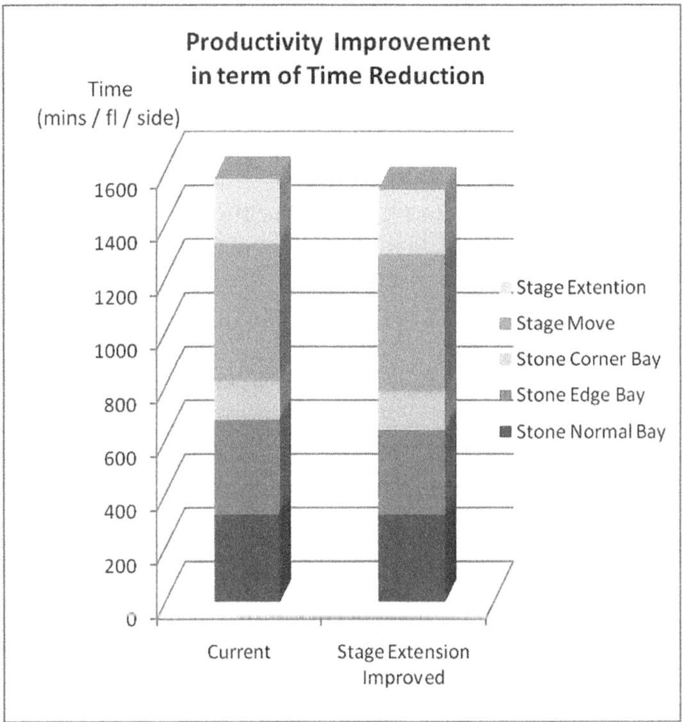

Figure A-43. Productivity Improvement with New Platform Extension

Source: Jiang et al. (n.d.), Figure 43; reproduced with permission.

$$\begin{aligned}&\text{new productivity index}\\&\quad=(O_2/I_1)_t/(O_1/I_1)_t\times 100\\&\quad=(6.10\text{ panels per labor day}/\\&\quad 5.95\text{ panels per labor day})\times 100\\&\quad=102.52\%\end{aligned} \tag{A-11}$$

The small productivity increase does not mean that this modification is not worth implementing on this project. Given the cost of the total project, which was in the millions of dollars, the resulting 2.52% increase in productivity could generate significant additional profit. Also, this modification would benefit the crew because Worker 3 could perform his work in a safer manner.

Increase Output (O_3)—Recommendation 3 was to increase the number of bay installations before moving the platform from two bays per move to three bays per move. This method could reduce the total installation time:

$$\begin{aligned}&\text{total installation time}=[(256\times 4)+(240\times 2)]/2\\&\quad=752\text{ min per floor}\\&\quad\text{total revised installation time}\\&\quad=[(256\times 4)+(240\times 2)]/3\\&\quad=501\text{ min per floor}\end{aligned}$$

Therefore, it reduces the installation time as follows:

$$\begin{aligned}&\text{total installation time reduction}\\&= 752\text{ min} - 501\text{ min}\\&= 251\text{ min per floor}\end{aligned}$$

Normally, it takes 1,572 min to finish a floor; thus,

$$\begin{aligned}&\text{new total time per floor} = 1{,}572\text{ min} - 251\text{ min}\\&= 1{,}321\text{ min per floor, or per 78 panels}\end{aligned}$$

$$\begin{aligned}&\text{average time per panel} = 1{,}321\text{ min}/78\text{ panels}\\&= 16.94\text{ min per panel}\end{aligned}$$

$$\begin{aligned}&\text{productivity rate per day} = 1/16.94\text{ min per panel}\\&\times 60\text{ min per crew hour}\\&= 3.54\text{ panels per crew hour}\\&= 3.54\text{ panels per crew hour} \times 8\text{ hours per day}\\&= 28.3\text{ panels per crew day,}\\&\text{or } 7.08\text{ panels per labor day}\end{aligned}$$

$$\begin{aligned}&\text{new productivity index}\\&= (O_3/I_1)_t/(O_1/I_1)_t\\&= (7.08\text{ panels per labor day}\\&/5.95\text{ panels per labor day}) \times 100\\&= 119.0\%\end{aligned}$$

Figures A-44 and A-45 show the resulting time and time reduction.

A.9 Work Improvement Conclusions

Implementing the three recommendations cited previously would increase the productivity index for the stone-panel installation. The following shows the increases that would result from implementing the recommendations:

Reduce Input (I_2), Increase Output (O_2), and Increase Output (O_3)

$$\begin{aligned}&\text{new number of panels per crew day}\\&= 1\text{ crew}/[(1{,}572\text{ min per floor installation time} - 39\text{ min}\\&\text{time reduction for moving platform per floor}\\&- 251\text{ min time reduction for installation})/(78\\&\text{panels} \times 8\text{ h/crew day})] \times 60\text{ min/hour}\\&= 29.2\text{ panels/crew day}\end{aligned}$$

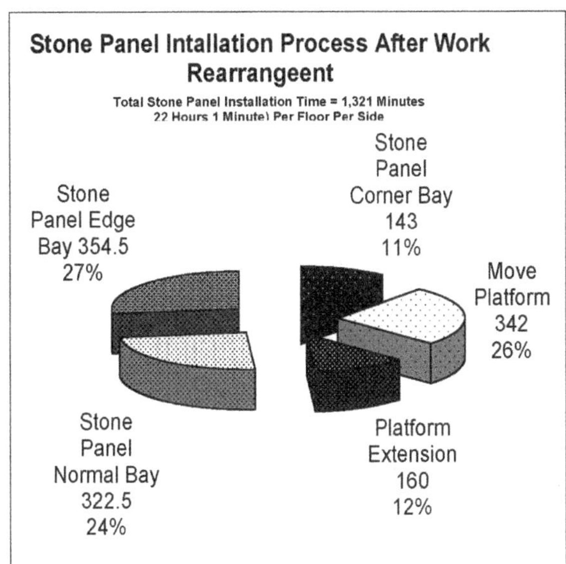

Figure A-44. Work Breakdown with Platform Extension and Work Rearrangement

Source: Jiang et al. (n.d.), Figure 44; reproduced with permission.

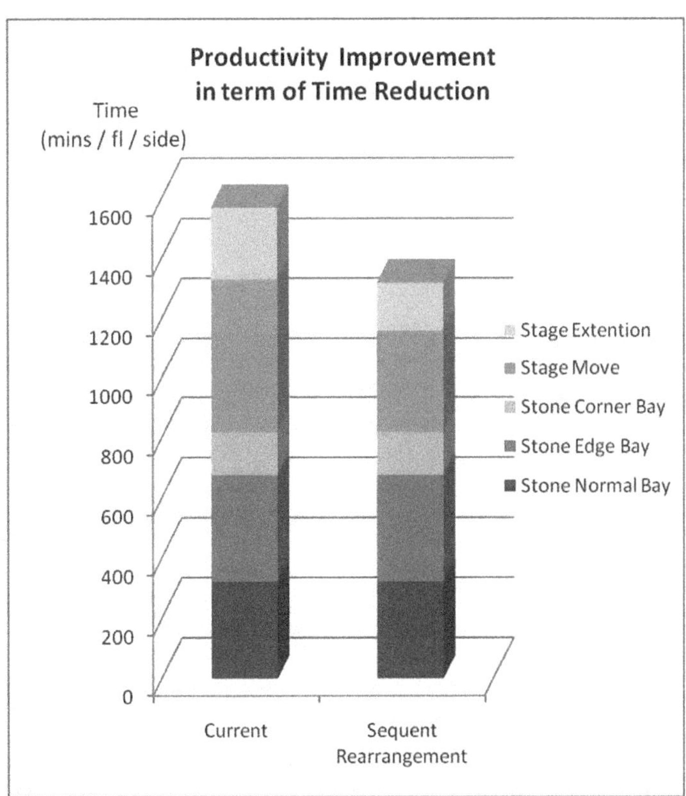

Figure A-45. Time Decrease after Implementation of Recommendations

Source: Jiang et al. (n.d.), Figure 45; reproduced with permission.

$$\text{new worker requirement} = (29.2 \text{ panels/crew day}) / (3 \text{ workers/crew}) = 9.73 \text{ workers}$$

$$\text{new number of panels per labor day} = (O_4 / I_2)_t$$
$$\text{improved project productivity index} = (O_3 / I_2)_t / (O_1 / I_1)_t = (9.73 \text{ workers}/5.95 \text{ workers}) \times 100 = 163.5\%$$

Figures A-46 and A-47 summarize the improved times for the stone-panel installation process. The recommendations could potentially increase productivity more than the 10% goal for this project. However, they are based on assumptions and a hypothetical analysis. In this case study, the output is the number of stone panels installed and the inputs are time, worker performance, tools, and machines.

A.10 Summary

This case study demonstrated how to develop and implement a productivity improvement investigation using the stone-panel installation process for a high-rise building. It illustrated how productivity team members could conduct a productivity improvement study by collecting data, analyzing the data, and preparing recommendations for increasing the productivity of the stone-panel installation process. Numerous photos were included in the case study to provide visual representations of the work-

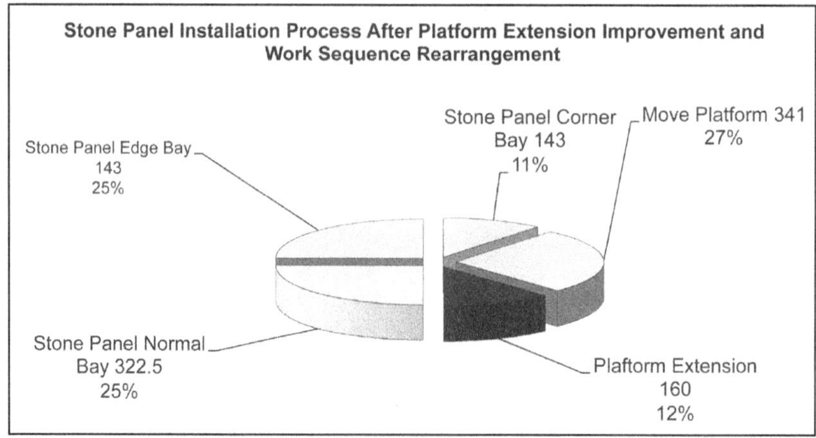

Figure A-46. Work Breakdown with Platform Extension and Improved Work Sequence Rearrangement

Source: Jiang et al. (n.d.), Figure 46; reproduced with permission.

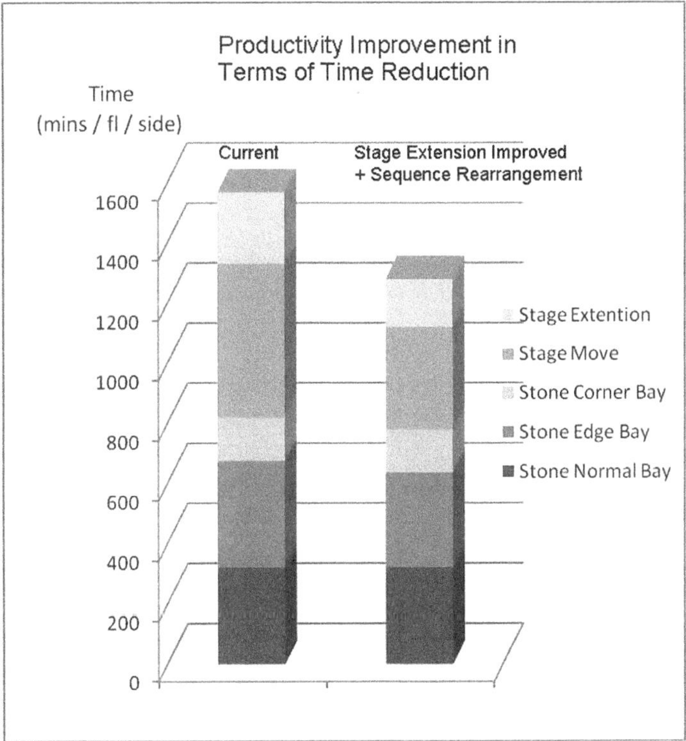

Figure A-47. Time Decrease after Platform Extension Improvement and Work Sequence Rearrangement

Source: Jiang et al. (n.d.), Figure 47; reproduced with permission.

flow processes and crew configurations originally used during the stone-panel installation process. The productivity improvement team members used the information obtained during the case study to develop revised crew balance charts and workflow processes that demonstrate how productivity could be improved on this portion of the project.

A.11 Key Terms

column cover
column cover and spandrel systems
curtain walls
glazers
glazing infills
mullion
spandrel
stick systems
stonemasons
transoms
unit and mullion systems
unit systems
welders

References

Conrad, K. (2009). *River Park Tower II project.* March 29. <http://www.bizjournals.com/sanjose/stories/2009/06/29/newscolumn1.html?page=all> (April 2014).

Hong Kong Polytechnic University. (2008). *Curtain wall: Stick system,* Department of Civil and Structural Engineering. <http://www.cse.polyu.edu.hk/~civcal/wwwroot/superstr/curtain_wall/system.htm> (Sept. 1, 2008).

Jiang, G., Mukherjee, A., Raksamata, C., Singh, I., Srisakorn, S., and Tefera, M. (n.d.). *Curtain wall installation: River Park Tower II project,* Course project, San Jose State University, San Jose, CA.

Index

Page numbers followed by *e*, *f*, and *t* indicate equations, figures, and tables, respectively.

About the Author

J. K. Yates, Ph.D., is the former dean of the College of Engineering Technology at Ferris State University in Big Rapids, Michigan, one of the largest colleges of engineering technology in the United States. The author was formerly the department head and the Joe W. Kimmel Endowed Distinguished Professor of Construction Management in the Department of Construction Management at Western Carolina University in Cullowhee, North Carolina. Dr. Yates was previously a professor and department chair in the Department of Construction Management and Engineering at North Dakota State University; in charge of the Construction Engineering and Management focus area at Ohio University; and program coordinator for the Construction Engineering program in the Civil and Environmental Engineering Department at San Jose State University in California. Dr. Yates was also a professor at the Polytechnic School of Engineering at New York University (formerly Polytechnic University and Brooklyn Polytechnic) and at Iowa State University and served as a visiting professor at the University of Colorado for one year.

Dr. Yates received a bachelor of science degree in civil engineering from the University of Washington with a minor in archeology/anthropology and a Doctorate of Philosophy (Ph.D.) degree in civil engineering from Texas A&M University, with minors in global finance and management, global political science, business analysis, and construction science.

Dr. Yates has worked for several of the top-ranked domestic and global engineering and construction firms, including Bechtel, URS, CH2M Hill, and Williams Company, and served as a consultant during legal cases and on international contracts. Dr. Yates is the author of eight books and numerous refereed journal articles and is a member of ASCE, the Project Management Institute, and the American Association of Cost Engineers International. Dr. Yates received the Distinguished Professor Award from the Construction Industry Institute (CII) in 2010 and from the Polytechnic School of Engineering at New York University in 1994, was named the Associated General Contractor's Outstanding Construction Professor in America in 1997, was one of the *Engineering News Record*'s Those Who Made Marks on the Construction Industry in 1991, and received the Ron Brown Award for industry/academic collaborations with the Hewlett Packard Foundation from President Bill Clinton in 2001. Dr. Yates has traveled for work and pleasure to 25 countries and worked in Bontang, East Kalimantan, on the island of Borneo in Indonesia.